Ecological Aspects of Used-water Treatment

Ecological Aspects of Used-water Treatment

Edited by

C. R. CURDS

British Museum (Natural History), London, England

and

H. A. HAWKES

The University of Aston in Birmingham, Birmingham, England

Volume 1—
The Organisms and their Ecology

1975
ACADEMIC PRESS
London New York San Francisco
A Subsidiary of Harcourt Brace Jovanovich, Publishers

ACADEMIC PRESS INC. (LONDON) LTD.
24/28 Oval Road,
London NW1

United States Edition published by
ACADEMIC PRESS INC.
111 Fifth Avenue
New York, New York 10003

Library of Congress Catalog Card Number: 75 19628
ISBN: 0 12 199501 8

PRINTED IN GREAT BRITAIN BY
WHITSTABLE LITHO, STRAKER BROTHERS LTD.

Contributors

R. A. BAKER Department of Pure and Applied Zoology, University of Leeds, England.

KATHRYN BENSON-EVANS Department of Botany, University College, Cathays Park, Cardiff, Wales.

R. F. CROWTHER Severn-Trent Water Authority, Upper Tame Division, Tame Valley House, 156–170 Newhall Street, Birmingham B3 1SE, England.

C. R. CURDS Department of Zoology, British Museum (Natural History), London SW7 5BD, England.

MARGARET DOOHAN* Department of Zoology, Royal Holloway College, Englefield Green, Surrey, England.

N. HARKNESS Severn-Trent Water Authority, Abelson House, 2297 Coventry Road, Sheldon, Birmingham B26 3PR, England.

M. A. LEARNER Department of Applied Biology, University of Wales, Institute of Science and Technology, Cathays Park, Cardiff CF1 3NU, Wales.

E. B. PIKE The Water Research Centre, Stevenage Laboratory, Elder Way, Stevenage, Hertfordshire, England.

F. SCHIEMER Limnologisches Institut, Österreichische Akademie der Wissenschaften, Berggasse 18/19, Vienna A-1090, Austria.

J. F. DE L. G. SOLBÉ Water Research Centre, Stevenage Laboratory, Elder Way, Stevenage, Hertfordshire, England.

T. G. TOMLINSON 2 Deard's End Lane, Knebworth, Hertfordshire, England.

I. L. WILLIAMS Applied Hydrobiology Section, University of Aston in Birmingham, Birmingham B4 7ET, England.

P. F. WILLIAMS Department of Botany, University College of South Wales and Monmouth, Cathays Park, Cardiff, Wales.

* *Present name and address:* MARGARET FRAYNE, Avery Hill College of Education, Bexley Road, Eltham, London SE9 2PQ.

Preface

Since his earliest days man has used water for cleansing and processing food and other materials. His increasing ability to exploit and process natural resources and to synthesize complex organics has added to the variety of wastes he produces and the amount of water he uses. With the change from an agrarian-based society to an industrial one and the consequent concentration of populations, the problem of waste disposal became acute. The subsequent introduction of the water-carriage system of sewage disposal further increased the use of water.

Having used water for his various domestic and industrial purposes, man has regarded it as "waste water" and simply recognized the problem as being one of disposal with minimal effect on the environment—pollution prevention. A more modern concept is to regard this used water as a valuable natural resource to be returned to supplement the hydrologic cycle. Treatment processes in this context have a dual purpose of pollution prevention and water reclamation. In keeping with this philosophy of "recovering" rather than "disposal", the term "used water" has been adopted in this work as being more appropriate than the more commonly used "waste water".

Used-water treatment by biological oxidation plants may be regarded as the environmental control of the activity of populations of appropriate organisms. Therefore, to those engaged in the design, operation and development of the processes, an understanding of the organisms, their activities and general biology is essential. A study of the environmental requirements of the organisms involves us in autecology; a study of the interaction of populations feeding on the organic matter in the treatment plant involves us in synecology. These different ecological aspects, essential to an understanding of the used-water treatment processes, form the theme of this work.

Ecological studies on treatment plants have been carried out since the turn of the century; some were instigated by academic interests, others by the need to solve operating problems and nuisances and yet others to improve efficiency. The results of such investigations are to be found in literature representing a wide range of disciplines. In this present work these scattered results have been collected together by the contributing authors, each dealing with a specific taxon or topic.

In this volume (Volume 1—The Organisms and their Ecology), different chapters deal respectively with the organisms of different taxa present in treatment plants, their role and ecology. In a companion volume (Volume 2—The Processes and their Ecology), the different used-water processes are first described. Subsequent chapters deal with microbial physiology, kinetics and mathematical modelling, and synecology of the processes, as a guide to their design and operation.

June, 1975

C. R. C.
H. A. H.

Contents of Volume 1

Contents of Volume 2—
The Processes and their Ecology

1

Aerobic Bacteria

E. B. Pike

The Water Research Centre
Stevenage Laboratory
Elder Way
Stevenage
Herts
England

I. Introduction

Bacteria constitute the major part of the biomass and the basic trophic level in all stages of biological sewage treatment, apart possibly from aerobic oxidation ponds (lagoons) for tertiary treatment of effluents, in which the activities of algae, which are actively photosynthetic under daylight conditions, may predominate.

The predominant bacteria are saprophytic, being responsible for the degradation and ultimate mineralization of organic compounds in the wastes. However, some bacteria are chemosynthetic autotrophs (chemolithotrophs), notably the genera *Nitrosomonas* and *Nitrobacter*, respectively oxidizing ammonia and nitrite, and these are regularly found in plants in which the rate of treatment and the specific growth rate of the sludge biomass is low enough to enable them to become established. *Thiobacillus* species which oxidize thiosulphate to sulphate and oxidize thiocyanate occur in plants treating carbonization liquors from gas works (Hutchinson and White, 1964) and have also been recorded in biological filters (Hotchkiss, 1924a, b). Photosynthetic bacteria, typified by members of the Athiorhodaceae, are invariably anaerobic, but have been recorded in oxidation ponds treating raw sewage.

The main reason for the success of bacteria, in comparison with other protists, in degrading waste materials is directly related to their small size and their relatively large surface area to volume ratio which increases their capacity for exchanging nutrients and catabolites with their suspending fluid. Thus, it is well known that many heterotrophic bacteria can exhibit minimum doubling times of about 20 minutes in pure culture under optimum conditions for growth whereas values of around 10 h are displayed by many protozoa and mammalian tissue cells (Maynard Smith, 1969), and a minimum value of 1·6 h was found in Curds and Cockburn (1971) for the ciliated protozoon *Tetrahymena pyriformis*, feeding upon *Klebsiella aerogenes* in two-stage continuous culture.

There has been considerable interest in the kinetics and nature of substrate removal in sewage treatment processes. Such information is of direct use for constructing predictive mathematical models of sewage treatment. It should be emphasized also that a further desirable requisite is that predictive models should also recognize the behaviour of the micro-organisms of which the active biomass (or suspended solids) is comprised. This type of study is essentially synthetic in that a knowledge of the complete process is built up from knowledge about the behaviour of its components. The information obtained is invaluable, because it can often be directly applied for design of processes, but is incomplete, because of simplifying assumptions which have to be made and which are inevitable when one is considering utilization of a heterogeneous substrate by a community of diverse micro-organisms, in which many of the bacteria appear to be non-viable, although their enzymes remain active (Wooldridge and Standfast, 1933, 1936). Another complication is that plant operation parameters fluctuate.

The complementary approach, using the methods of ecology, is essentially analytical, in that one observes the complete system in operation and attempts to discover the ways in which the ecosystem reacts in response to applied, independent factors. One of the difficulties of synecological studies on microbial communities—as indeed with communities of higher organisms—results from the difficulty of controlling all factors except the one being studied and from the existence of interactions between species, caused alternatively by predation, or excretion of toxic metabolites and by the dependency of certain populations upon catabolites excreted by others. Since the advent of electronic computers, it has become practicable to analyse interactions between many observed variables in an ecosystem by multivariate statistical methods and to obtain a generalized quantitative description of community structure or behaviour. Such methods have been widely used in ecological studies of plant and animal communities, but have yet to be applied to any extent in the study of microbial communities. It might be supposed, however, that microbial communities, such as those present in used-water treatment processes, would be very suitable for ecological studies because of the rapidity of growth responses and adaptation to new physiological or nutritional conditions displayed by micro-organisms, in comparison with responses of higher organisms. It has been suggested that the activated-sludge process is an ideal model for quantitative ecological studies because of the powerful selective pressures which exist, the interactions and interdependencies between the different micro-organisms and the inherent regulatory mechanisms (la Rivière, 1972).

II. Ecological Methods in the Study of Bacteria in Used-water Treatment Processes

The ecology of bacteria in sewage treatment—as indeed in other habitats—can be studied in three main ways: by microscopic, cultural and metabolic techniques. Microscopic and cultural methods have most use in the estimation of total and viable cell numbers, in taxonomy and identification of bacteria and in studies of population behaviour. Metabolic methods are inherently more suitable for estimating biomass, kinetic constants and for measuring substrate turnover. A fourth approach is the investigation of energy flows between the various trophic levels in the food chains of sewage treatment. So far there has been little, if any, study in this way, although the approach is theoretically attractive, because a major aim of waste treatment is that there should be minimal retention of energy in the ecosystem for production of cellular materials as sludge and that as much energy as possible should be dissipated in the oxidation of organic pollutants to CO_2 and water.

TABLE I. Numbers of total and viable bacteria at different stages of sewage treatment (reproduced from Pike and Carrington, 1972). Values are geometric means.

Source (and number) of samples[b]	Bacterial counts[a]				Viability (per cent)	Total suspended solids (mg/l)
	In samples (No./ml)		In suspended solids (No./g)			
	Total	Viable	Total	Viable		
Settled sewage (46)	$5{\cdot}6 \times 10^{8}$	$6{\cdot}3 \times 10^{6}$	$3{\cdot}0 \times 10^{12}$	$3{\cdot}4 \times 10^{10}$	1·1	190
Activated sludge mixed liquor, conventional rate (18)	$5{\cdot}9 \times 10^{9}$	$4{\cdot}9 \times 10^{7}$	$1{\cdot}3 \times 10^{12}$	$1{\cdot}1 \times 10^{10}$	0·83	4600
Activated sludge mixed liquor, high rate (24)	$1{\cdot}4 \times 10^{10}$	$2{\cdot}4 \times 10^{8}$	$3{\cdot}0 \times 10^{12}$	$5{\cdot}0 \times 10^{10}$	1·7	4800
Filter slimes (18)	$6{\cdot}2 \times 10^{10}$	$1{\cdot}5 \times 10^{9}$	$1{\cdot}3 \times 10^{12}$	$3{\cdot}2 \times 10^{10}$	2·5	54 000
Secondary effluents (16)	$5{\cdot}4 \times 10^{7}$	$1{\cdot}1 \times 10^{6}$	$1{\cdot}9 \times 10^{12}$	$4{\cdot}1 \times 10^{10}$	2·1	28
Effluents, high-rate activated-sludge plants (24)	$4{\cdot}8 \times 10^{7}$	$1{\cdot}4 \times 10^{6}$	$3{\cdot}3 \times 10^{12}$	$1{\cdot}0 \times 10^{11}$	3·0	14
Tertiary effluents (11)	$2{\cdot}9 \times 10^{7}$	$6{\cdot}6 \times 10^{4}$	$3{\cdot}0 \times 10^{12}$	$6{\cdot}8 \times 10^{9}$	0·23	9·7

[a] Total counts obtained with Helber counting chamber, viable counts by plate-dilution frequency method (Harris and Sommers, 1968) on CGY agar, incubation for 6 days at 22°C (Pike *et al.*, 1972); viability expressed as percentage ratio of viable count to total count.

[b] Samples from sewage works and laboratory pilot plants; high-rate plants are pilot scale, working at loadings of 0·46–2·5 kg BOD removed/kg MLSS day (Boon and Burgess, 1972), secondary effluents from nine filters and seven activated-sludge plants, tertiary effluents from ten lagoons, one grass plot.

A. Microscopic Methods

2. Counts of Total Bacteria

Total cell counts of bacteria are usually obtained microscopically by using a bacterial counting chamber, such as the Helber chamber, which consists of a microscope slide, having a central chamber machined to a depth of 0·02 mm below the level of the upper surface of the slide. The floor of the chamber carries an engraved graticule, 1 mm^2 in area, subdivided into 400 squares. It has been noted (Postgate, 1969; Pike and Carrington, 1972) that liquid depths can be up to some 50 per cent greater than the nominal value, implying that chamber depths should be measured each time the chamber is used, either by interferometry (Postgate, 1969) or by observing movement of the fine adjustment micrometer on the microscope, while focusing on the bottom surface of the chamber and the under surface of the cover glass. Phase-contrast illumination is preferable to bright-field illumination. In the author's experience, the thick cover glasses supplied with the chambers tend to blur the image formed using the high power objective (×40); the definition—and thus the differentiation of bacteria from other particles—can be improved by using ordinary No. 1½ thickness cover glasses, although these affect chamber depth.

It is well known that total counts of bacteria by microscopy are much higher than corresponding viable counts, and there are at least two reasons for this. Firstly, it is often difficult to distinguish between bacteria and debris when both are near the limits of resolution of the light microscope, and secondly, many of the bacteria may be either dead or incapable of growing under the conditions of the viable counting method. Table I gives total and viable bacteria at different stages of sewage treatment. Numbers of total bacteria in the dry suspended solids are of the order of 10^{12}/g at all stages of treatment, confirming that the particles counted were of bacterial dimensions, although not proving that the particles were exclusively bacteria.

2. Cytological Methods

The structural components of individual bacteria and of their aggregations, in for example the flocs of activated sludge and in filter slimes, can be demonstrated by suitable cytological methods, such as those described in the monograph by Bisset (1955). Normal bacteriological staining methods, such as Gram's method, involve drying, heat-fixation and chemical treatments of the smears, which cause serious distortion of the cell structures. This can be overcome by allowing smears of bacteria to dry in the air without application of heat. It is convenient to prepare smears on cover glasses; these can be readily treated in small beakers or watch glasses containing the staining reagents. Stained smears should be mounted in water; although this renders them impermanent, the shrinkage resulting from dehydration and mounting in the usual mountants is avoided. If thin cover-glasses (No. 1 or 1½) are used, and the edges are sealed to the slide with wax cement, it is quite possible to observe the specimens under the oil immersion (2 mm) objective.

Some of the staining methods which have been found useful in studying the morphology of activated-sludge flocs are described below in the section dealing with flocculation in this process. There is no reason why these methods should not also be used for studying bacteria in other waste-treatment communities.

B. Cultural Methods

These are indispensable for descriptive, taxonomic and quantitative studies in which viable bacteria are isolated, identified and counted. However, there are serious discrepancies between total counts of bacteria in sewage-treatment processes by microscopy and viable counts obtained by cultural methods, the latter being about two orders of magnitude less (Table I; Ware, 1961; Pike and Carrington, 1972). This apparently low viability is also characteristic of other microbial ecosystems such as soil (Babiuk and Paul, 1970), the rumen (Annison and Lewis, 1959) and fresh water (Jannasch, 1958, Collins, 1963). It has already been noted that total counts may be higher than the true value, if debris of bacterial size is confused with bacterial cells; it is also clear that no single medium or set of incubation conditions will support growth of all types of bacteria present. Moreover, it is typical for substrate levels in sewage-treatment processes—measured on effluents or interstitial liquors—to be low and for microbial cell concentrations to be high. It can be deduced that processes are typically operated in the stationary or decline phases of microbial growth, and that because nutrient levels are insufficient to meet the needs of cell growth and maintenance energy (Pirt, 1965) requirements, a proportion of the biomass is dead or non-viable, although enzymatic activities persist. Another factor may be that the methods involved in culture could cause "substrate-accelerated death" (Postgate and Hunter, 1963), because bacteria taken from an environment of starvation are diluted—and thereby washed—in a non-nutrient diluent and transferred to a rich growth medium. These considerations imply that cultural methods cannot be used as estimates of biomass or of microbial activity, although this does not prevent their being used for measuring rates of increase or decline in numbers of those viable bacteria for which they are selective. A wide range of media and cultural techniques are available for culturing specific groups of bacteria from different habitats. Useful collections of isolation and cultural techniques for most genera likely to be encountered in sewage treatment, including autotrophs, are detailed in Table II.

The usual culture methods involve a period of incubation, which implies that counts cannot be obtained in less than a working day. Some attempts have been made to devise rapid methods, giving results within a few hours, for long enough to produce micro-colonies which can be counted microscopically, as in the method of Jannasch (1958) for planktonic bacteria. Winter *et al.* (1971) describe a similar rapid membrane filtration method for counting food bacteria. The slide culture method of Postgate (1969) for determining bacterial viability is of this type, permitting estimation of the proportion of cells in a sample that are viable, although not their absolute numbers.

Although there is no universal culturing technique, it is very desirable to have available methods giving rapid and optimal counts of viable heterotrophic bacteria. A variety of media have been proposed or used for counting aerobic heterotrophs in activated sludge, such as nutrient agar (Allen, 1944), nutrient agar with skim-milk (Jasewicz and Porges, 1956), tryptone/glucose-extract/skim-milk agar (Adamse, 1968) or sewage agar (Dias and Bhat, 1964). Prakasam and Dondero (1967a, b) and Lighthart and Oglesby (1969) showed that none of various media tested would adequately support growth of all nutritional types of bacteria encountered. Pike *et al.* (1972) found that optimal counts of aerobic heterotrophic bacteria were obtained on casitone/glycerol/yeast extract agar plates (CGY agar, containing Difco casitone, 5 g; glycerol, 5 g; Oxoid yeast extract, 1 g; Oxoid agar No. 3, 13 g; water, 1 litre; pH 6·7 after sterilizing by autoclaving for 15 min at 121°C), inoculated by spreading and incubated for 6 days at 22°C.

The bacteria in sewage ecosystems are typically aggregated particularly in sludges and filter slimes, and for accurate and reproducible counts must be rendered completely and randomly dispersed before inoculation of plates or tubes of media. All viable counting methods require the bacteria to be present as randomly dispersed, individual cells. Various homogenization procedures have been recommended for dispersing activated sludge bacteria, such as forcing the suspension through a cream-maker (Allen, 1944), ultrasonic treatment with a transversely vibrating filament (Williams *et al.*, 1970) or a stretched wire (Williams *et al.*, 1971), or treatment with the Silverson homogenizer (Gayford and Richards, 1970). Similar degrees of deflocculation, measured by comparing the maximal increase in count obtained during treatment with that before, were

TABLE II. Key references to isolation and cultivation methods for different genera or physiological types of bacteria found in used-water ecosystems

Authors, including editors of collective works	Genera or types of bacteria considered (and authors of specific papers)
Skerman (1967)	Guide to identification of genera in Bergey's Manual (Breed *et al.*, 1957), also contains isolation methods
Collins (1963)	Methods for isolating heterotrophic and autotrophic bacteria in natural waters
Norris and Ribbons (1969)	Individual papers on isolation selection and growth of various micro-organisms, including autotrophs (V. G. Collins, pp. 1–52), phototrophic bacteria (N. G. Carr, pp. 53–77), psychrophiles and thermophiles (T. D. Brock and A. H. Rose, pp. 161–168), myxobacteria (J. E. Peterson, pp. 185–210), bacteriophages (E. Billing, pp. 315–329)
Aaronson (1970)	A compendium of isolation methods for many species from different habitats
Lochhead and Chase (1943)	Media for cultivation of bacteria of different nutritional requirements. Original media modified for activated-sludge bacteria by Dias and Bhat (1964) and Prakasam and Dondero (1967b)
Shapton and Gould (1969)	Individual papers on isolation of aerobic bacteria, including staphylococci (A.C. Baird-Parker, pp. 1–8; R. J. Gilbert, M. Kendall and B. C. Hobbs, pp. 9–15), salmonellae (J. H. McCoy and G. E. Spain, pp. 17–27; D. L. Georgala and M. Boothroyd, pp. 29–39), coli-aerogenes, *E. coli*, faecal streptococci, *Clostridium perfringens*, actinomycetes by membrane filtration (N. P. Burman, C. W. Oliver and J. K. Stevens, pp. 127–134)
Baumann (1968)	Isolation methods for *Acinetobacter* spp.
Unz and Farrah (1972)	Isolation medium for *Zoogloea ramigera*
Mülder (1964a)	Isolation and examination of *Arthrobacter*

TABLE II. (*Contd.*)

Authors, including editors of collective works	Genera or types of bacteria considered (and authors of specific papers)
Lechevalier (1964)	Isolation methods, illustrated key for actinomycetes
Starr and Seidler (1971); Starr and Huang (1972)	Both detail methods for isolating and maintaining *Bdellovibrio* spp.
Report (1969)	Official methods for bacteriological examination of water supplies in United Kingdom. Contains isolation and cultural methods for bacteria of public health significance
Scarpino (1971)	Review of methods for bacterial and viral analysis of water
Berg (1971); Shuval and Katznelson (1972)	Some references and procedures for concentrating and isolating viruses from natural and polluted waters

obtained by Pike *et al.* (1972) when activated sludges were treated by shaking with glass beads (Ballotini, Grade 11) or by ultrasonication, using either a high intensity probe or immersion of a tube of sample in an ultrasonic cleaning tank; the last mentioned was simplest and was least lethal in its effect on the bacteria. The deflocculation can be assisted by suspending the sludge during treatment in 5 mg/l sodium tripolyphosphate solution (Pike *et al.*, 1972), or in a solution of sodium pyrophosphate and Lubrol W (both 0·01 per cent, w/v) (Gayford and Richards, 1970). The procedure adopted by the author for sewage, activated sludge and effluents is to suspend for 1 min a boiling tube, containing sample (3 ml) and sodium tripolyphosphate diluent (27 ml), in an ultrasonic cleaning tank (Model KS100/101, 100 W, 45 kHz; Kerry Ultrasonics Ltd, Wilbury Way, Hitchin, Herts) filled with water. This treatment, applied for 2–3 min, is usually successful for deflocculating slimes from high-rate biological filters, but slimes from conventionally loaded filters are usually highly resistant and for these the author recommends shaking with glass beads in the Braun tissue homogenizer for 10–30 s.

C. Metabolic Methods

These can be used either to estimate the biomass of natural populations or to measure their activities against specific substrates, by measuring the uptake of the substrate itself, or associated changes such as oxygen consumption or dehydrogenase activity. There are inherent difficulties when mixed cultures as opposed to pure cultures, are being studied, in that the changes noted will be caused by the resultant behaviour of the individual members of the microbial community, which may interact. By adding a substrate, one is also altering the system. If the substrate is added in excess of limiting concentrations, the maximum uptake rate is dependent only on population size and can be used in relative measurements of biomass (El-Shazly and Hungate, 1965).

Some of the difficulties of studying mixed bacterial communities were described by Stumm-Zollinger (1968), who attempted to synthesize artificial populations from mixed laboratory cultures of bacteria from natural waters and activated sludge by growing them upon a complex non-selective medium and upon a medium containing mineral salts, glucose and galactose. These did not simulate natural populations adequately and the community structure of natural populations could not be maintained unaltered in the laboratory, because acclimatization to a substrate led to rapid enrichment of one or a few strains. Stumm-Zollinger partially overcame this by synthesizing her populations for metabolic experiments by mixing strains of bacteria, after acclimatization to the substrates used, in the proportions and kinds found in the population being simulated.

Difficulties of this kind can also be overcome by two-stage continuous culture in which individual strains are cultured separately in parallel vessels of the first stage and mixed, to synthesize an artificial community in the second stage. Such artificial communities include the four-member, photosynthetic "microcosm" of Taub (1971), comprising *Chlamydomonas reinhardtii, Escherichia coli, Pseudomonas fluorescens* and *Tetrahymena vorax*, and the two-stage systems used at the Water Pollution Research Laboratory (WPRL) to study growth of *Tetrahymena pyriformis* when feeding upon *Klebsiella aerogenes* (Curds and Cockburn, 1971; Department of the Environment, 1971), and the interacting growth of two phenol-degrading bacteria (Jones and Carrington, 1972).

A comprehensive review by Brock (1971) describes methods for measuring microbial growth rates and biomass in natural habitats. Direct microscopic methods can be used for biomass estimation, if the organisms are of reasonably constant size. Thus, the data (Pike and Carrington, 1972) of Table I suggest that the suspended solids of waste-treatment systems contain some $1–3\times10^{12}$/g of particles considered to be bacteria under the microscope. A total count of $1{\cdot}4\times10^{12}$/g of suspended solids would be given by rod-shaped cells of typical bacterial size, 1·5 μm long, 0·7 μm diameter, density 1·1 g/cm^3, and these figures are in general agreement with data presented by Strange (1972) for stationary-phase cells of *Aerobacter aerogenes* (NCTC 418), in which 1 g of cell mass contained either $2{\cdot}9 \times 10^{12}$ cells (mannitol-limited medium), or $4{\cdot}2 \times 10^{12}$ (tryptone–glucose medium). Difficulties arise when considering filamentous bacteria and large, aggregated flocs, perhaps with projecting filaments. Projected area methods are available (Brock, 1971; Bott and Brock, 1970) and have been used in studies of bulking activated sludge (Sladká and Zahrádka, 1970; Pipes, 1967, 1968), but of necessity either neglect the third, depth dimension, or attempt to estimate it from assumptions which are only truly valid for regular-shaped particles. A related difficulty applies in the estimation of the biomass of percolating filters, where not only the thickness but also the area of the film may be impossible to determine experimentally.

The difficulty of differentiating bacteria from inert debris can to some extent be overcome if selective staining or other optical techniques are available. In a study of soil biomass (Babiuk and Paul, 1970), fluorescein isothiocyanate staining and fluorescence microscopy was used to estimate bacterial numbers. In theory, it should be possible to use fluorescent enzyme substrates, such as fluorescein diacetate, to distinguish metabolizing and inactive cells by microscopy (Rotman and Papermaster, 1966), although neither this method nor the fluorescent dyes, acridine orange, phenol aniline blue and erythrosin, were able to differentiate soil bacteria from debris (Babiuk and Paul, 1970). Casida (1968) has advocated using Kodak Infra-Red Aerofilm type 8443A and conventional microscopy for selectively demonstrating and recording bacteria in material; the bacteria appear red against a false-colour background.

The metabolic activity of whole activated-sludge solids is much greater than would be supposed, considering the activities of pure culture isolates from the sludge and their numbers originally present in the sludge determined by viable counts (Department of the Environment, 1971). This arises from the low viability of the sludge bacteria and because the enzymes of the non-viable cells tend to retain their activity. In experiments at WPRL the specific uptake rates of glucose, stearate (as sole carbon source) and acetate by isolates from activated sludge were much greater than for the sludge from which they were isolated, and indicated that these metabolic types comprised respectively 8–13%, 14–28% and 5–10% of the total biomass.

Work at WPRL (Ministry of Technology, 1970) has indicated that the content of deoxyribonucleic acid (DNA) or the dehydrogenase activity towards triphenyl tetrazolium chloride (TTC) are not suitable for estimating the active biomasss of activated sludge, because the former may also be associated with dead cell material and because dehydrogenase activity, measured by production of formazan from TTC, did not correlate well with oxygen uptake. However, in one situation, a more recent study (Irgens, 1971) showed that DNA content of volatile sludge solids could be used to estimate biomass, represented by viable bacteria in activated sludge, Patterson *et al.* (1969) considered that the use of TTC to measure dehydrogenase activity presented difficulties in technique and in interpretation of results, because of its interference with normal cell metabolism, and recommended the use of other electron acceptors, or the measurement of oxygen uptake rates, which measure cellular activity directly.

Adenosine triphosphate (ATP) has been advocated as a measure of biomass because it appears to be non-conservative, associated only with living cells, and because its concentration appears to bear a reasonably constant ratio to organic carbon content in many cells. ATP is conveniently analysed by the firefly "cold light" reaction in which the specific reaction of ATP extracts with the luciferin and luciferase enzyme complex (extracted from firefly lanterns) results in emission of light energy quanta directly proportional to the amount of ATP added. The light evolved can be measured by liquid scintillation counters or more conveniently by a commercial "Luminescence Biometer" (Dupont Instruments Ltd, Wilbury Way, Hitchin, Herts) which permits rapid assays to be made (Anon, 1970). The method has been used for assay of bacterial contamination in foodstuffs (Sharpe *et al.*, 1970), for estimating bacterial numbers in oceanic waters (Holm-Hansen and Booth, 1966) and for viable biomass in activated sludge (Patterson *et al.*, 1969, 1970; Brezonik and Patterson, 1971). Although one might suppose that the relationship between ATP and biomass would be complicated by the heterogeneous populations in activated sludge, for example by protozoa and metazoan animals in addition to bacteria, Patterson *et al.* (1969, 1970) claim that ATP can be used for quantitative estimation of microbial biomass and as a specific indicator of viability, and that it is non-conservative. In further experiments upon experimental plants treating dried milk solids, Brezonik and Patterson (1971) found that ATP could be used to measure biomass under constant environmental conditions; the ATP content of volatile suspended solids was affected by sudden temperature changes, anaerobiosis, varying substrate loadings and by addition of toxic chemicals, and could therefore be used as a monitor of these conditions within a plant.

D. Taxonomic Methods

1. The Classification of Bacteria

There are three stages in taxonomy—classification, nomenclature and identification: firstly, unknown organisms are studied intensively and arranged into convenient groupings reflecting their mutual similarities or dissimilarities; then one may wish to recognize the groupings by giving them names; and finally one will wish to be able to identify new isolates. Before one can do this it is necessary to refer to the original data and select those characteristics which have special diagnostic value so that an identification key can be constructed.

The classifications of the macroscopic members of the plant and animal kingdoms, as opposed to micro-organisms, are traditionally based upon supposed natural relationships. Morphology and mode of sexual reproduction are key features and the schemes of classification are largely based upon subjective interpretation of evolutionary evidence, phylogeny and comparative anatomy. The resulting classifications are therefore hierarchical, resembling a "family tree" in structure. However, it should be realized that any classification is merely a convenient way of summarizing information in a mnemonic form about a collection of organisms, and that to be useful to the worker, it should also reflect the prime interest of the worker. It is not necessary for a classification to be based upon evolutionary principles, nor is it necessary for it to be hierarchical, and this is quite pertinent when the classification of bacteria is considered.

Early bacteriologists soon encountered difficulties when attempting to classify bacteria and this state of confusion still exists. There is no general agreement upon what differences suffice, for example, to delineate bacterial species. Because many bacteria appear to be morphologically simple and apparently reproduce by binary fission (although processes for exchange of genetic information between different cells do occur), their classification has been based upon other criteria than are used for higher organisms, for example physiology, biochemical abilities, antigenic analysis (which reflects differences in the chemical nature of the surface structures), cell wall composition and the ratio of guanine and cytosine to total bases ("GC ratio") in the intracellular DNA. In the face of many data for such fine distinctions, certain critical problems appear—how to decide which criteria are most important for splitting the isolates into taxonomic groups, how to define the relative order of rank of the groups (for example, species, genera, families, orders, classes, phyla) and what to do about aberrant isolates which do not fit comfortably into any group, or isolates which present a spectral range of characters and which cannot therefore be split into groups. There has thus been a growing school of opinion which would rely upon an objective approach ("numerical taxonomy") by using rigidly defined and formal statistical procedures to discover the relationships between isolates. The principles of numerical taxonomy (Sneath, 1957a, b, 1962; Sokal and Sneath, 1963) involve the study of as many characters of the isolates as possible; the data may be coded as simple binary ("present or absent") or ordinal scores, or as continuous numbers, and are then fed into a computer, which is instructed to calculate similarity or dissimilarity coefficients between all pairs of isolates; the isolates are then grouped by some defined method of clustering. These methods have the great advantage of eliminating personal bias and it may be considered that they are capable of producing true natural divisions. However, the results must reflect the quality of input; if the isolates are all more or less similar, or if the characters studied have no diagnostic value—which is unlikely if many are studied—this will be reflected in the output. The clustering procedure will to some extent distort the relationships, because the output is usually in the form of a two-dimensional dendrogram ("family tree") reduced from what is essentially represented by the data as a multi-dimensional distribution of the isolates as points in hyperspace. Clusters of like strains will fuse with other clusters at different levels of similarity, but any attempt to relate the levels at which they do fuse to, say, species or generic level must necessarily be arbitrary and artificial. Nevertheless the results of numerical taxonomy upon known strains of bacteria have been generally consistent with conventional classification and the methods are serving to reinforce existing taxonomic knowledge.

Readers who are interested in the problems and methods of bacterial classification are referred to the symposium edited by Ainsworth and Sneath (1962, pp. 456–463). Although any attempt to construct classification upon evolutionary principles must be speculative, the monograph by Bisset and Moore (1963) has the advantage of giving a concise, comprehensive and readable account of the entire class of bacteria to be found in nature. Bacteria of medical and public health significance are described in great detail in *Topley and Wilson's "Principles"* (Wilson and Miles, 1975). Various taxonomic studies have been made upon bacteria in waste treatment and these are discussed in the sections concerned with the various processes. Numerical taxonomy has been employed by Toerien (1970a, b) for describing populations in anerobic digesters and by Lighthart and Oglesby (1969) for organisms from raw sewage, activated sludge and secondary effluent. Many programs and algorithms for numerical taxonomy are now available; the author has used a suite of programs, CLUSTAN IA for IBM 360 and other computers (Wishart, 1969).

2. Bacterial Nomenclature

In general the nomenclature for bacteria follows that adopted for higher plants and animals as agreed in the *International Code of Nomenclature of Bacteria and Viruses* (1959), summarized in Ainsworth and Sneath (1962), and in the *International Code of Nomenclature of Bacteria*

(1966). The principles of bacterial nomenclature are described by R. E. Buchanan in *Bergey's Manual of Determinative Bacteriology* (Breed *et al.*, 1957, pp. 15–28). Adherence to the rule of observing the first validly published species name, and dissension over what constitutes valid publication in specific instances, has led to much confusion and the changing of well established names for those which are inappropriate or less well known. Problems of taxonomy are discussed in the *International Journal of Systematic Bacteriology*. There has been much controversy, for example over the validity and identity of *Zoogloea ramigera;* this species had been largely discarded by bacteriologists but was revived by Butterfield (1935) from Gram-negative, rod-shaped bacteria with a polar flagellum, isolated from sludge flocs and capable of flocculating in pure culture. The characters of zoogloeal growth and floc-formation have not been regarded as valid criteria for re-instating this species, particularly as many other bacteria can be induced to flocculate (McKinney and Horwood, 1952; McKinney and Weichlein, 1953; McKinney, 1956). However, Unz (1971) has recently proposed and described a new neotype strain.

Bergey's Manual (Breed *et al.*, 1957) is less a taxonomic or identification scheme for bacteria than an extremely comprehensive list of bacteria which are deemed to have been adequately described in the literature, together with their characteristics. The latest (8th) edition has appeared since the time of writing.

3. Bacterial Identification

The identification of unknown strains isolated from natural material is a major concern of a practical microbiologist and one that presents difficulties, if only because the preceding branches of taxonomy are deficient. Identification presupposes that there are reliable diagnostic criteria available for differentiating between isolates. Because bacterial species cannot be rigidly defined, largely because of aberrant forms and intergrading, quite apart from errors in the application of the criteria, there are no absolutely reliable diagnostic criteria. It is, moreover, the case that workers or laboratories involved in studying particular groups of bacteria have evolved their own identification methods by experience. Diagnostic methods specifically for bacteria encountered in waste treatment are urgently needed.

The monograph by Skerman (1967) is a valuable diagnostic key and compendium of methods for isolating and identifying the genera described in *Bergey's Manual* (Breed *et al.*, 1957). The volumes edited by Gibbs and Skinner (1966) and Gibbs and Shapton (1968) are collections of papers describing methods adopted at different laboratories for identifying various groups of bacteria. Of these, the key of Hendrie and

Shewan (1962) was found a valuable aid by Benedict and Carlson (1971) for preliminary assignment of isolates to genera in their study of aerobic bacteria in activated sludge. Farquhar and Boyle (1971a, b) describe simple microscopic and histological tests and provide a key for identifying filamentous bacteria in activated sludge. Cowan and Steel (1966) provide a manual with test methods and simple keys for identifying bacteria of medical importance.

E. The Study of Microbial Populations

It will be seen from Section III of this chapter that the bacterial communities in sewage treatment are complex and that a variety of genera are present in addition to the dominant members, which are usually heterotrophic Gram-negative rod-shaped bacteria with polar flagella. It is also obvious that conditions of plant operation, climatic factors and the nature of the used water will affect the composition of the microbial community and that, because bacteria can display relatively rapid adaptation to new substrates and are potentially capable of rapid multiplication, compared with higher forms of life, such changes in the community will closely follow and reflect these applied stimuli. The major difficulties in studying imposed changes in bacterial communities lie firstly in being able to collect a large amount of information about the numbers or active mass of the populations likely to be affected. It may not be known at the start of an experiment which populations will be affected, and this implies that some discretion should be exercised in the choice of populations and that one should choose as many as can be conveniently studied. The data will be voluminous, and the second problem lies in being able to summarize the information so as to identify and retain the main trends, while discarding those parts of the information lacking significance or arising from random experimental errors.

The classical method of describing bacterial communities is by isolation and identification of a sufficient number of randomly selected isolates. This is extremely time-consuming, the number of samples that can be examined is small and there is a very real danger of ignoring minority species, which may nevertheless have value as indicators of imposed stimuli. The results can be only qualitative or at best partly quantitative. The labour of screening many isolates can be lightened by using replica-plating, as in the studies of Prakasam and Dondero (1967b, c, 1970) and Lighthart and Oglesby (1969), or multi-point inoculator methods (Hill, 1970; Goodfellow and Gray, 1966, Watt *et al.*, 1966) for inoculating the different test media used for identification, but this is still a three-stage procedure, in that the strains must first be isolated, a master culture prepared and the test media inoculated. Pike and Carrington (1972; Department of the Environment, 1972) have adopted a more direct, single-step approach in which a variety of diagnostic and selective media are used to count about 30 different physiological types simultaneously for each sample. Enumeration is by a modification of the plate-dilution frequency method of Harris and Sommers (1968); five drops (0·02 ml) of homogenized sample, at each of four dilutions in descending decimal series, are used to inoculate either two Petri dishes of solid medium, or the respective four sets of five wells of a divided Petri dish (Repli-Dish, Dyos Plastics Ltd, Surbiton, Surrey) if a liquid medium is used. Different media are used for enumerating different populations. The count can be estimated by referring the pattern of positive reactions, after incubation, to standard "most probable number" (MPN) tables. McCrady's tables for enumerating faecal indicator bacteria in water (Cruickshank *et al.*, 1965, pp. 963–966; Report, 1969), which are designed to be used for counting using inoculation of five aliquots each of 10, 1 and 0·1 ml, may easily be adapted for the different volumes inoculated in the plate-dilution frequency method. In the original method of Harris and Sommers (1968) tables are given for inoculation with eight drops of diluted sample at each of six dilutions in descending decimal or quadrupling series.

Quantitative and semi-quantitative (i.e. at least in a binary, presence–absence form) data are suitable for analysis on the computer by multivariate statistical methods such as correlation analysis (Thiel *et al.*, 1968; Department of the Environment, 1971; Pike and Carrington, 1972), cluster analysis (Lighthart and Oglesby, 1969; Sokal and Sneath, 1963; Prakasam and Dondero, 1970; Curtis and Curds, 1971) and principal component analysis (PCA) (Toerien *et al.*, 1969; Sundman, 1970; Sundman and Gyllenberg, 1967; Pike and Carrington, 1972; Department of the Environment, 1973). All these methods are suitable for giving a less complex description of an ecosystem reacting to an applied stress than in the original data by providing a formalized statistical approach to the search for pattern and by retaining only the most significant information. The methods are indeed also suitable for numerical taxonomy, except that in this context, samples of a community are compared and not individual isolates.

Although there are various practical problems in the field of sewage treatment and pollution control which are of a microbiological nature and which could benefit from ecology study, there have so far been few applications of the above methods for studying such ecosystems. Noteworthy is work at the National Institute for Water Research, Pretoria on description of isolates in the non-methanogenic phase of anaerobic digestion by cluster analysis (Toerien, 1970a) and PCA

(Toerien, 1970b), and on the effects of progressive nitrogen limitation on chemical and biological characteristics of anaerobic digesters by correlation analysis (Thiel *et al.*, 1968) and PCA (Toerien *et al.*, 1969); the use of cluster analysis by Prakasam and Dondero (1970) to show differences in populations of 11 laboratory activated sludges acclimatized to different aromatic compounds, and by Curtis and Curds (1971) in investigation of "sewage fungus" communities in polluted rivers; and current work at WPRL (Pike and Carrington, 1972; Department of the Environment, 1972, 1973).

Taxometric methods for describing bacterial populations in activated sludge have been used by Prakasam and Dondero (1970), Lighthart and Oglesby (1969) and by the author (Department of the Environment, 1972, 1973; Pike and Carrington, 1972). Prakasam and Dondero (1970) characterized 11 sludges, each one acclimatized to a different aromatic compound and each defined by its content of various bacterial types with different degradation patterns towards the 11 aromatic compounds. The two methods used, a confidence interval method and Mountford's similarity index, gave somewhat different results but indicated that the populations of the sludges were quite dissimilar. The activity of seven of the sludges correlated well with the population structure.

Cluster analysis and principal component analysis have been used at WPRL to assist in the description and comparison of populations in different stages of sewage treatment (Department of the Environment, 1973), in sewage undergoing aeration for different mean retention times in a system of six chemostats in series (Department of the Environment, 1972), and in two high-rate activated sludge plants operated at different specific rates of wastage of sludge solids (Department of the Environment, 1973; Pike and Carrington, 1972).

III. The Microbial Ecology of Sewage Treatment Processes

A. Crude and Settled Sewage

1. *The Bacterial Flora of Faeces and Sewage*

Human excreta comprise about 135 g dry weight of faeces and 1400 ml urine/day (Elliot and Rowe, 1971). In adult faeces, the predominant bacteria are the obligate anaerobes, *Bacteroides* spp., present in numbers of about 10^{10}/g wet material; coli-aerogenes, streptocci and lactobacilli are regularly present at levels of 10^{6}–10^{8}/g. while *Veillonella* spp., *Streptococcus salivarius, Bacteroides melanogenicus* and staphylococci are present in lower numbers. Anaerobes outnumber aerobes by a factor of 40 and total bacteria by microscopy constitute about 30% of the biomass (van Houte and Gibbons, 1966). The following geometric mean counts/g (dry weight) for adult faeces were obtained by Mata *et al.* (1969) in a study of the faecal microflora in healthy persons in a Guatemalan highland village: coli-aerogenes bacteria, 4×10^8; enterococci, 8×10^7; anaerobic lactobacilli, 10^9; *Bifidobacterium* spp., $2{\cdot}5 \times 10^9$, micro-aerophilic lactobacilli, 4×10^8. The ranges of the counts for individual groups of bacteria covered some two to three orders of magnitude. The coliform contents of faeces from farm animals and poultry and their daily *per capita* loads have been summarized by Geldreich *et al.* (1962).

Normal urine, as voided, contains few bacteria and the presence of large numbers ($>10^5$/ml) of bacteria (bacteruria) indicates infection of the urinary tract. In the study of Elliot and Rowe (1971), 63 (77 per cent) of the 82 strains, isolated from 209 urine samples, were Gram-negative bacteria, including *Escherichia coli* (18 strains), *Aerobacter aerogenes* (10), *Proteus* spp. (8), *Pseudomonas aeruginosa* (17). The 19 (23%) Gram-positive strains were coagulase-negative staphylococci.

The bacterial content of sewage is subject to wide fluctuations with time. Much of the variability is caused by daily fluctuations in patterns of excretion and water use and by incursions of storm water in combined sewerage systems. Long-term seasonal changes may also occur, but are comparatively small in comparison with diurnal fluctuations, in which the count of coli-aerogenes bacteria, generally of the mean order of 10^5/ml in sewage, may fluctuate over nearly two orders of magnitude. Allen *et al.* (1949) recorded that numbers of total viable bacteria growing at 20°C, coli-aerogenes bacteria and *Streptococcus faecalis*, in settled sewage at two works were higher in the warmer than in the colder months of the year, although the extent of these changes was of about the same magnitude as short-term changes lasting a few weeks. In the case of settled domestic sewage from a population of 1700, average counts of total viable bacteria at 20°C, coli-aerogenes, *Escherichia coli* and *S. faecalis*, for April–October were respectively 2·6, 2·1, 1·4 and 1·8 times the respective values for the preceding November–April. The authors considered that the changes could be accounted for by multiplication of the bacteria in the sedimentation tanks during the warmer months of the year. Tomlinson *et al.* (1962) recorded larger changes in settled domestic sewage from Stevenage, Herts. Total viable bacteria growing at 30°C fluctuated over a range of about six-fold, with peak counts in March–April, June–July and October–November, whereas coli-aerogenes, initially 10^5/ml in January, rose to a maximum of 3×10^6/ml in May and thereafter fell to 4×10^5/ml in December. Small changes occurred over the summer months in the daily loads of coli-aerogenes bacteria discharged in crude sewage to a sea outfall at Sidmouth, Devon (Site C) (Gameson *et al.*, 1967); in early September the average daily loads were about three times those in May. As this is a typical medium-size seaside resort (permanent population 11 000) it is possible that a certain proportion of the change, but not all, could be attributed to the influx of tourists. The data given by Harkness (1966) in his Table I of monthly counts of coli-aerogenes, *E. coli* and total viable bacteria (37° and 22°C counts) in crude sewages at two works in Birmingham, show considerable increases in August and September, and here, there is a suggestion either of additional bacterial multiplication in the sewers, or of an increase in the human rate of excretion of faecal bacteria, because of the general efflux of holidaymakers from this region during these months, or, more likely, it is due to the absence of large volumes of industrial wastes.

2. *Pathogenic Organisms in Faeces and Sewage*

A wide variety of enteric pathogens can be detected in sewage. As the sensitivity and reliability of the methods used for detection are improved, so the chances of successful isolation or the number recovered will be increased. The main significance of finding an enteric pathogen in the used water of a community is the demonstration that a proportion of the community is infected, between the extremes of displaying acute, active infection and the low-grade, symptomless carrier state. Pathogenic organisms, particularly bacteria of the Salmonella group associated with food-poisoning, can originate from abattoirs, poultry-processing wastes, effluents from intensive animal and poultry-rearing units, and from rodents in the sewers. Large numbers of poliovirus particles can be demonstrated in sewage following successful mass vaccination of a community with live, attenuated vaccine. Children are commonly found to be symptomless excreters of poliovirus and other enteric viruses; the prevalence of poliovirus excreters varies with such factors as standard of living, season of the year and climate, being highest in the summer and autumn in temperate climates (Kollins, 1966).

The range of pathogens which can be encountered in sewage has been considered generally by Pike and Gameson (1970) and Hawkes (1971); the reader is referred to these for further details. Those which are common in temperate climates include bacteria of the genera *Salmonella* (typhoid, paratyphoid, food poisoning), *Shigella* (bacterial dysentery), *Mycobacterium* (tuberculosis, Johne's disease of cattle) and *Leptospira icterohaemorrhagiae* (Weil's disease, leptospiral jaundice). The greater part of the viruses which can be detected in used waters are of human origin and particularly comprise those excreted in large numbers in faeces, such as poliovirus, Coxsackievirus groups A and B, Echovirus, infectious hepatitis virus, reovirus and adenovirus (Grabow, 1968). There has been insufficient evidence to prove that poliomyelitis is water-borne in origin, and Mosley (1967) has concluded that this disease should be regarded as one in which water transmission is of rare occurrence, at least in epidemic outbreaks. The only virus disease for which the water route has been proven epidemiologically is infectious hepatitis, as was demonstrated by the mass epidemic at Delhi in 1955–1956, following from flooding of the water intake with sewage–contaminated river water and a breakdown of the chlorination procedure (Dennis, 1959; Kollins, 1966). Water-borne parasitic animal species include *Entamoeba histolytica* (amoebic dysentery), *Schistosoma* spp. (Bilharzia, swimmer's itch), *Taenia* spp. (tapeworms), *Ascaris* spp. and *Enterobius* spp. (roundworms) and the soil amoebae *Hartmanella-Acanthamoeba* spp. and *Naegleria* spp., which have been recently incriminated as the cause of occasional outbreaks of meningitis in swimmers (Culbertson, 1971; Newton, 1972).

The relationship of infection to personal contact with, or ingestion of, pathogenic organisms in a polluted environment is extremely complex, reflecting the fact that the state of health of an individual is the resultant of two opposing processes—the innate and acquired immunity of the individual and the invasiveness and toxigenicity of the pathogen. The relationships of faecal pollution of waters to public health have recently been considered generally by Moore (1971), and in relation to pollution of coastal and recreational waters by Moore (1970a, b) and Pike and Gameson (1970). Factors affecting the dissemination of faecal bacteria and pathogens in aquatic environments have been discussed by Gameson *et al.* (1970) (coli-aerogenes bacteria in coastal waters), McCoy (1971) (*Salmonella* spp. in estuarial waters) and Hawkes (1971). The special public health problems presented by the dissemination of enteroviruses in polluted waters have been considered in a symposium book (Berg, 1967), in a review by Grabow (1968) and by the Committee on Environmental Quality Management of the Sanitary Engineering Division, American Society of Civil Engineers (Committee, 1970).

Strains of *Escherichia coli* which are resistant to antibiotics have become common, probably as a result of the widespread use of antibiotics in therapy and of the medication of the foodstuffs fed to intensively reared farm animals and poultry. Many of these antibiotic-resistant strains have also been shown to possess R-factors enabling them to transfer, *in vivo* or *in vitro*, their resistance to other enterobacteria including *Salmonella* spp. As yet, the epidemiological and ecological implications have not been evaluated fully. Amongst faeces of healthy adults and children under five years, none of whom were receiving antibiotics, resistant strains of coli-aerogenes bacteria were carried by a higher proportion of the children (67%) than adults (46%), and, among rural adults, by 63% of those in close contact with farm animals compared with 29% of those in other occupations. Transmissible R-factors were found in 61% of resistant strains (Linton *et al.*, 1972). Numbers of antibiotic-resistant *E. coli* in rivers have been related to the degree of pollution by sewage (Williams Smith, 1970); a high proportion of isolates from rivers and coastal waters which were resistant to various antibiotics were able to transmit resistance to *Salmonella typhi* or *Salmonella typhimurium* (Williams Smith, 1970, 1971). The claim by Williams Smith (1971) that R^+ *E. coli* found in sea water provided evidence of contamination by human sewage has been contested by Regnier and Park (1972) on the grounds that antibiotic-resistant forms constitute a roughly constant proportion of coliform bacteria, from whatever source they were found by Williams Smith (1970, 1971).

B. The Activated-sludge Process

1. General Outline of the Process

The activated-sludge process is probably the earliest industrial application of the continuous fermenter with feedback, having been first described by Arden and Lockett (1914) following experiments with Manchester sewage (see also Lockett, 1954a, b). The feedback of most of the cell yield from the sedimentation tank encourages rapid adsorption and oxidation of pollutants in the incoming sewages and also serves to stabilize operation over the wide range of dilution rates and substrate concentrations imposed by diurnal and other fluctuations in the flow and strength of the sewage. Stability is also provided by the continuous inoculation of the aeration tank with micro-organisms in the sewage and air flows, which are ultimately derived from human and animal excreta, soil run-off, water and dust. The correct operation of the process requires that the concentrations of substrate and cells should be low, in order that the required quality of effluent may be maintained, and that the yield of cells removed from the system as waste sludge—which presents disposal problems—should also be minimal. Thus, dilution rates, specific growth rates of the sludge micro-organisms and nutrient levels are also low in comparison with those of other fermentation applications. Operating conditions also aim to produce compact, flocculant growth, able to settle rapidly. Under certain conditions, growth of filamentous micro-organisms produces a "bulking sludge" condition. In addition, if aeration is continued well into the nitrification stage and is followed by retention in the settling tank long enough for denitrification to occur, sludge flocs rise to the surface of the aeration tank, carried by entrained nitrogen gas bubbles.

In the conventional process, as used in Britain, treating wastes with a 5-day biochemical oxygen demand (BOD) of about 250 mg/l, the mixed liquor-suspended solids content—an approximate measure of the biomass—is limited to about 6000 mg/l or less by the ability of the settling stage to concentrate and return sludge. In theory, it would be possible to increase the efficiency of activated-sludge treatment, or to treat stronger wastes within a given volume or retention time, were it possible to increase the concentration of cells returned (Pirt and Kurowski, 1970; Hardt *et al.*, 1970). Strong wastes, as from the dairy, food and fermentation industries or farm slurries, can be given initial treatment by so-called "dispersed growth" processes typified by the Pasveer oxidation ditch, in which the sludge is largely retained within the system and is not fed back (Scheltinga, 1969).

At the present time, there is much interest in the use of activated-sludge treatment, particularly for larger installations, because of the greater treatment capacity provided for a given size of plant than with alternative secondary treatments such as biological filtration. This is a direct result of the feedback mechanism. In its turn, this has given rise to the recent great interest in the kinetics of nutrient uptake, the biochemistry of bacteria involved in the process, and the organisms and conditions giving rise to flocculation and sludge bulking. Apart from these aspects, the ecology of the bacteria remains largely unexplored.

2. *Selective Pressures*

The activated-sludge community is specialized and shows a lower diversity than that of a percolating filter. The dominant members are heterotrophic bacteria, present in the sludge flocs and dispersed in the liquor. These, with saprobic protozoa, form the basic trophic level. The ecology of the process has been discussed by Hawkes (1963), Pipes (1966) and Pike and Curds (1971).

Selective pressures in the process arise from the intrinsic nature of the reactor, its operation and the nature of the sewage.

In practice, the conditions in the aeration unit diverge from those of the theoretical completely-mixed reactor with feedback (Herbert, 1961) and tend towards those of the tubular, piston-flow reactor, particularly in plants where the flow is through long, deep aeration channels, or through a series of individual aeration pockets. Oxygen uptake is most rapid in the initial stages of aeration, and causes a gradation in dissolved oxygen concentration along aeration channels, which can be eliminated by "tapered aeration" or "step feeding" designs.

Under steady conditions, the growth rate of sludge micro-organisms is equivalent to the specific sludge wastage rate (biomass removed in unit time/total biomass in the system) (Curds, 1971). High specific sludge wastage rates and dilution rates in the aerator encourage high growth rates, suppress higher trophic levels and select for faster-growing bacteria, although nutrient removal is less complete. Slow-growing bacteria such as *Nitrosomonas* spp. cannot become established in the system unless they receive a minimum effective period of aeration which is roughly equal to the fractional increase in the concentration of activated sludge during aeration divided by the growth rate constant of the organism in the mixed liquor (Downing *et al.*, 1964). Similarly, because free-swimming bacteria are constantly removed by protozoal grazing and by discharge in the effluent, they will be required to grow faster than flocculating bacteria to remain established. Thus, the process inherently selects for flocculating bacteria.

In plants in which piston-flow conditions are approximated it is unlikely that relative changes in numbers of different micro-organisms could be detected, because the specific growth rates of the sludge microflora are longer than the sewage dilution rates. However, in the case of multi-stage activated-sludge plants, each stage with its own settling tank and return line, the computer simulation by Curds (1971) suggests that the major changes would occur in the initial stages. In his simulation for five completely-mixed reactors in series, most of the substrate had been removed in the first and second stages and concentrations of flocculating sludge bacteria did not increase appreciably in subsequent stages. In a reactor with attached ciliates, most of the dispersed bacteria were removed by the first two stages, but in a reactor with ciliates attached to sludge flocs, these were actively consuming dispersed bacteria even in the fifth stage.

Selection will also be caused by dissolved oxygen concentrations, pH value, excess or deficiency of certain nutrients (discussed later in connection with sludge bulking) and by presence of toxic wastes. The short- and long-term effects of various inhibitors upon the rate of oxidation of ammonia by *Nitrosomonas* spp. has been studied by Tomlinson *et al.* (1966) in experimental fill-and-draw activated-sludge plants. Short-term effects in the vessels appeared to be similar to those reported for pure cultures of *Nitrosomonas;* the most powerful inhibitors were homologues of thiourea, thio-acetamide and dithio-oxamide, together with cyanide ion, which, unlike the former compounds, had little effect upon activity of *Nitrobacter* spp. in activated sludge. Copper, mercury and chromium affected *Nitrosomonas* very much less in activated sludge than in pure culture. Immediate, or short-term effects were not a good guide in general to long-term effects, since there was evidence of acclimatization of the plants to the powerful organic inhibitors, particularly to thiourea, whereas the long-term effects of heavy metals appeared to be more severe than expected from short-term experiments. More recent experiments (Department of the Environment, 1971, p. 117) with small activated-sludge plants showed that nickel, hexavalent chromium and a mixture of these with copper and zinc each adversely affected the chemical oxygen demand of the settled effluent when added to sewage at 5 mg/l concentration; when the mixed liquor-suspended solids concentration was reduced from 4000 to 2000 mg/l, nitrification was almost completely inhibited.

There is apparently no information on the effect of seasonal factors upon the microbial flora of activated-sludge plants, although plants which produce partial nitrification during warmer months of the year may cease to nitrify altogether in the winter. This phenomenon is a direct consequence of the effect of temperature upon the maximum specific growth rate of the *Nitrosomonas* spp. in the plant, possibly aggravated slightly by an increased rate of sludge production (Downing *et al.*, 1964).

3. *Study of Bacterial Populations in Activated Sludge*

(*a*) *Culture and enumeration.* Various technical difficulties impede the study of bacterial populations and were the subject of a study by Pike *et al.* (1972). Before viable or total microscopic counts can be made, mixed liquor samples must be homogenized to release individual bacteria from the flocs with minimum lethal effect. Various combinations of homogenization methods and suspending medium have been advocated (Allen, 1944; Gayford and Richards, 1970; Pike *et al.*, 1972; Williams *et al.*, 1970; Williams *et al.*, 1971) and these generally bring about increases in count of up to two orders of magnitude over counts obtained before treatment.

Work at WPRL (Table I; Pike and Curds, 1971; Pike and Carrington, 1972) and by Sladká and Zahrádka (1970, p. 45) shows great discrepancy between total microscopic counts of bacteria in mixed liquors and viable counts on the same material. The discrepancy and the apparent viability (viable count/total microscopic count) decreases as the rate of loading of the sludge and as the specific sludge wastage rate increases (Pike and Carrington, 1972). This, coupled with the observation that greatly decreased viability occurred in continuous cultures of *Aerobacter aerogenes* at very low dilution rates (Tempest *et al.*, 1967), lead one to suppose that many of the bacteria in mixed liquor may be moribund or dead, since the process is typically operated at the stationary or decline phases of growth under low nutrient conditions.

A variety of media have been used in taxonomic or population studies, such as nutrient agar (Allen, 1944), nutrient agar with skim-milk (Jasewicz and Porges; 1956), tryptone/glucose-extract/skim-milk agar (Adamse, 1968) and sewage agar (Dias and Bhat, 1964), although the experiments of Prakasam and Dondero (1967a, b) and Lighthart and Oglesby (1969) showed that of the variety of media tested, none could support growth of all types of bacteria encountered. Where much routine examination of mixed liquors is envisaged, there is a need for a general-purpose medium giving maximal counts within a reasonable incubation period; Pike *et al.* (1972) recommended use of CGY agar, inoculated by the spread plating or plate dilution-frequency (Harris and Sommers, 1968) methods and incubated for 6 days at 22°C.

Various workers have attempted to classify isolates by their nutritional requirements, using the methods of Lochhead and Chase (1943) for soil bacteria. Dias and Bhat (1964) found that isolates requiring neither amino acids nor vitamins for growth in a medium containing glycerol, succinate and NH_4NO_3, comprised only 24% of 110 from raw sewage and 8% of 150 from activated sludge. Fifty-five per cent of 71 *Arthrobacter* strains isolated by Adamse (1970) from a dairy waste plant needed added vitamins for growth with inorganic nitrogen and only 17% for growth with amino acids. Activated-sludge extract agar was found by Prakasam and Dondero (1967a, b) to give counts generally higher than those on the other media tested; the efficacy of an extract depended upon its source and that of the sample tested. Approximately half of the 127 isolates from this medium failed to grow on any of their defined media containing glucose, amino acids, vitamins and mineral salts. On the other hand, Pike *et al.* (1972) found that counts of activated-sludge bacteria on unsupplemented CGY agar were higher than on CGY agar in which 10%–100% of the water in the formulation had been replaced by soil extract, by liquors from settled sewage or activated sludge or by filter-sterilized mixed liquor.

(*b*) *Taxonomic studies of activated-sludge micro-organisms.* The majority of bacteria described in activated sludge belong to Gram-negative genera (Table III). This table may be compared with the Table I of Pipes (1966); a cursory examination of the latter could suggest that the principal genera are *Achromobacter, Alcaligenes, Bacillus, Flavobacterium, Micrococcus* and *Pseudomonas.* However, it is clear that the authors detailed in Table III used a variety of media and methods for isolation, most of them selective for the more fastidious heterotrophic bacteria, or even for flocculating or zoogloeal bacteria alone (Butterfield, 1935; van Gils, 1964; Unz and Dondero, 1970) and that many of the plants examined were not typical of conventional sewage works (Table IV).

TABLE III. The principal genera which have been recorded in taxonomic and other studies of activated-sludge bacteria. Adapted from Pike and Curds (1971) and Pike and Carrington (1972)

Genus[a]	References[b]
Pseudomonas	Allen (1944), Jasewicz and Porges (1956), van Gils (1964), Adamse (1968), Lighthart and Oglesby (1969), Tezuka (1969), Unz and Dondero (1967, 1970), Ministry of Technology (1968), Benedict and Carlson (1971)
Comamonas	Dias and Bhat (1964)
Lophomonas	van Gils (1964)
Nitrosomonas	Loveless and Painter (1968)
Zoogloea	Butterfield (1935), Dias and Bhat (1964), Tezuka (1969)
Sphaerotilus	Harkness (1966), Austin and Forster (1969), Farquhar and Boyle (1971a, b)
Large filamentous micro-organisms	Cyrus and Sladká (1970), Sladká and Zahrádka (1970), Farquhar and Boyle (1971a, b)
Azotobacter	Dias and Bhat (1965)
Chromobacterium	Allen (1944)
Achromobacter	Allen (1944), van Gils (1964), Austin and Forster (1969), Lighthart and Oglesby (1969)
Flavobacterium	Adamse (1968), Austin and Forster (1969), van Gils (1964), Jasewicz and Porges (1956), Lighthart and Oglesby (1969), Tezuka (1969), Ministry of Technology (1968), Unz and Dondero (1970), Benedict and Carlson (1971)
Coli-aerogenes bacteria	Dias and Bhat (1965), Austin and Forster (1969)
Alcaligenes	Dias and Bhat (1965)
Micrococcus, *Staphylococcus*	Allen (1944), Jasewicz and Porges (1956)
Bacillus	Jasewicz and Porges (1956), Benedict and Carlson (1971)

[a] Names of organisms as used by authors in reference list.

[b] Further references are given by Pipes (1966).

(*Continued overleaf*)

Table III (*Contd.*)

Genus[a]	References[b]
Arthrobacter, coryneform bacteria	Dias and Bhat (1965), van Gils (1964), Adamse (1968, 1970)
Nocardia, *Mycobacterium*	Anderson (1968)
Bacteriophage, *Bdellovibrio*	Dias and Bhat (1965)

4. *Flocculation of Activated Sludge*

(*a*) *Cytology and floc structure.* Careful cytological observation (Fig. 1) shows that the flocs are generally composed of aggregations of small rod-shaped bacteria, surrounded by polymeric material, so giving the classical zoogloeal appearance (Butterfield, 1935). This appearance is not readily visible by phase-contrast microscopy but can be revealed by suitable vital staining techniques in which mixed liquor suspensions are mixed with a small quantity of stain and observed as wet-mounted preparations under the 2 mm oil-immersion objective. The author recommends Unna's borax methylene blue (G. T. Gurr Division, Searle Laboratory Services, High Wycombe, Bucks), which stains differentially the matrix material and bacterial cytoplasm, and lactophenol-cotton blue. These stains are often taken up at different rates by the bacteria and sometimes reveal that the flocs have a definite micro-colonial structure and are not an amorphous aggregation of bacterial cells. The cell wall structure of the bacteria can be demonstrated by staining techniques, specific for cell wall material. The tannic acid–crystal violet method (Bisset, 1955) appears to give better results with activated-sludge flocs than the method of Hale (1953). The matrix material itself can be stained with 0·5% (w/v) alcian blue, applied to smears for 30 min; the matrix is probably true capsular material and not soluble slime, since it can be detected in relief by mounting specimens in Indian ink (Cruickshank *et al.*, 1965, p. 658) or in 10% (v/v) skim-milk (Dondero, 1963; Unz and Dondero, 1970). It is possible that further information on the nature of the matrix material and its relation to cellular structure could be obtained by applying the specific capsular reaction (Tomcsik and Guex-Holzer, 1954a, b), although this remains to be tried.

Polymeric material can be extracted from activated-sludge solids by boiling or autoclaving with water; the gel extracted can be precipitated with acetone or ethanol (Department of the Environment, 1972; Wallen and Davis, 1972). It is not clear whether the gel material represents the zoogloeal matrix of the sludge flocs in which the cells of bacteria are embedded, or whether additional polymer is extracted from the cells themselves. In the former study, the acetone-precipitated fraction was shown to contain (in descending order of importance) glucose, galactose, fucose, mannose and arabinose; some sludges also yielded rhamnose, ribose and galacturonic acid. In the latter study, acid hydrolysates of the extracted material contained glucose, mannose, rhamnose, galactose hexosamine, maltose and lactose. Gels extracted from pure cultures of *Zoogloea ramigera* and other sludge isolates contained all or some of the sugars found in activated sludge. The polymer extracted from activated sludge by Forster (1971) was similar in composition to those above; at low sludge volume indices, the main ionogenic material of the sludge surface was glucuronic acid.

Recent work at Wageningen (Mülder *et al.*, 1971) has shown that the individual cells of flocculating strains of bacteria possess fibrils, visible in electron micrographs, which connect the cells in the flocs. The fibrils appeared to be cellulosic in nature and treatment of the flocs with cellulase caused their dispersion. Electron micrographs of activated sludge revealed a certain number of cells with fibrils, but cellulase treatment had only a slight effect upon floc structure.

Extracellular polymeric fibrils have also been observed by Marshall *et al.* (1971) in a halophilic marine pseudomonad, and were thought to be involved in the sorption of this organism to glass surfaces and to the formvar films on the grids used in the electron microscopy. In some instances bacteria were noted to have been removed by shearing forces, leaving attached polymeric material behind on the formvar surface as "footprints".

(*b*) *Mechanisms of flocculation.* Although substrate removal in aerobic processes can proceed swiftly and efficiently without flocculation, as in dispersed growth systems, flocculation is essential in the activated-sludge process, as currently operated, for bringing about efficient concentration of the sludge biomass in the settling tank and for ensuring a high quality effluent of low suspended-solids content. However, the gravitational settling characteristics of activated sludge limit the amount of concentration received by the sludge biomass, and so limit the biomass carried in the aeration tank and the rate of substrate removal. Unless alternative methods of biomass recovery, concentration and return, such as continuous filtration (Pirt and Kurowski, 1970) or centrifugation, can be devised, with equal or greater reliability and lower cost of operation than

TABLE IV. The nature of the activated-sludge plants and the isolation methods used by the authors of Table III in their taxonomic studies

Authors[a]	Details of plants	Methods of isolation
Adamse (1968, 1970)	(a) Laboratory plant at 25°C treating synthetic dairy waste, (b) oxidation ditch treating dairy waste	Desaga homogenization, plating on Oxoid glucose extract agar, incubation 3–4 days at 25°C
Allen (1944)	(a) Large sewage works, (b) laboratory fill-and-draw plant treating domestic sewage	Homogenization in cream-maker. Isolation from (a) on nutrient and caseinate agar, (b) nutrient agar or synthetic sewage agar. Incubation at 20°C for at least 10 days
Anderson (1968)	Normal sludge	Not given
Austin and Forster (1969)	Lubeck pilot plant treating 0·5–0·7 l/sec domestic and hospital waste	Hand homogenization. Isolation by "serial dilution subculturing techniques" Incubation at 30°C?
Benedict and Carlson (1971)	Domestic treatment plant and laboratory-scale unit treating synthetic sewage	Plating on two non-selective media, identification of 129 isolates
Butterfield (1935)	Source not specified, "normal activated sludge"	Picking of zoogloeal flocs, washing and serial dilution in standard lactose broth incubated at 22° and 37°C
Cyrus and Sladká (1970)	Municipal, industrial, laboratory plants	Microscopic descriptions
Dias and Bhat (1964)	(a) Laboratory plant, (b) pilot fill-and-draw plant, (c) aeration of sewage on rotary shaker	Dispersal on rotary shaker, isolation on sewage agar, incubated 10–15 days at 16–27°C, subculture on Difco PPYE broth
Dias and Bhat (1965)	(a) Laboratory plant, (b) aeration of sewage on rotary shaker	Selective technique used for coli-aerogenes bacteria, bacteriophages, *Bdellovibrio* and nitrogen-fixing bacteria

Farquhar and Boyle (1971a, b)	16 samples from full-scale works (13) and laboratory units (3)	Direct microscopic identification by morphological and histochemical tests, identification key given
van Gils (1964)	(a) Sewage works, (b) laboratory units at 20°C, continuous or fill-and-draw, fed domestic or synthetic sewage	Settled sludge, washed, resuspended in saline, blended, plated mainly on tryptone–glucose agar incubated 6 days at 25°C, cultures maintained on yeast extract–glucose agar slopes
Jasewicz and Porges (1956)	18-litre fill-and-draw plant treating 0·1% (v/v) milk at 30°C	Plating on nutrient agar +0·1% (v/v) skim-milk, incubated 3 days at 30°C
Lighthart and Oglesby (1969)	Sewage works examined over 3 seasons	Blended sample plated on to vitamin-enriched tryptone–glucose-extract agar, incubated 10 days at 15°–20°C
Ministry of Technology (1968)	Pilot scale fill-and-draw plant	Mickle disintegration, 30 isolates from one sample plated on yeast-extract agar
Sladká and Zahrádka (1970)	Experimental plant examining effects of varying loading, different sludge ages, two-phase aeration, extended aeration	Microscopic descriptions of populations
Tezuka (1969)	Sewage works treating mainly municipal waste	Plating on agar with mineral base +0·1% yeast extract and 0·2% polypeptone
Unz and Dondero (1970)	Not given	Dissection of individual cells from washed zoogloeas
Wattie (1943)	Sewage works and experimental	(a) Method of Butterfield (1935), (b) Manual shaking with glass beads, serial dilution in lactoce broth—tubes showing flocs after 48–120 h at 20°C selected for plating on to dilute nutrient agar, which was incubated 96 h at 20°C

[a] Further references given by Pipes (1966).

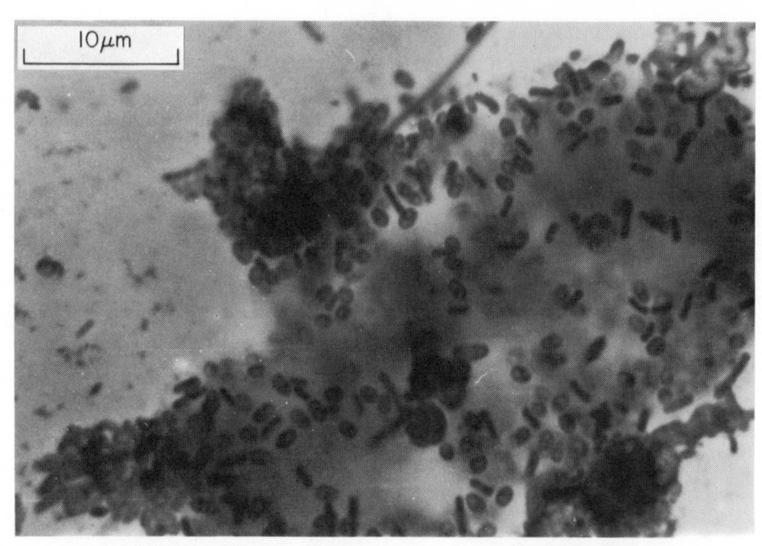

FIG. 1. Bacteria embedded in matrix of an activated-sludge floc in mixed liquor from a high-rate plant. Stained by tannic acid–crystal violet method to demonstrate bacterial cell walls and cross-septa. Water-mounted preparation, ×100 oil-immersion objective.

gravitational settling, then it is unlikely that the need for selection of flocculating organisms will be superseded in this method of waste treatment, which depends upon the principle of the continuous fermenter with feedback, or its full potential realized in practice.

Accordingly, much effort has been expended upon accounting for the phenomenon of flocculation. Several theories have been proposed, reviewed by Painter and Hopwood (1967, 1969), Coackley (1969), Sladká and Zahrádka (1970) and Campbell (1972), but without completely explaining the process. There seems to be general agreement that flocculation can be explained by the phenomena of colloid science, but there is disagreement over the actual mechanisms involved. It is generally recognized that the operation of plants at high dilution rates and high substrate concentrations tends to predispose to dispersed or filamentous growth, and that flocculation is encouraged by the opposite conditions (Tenney and Stumm, 1965; Sladká and Zahrádka, 1970).

An early view was that floc-formation was attributable to a specific organism *Zoogloea ramigera*, since strains resembling this species could be isolated by microdissection of bacteria from washed flocs, followed by serial dilution in liquid media (Butterfield, 1935; Heukelekian and Littman, 1939; Wattie, 1943). Those isolates produced flocculent growth when aerated in pure cultures. Later workers showed that a variety of bacteria could be isolated from activated sludge and that those able to flocculate in pure cultures belonged to a variety of genera (McKinney and Horwood, 1952; McKinney and Weichlein, 1953; McKinney, 1956; Dias and Bhat, 1964; Anderson, 1968).

McKinney (1952, 1956) proposed that the normal dispersed condition of bacterial cultures was caused by the mutual repulsion between negatively-charged cell surfaces, and that flocculation observed in old cultures resulted from their lack of energy. More recently, Peter and Wuhrmann (1971) suggested that bacterial suspensions resemble protected dispersoids and that, because flocculation could be brought about experimentally by adding optimum amounts of cationic polyelectrolytes, flocculation in activated-sludge treatment could result from naturally occurring polyelectrolytes, such as humic acids in the liquor. Tenney and Stumm (1965) similarly suggested that other naturally occurring polyelectrolytes were responsible, such as complex polysaccharides and polyamino acids excreted at the cell surface during the decline and endogenous growth phases.

Although accumulation of intracellular poly-β-hydroxybutyric acid (PHB) has been related to flocculation in pure cultures of activated-sludge isolates (Crabtree *et al.*, 1965), later work indicates that this may not be the only mechanism (Painter and Hopwood, 1969), or that there may be little causal connection. Dias and Bhat (1964) found that 51% of their isolates and 95% of *Zoogloea* isolates could form sudanophilic inclusions. Although Crabtree *et al.* (1965) reported that PHB comprised 12% by dry weight of activated-sludge solids, much smaller amounts, or none, were reported by van Gils (1964) and by Painter *et al.* (1968) in activated sludge. Experiments with *Zoogloea* sp. (Angelbeck and Kirsch, 1969) and *Flavobacterium* sp. (Tezuka, 1969) from sludge did not relate PHB levels with flocculation ability.

Flocculation in pure cultures of *Zoogloea* and other bacteria was shown to be independent of C:N ratio (van Gils, 1964; Tezuka, 1967) and to be induced by low pH values, whereas the effect of calcium and magnesium ions was conflicting, either causing flocculation (van Gils, 1964; Tezuka, 1967) or reversing it (Angelbeck and Kirsch, 1969; Tezuka, 1969).

The evidence given suggests that much requires to be known about the mechanisms of flocculation and that a great deal of evidence relates to pure cultures of *Zoogloea* and other bacteria, grown on laboratory media. There is a need for studies of the activated-sludge biocoenosis itself under different plant operating conditions, and for physico-chemical studies of the surface of sludge bacteria.

5. Bulking of Activated Sludge

In the growth condition termed "bulking sludge", loose, flocculent, cotton-wool-like growths of filamentous micro-organisms impede settling. Uncritical observation in the past ascribed the condition to *Sphaerotilus* spp., although it is now realized that bulking sludge is represented by various communities, some of which resemble those of "sewage fungus" conditions in polluted rivers and are dominated by various filamentous genera, such as *Sphaerotilus, Bacillus, Nocardia, Beggiatoa, Thiothrix* (Pipes, 1969), Vitreoscillaceae, *Leucothrix, Lineola* (Cyrus and Sladká, 1970), *Geotrichum* (Pipes and Jones, 1963; Schofield, 1971) and *Zoophagus* (Cooke and Ludzack, 1958; Pipes, 1965; Pipes and Jenkins, 1965). Mülder *et al.* (1971) distinguished two types of bulking in conventional aeration tanks and oxidation ditches in the Netherlands. The first type, in which *Sphaerotilus natans* was dominant, was characteristic of overloaded conditions in aeration tanks and sometimes in oxidation ditches. The second type, in which *Sphaerotilus* was absent, was very often encountered in underloaded oxidation ditches and was characterized by the sheath-former *Streptothrix hyalina*, and by a number of Gram-negative and Gram-positive non-sheath-forming bacteria. Pasveer (1969) has also recorded bulking in an underloaded oxidation ditch caused by a filamentous coliform organism.

Sludge bulking may also be associated with filamentous growth of the anaerobic hydrogen sulphide-oxidizing bacteria of the genus *Thiothrix* (Farquhar and Boyle, 1972). This organism can be encouraged when septic sewage, containing appreciable amounts of H_2S, is being treated; the organism can be recognized by the presence of yellowish refractile globules of sulphur which are deposited intracellularly in the filaments (Fig. 2) and which dissolve when the preparation is treated by irrigation with ethanol.

Excessive filamentous growths may also cause operational problems by trapping air bubbles, resulting in a stable condition of dense foaming, which can overflow from aeration and settling tanks. In two cases examined, both in surface-aeration plants, organisms resembling *Leucothrix* spp. were dominant (Fig. 3).

Bulking can seriously interfere with plant operation and cause poor effluent quality; because of this and because of the nuisance caused by sewage fungus communities in streams, there has been much research into the physiological conditions encouraging them, particularly the growth of *Sphaerotilus* spp. (Mülder, 1964b; Phaup, 1968; Curtis, 1969). *Sphaerotilus* can utilize a variety of carbohydrates and either inorganic nitrogen (in which case it also requires Vitamin B_{12}) or amino acid

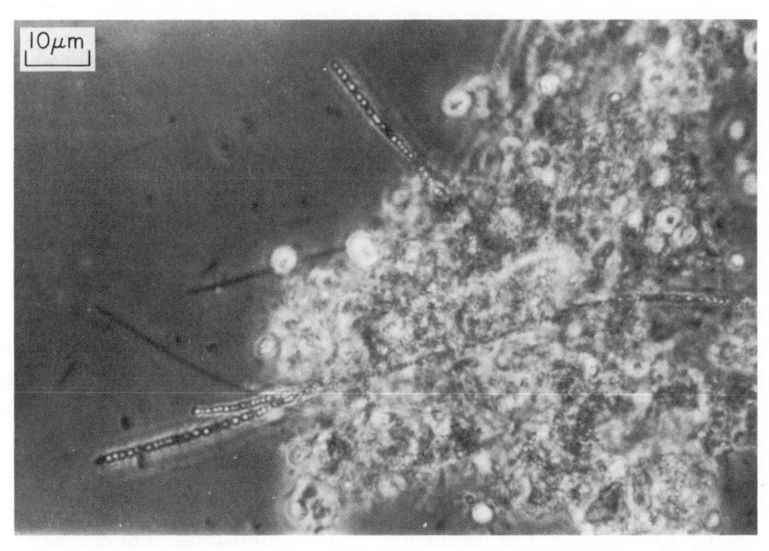

FIG. 2. Floc in mixed liquor from an activated-sludge plant treating septic sewage and affected by a bulking-sludge condition. Protruding filamentous bacteria, possibly *Thiothrix* sp., contain intracellular sulphur granules. Phase-contrast illumination, ×40 objective.

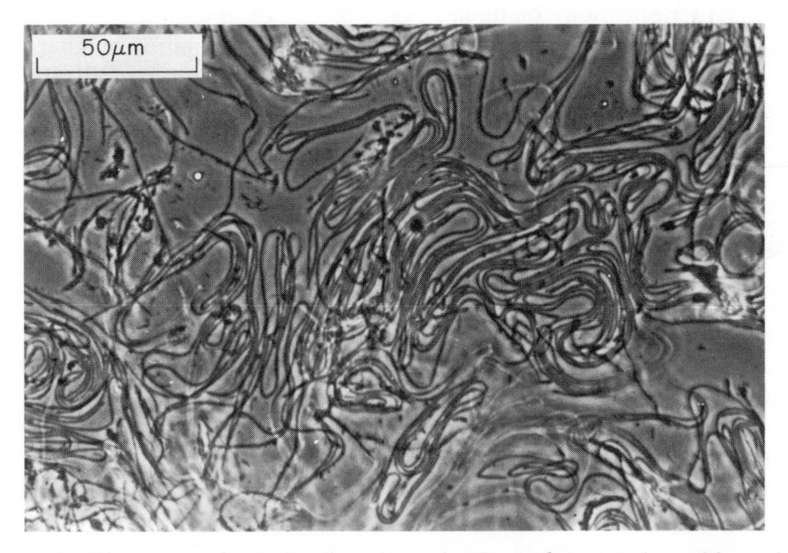

FIG. 3. Filamentous bacteria, thought to be *Leucothrix* sp., in stable surface foam accumulating on aeration pockets of a surface-aerated activated-sludge plant. Phase-contrast illumination, ×20 objective.

nitrogen. Bulking is apparently encouraged by low dissolved oxygen concentrations, the combined effect of high carbon to nitrogen and phosphorus ratios (Hattingh, 1963), or, primarily, by nitrogen deficiency (Dias *et al.*, 1968). High acetate concentrations of the order of 100 mg/l have been incriminated as the cause of bulking in small experimental plants (Ministry of Technology, 1968, p. 145). Bulking activated sludge and pure cultures of *Sphaerotilus* spp. have low DNA contents compared with normally-settling sludge and *Zoogloea ramigera* (Genetelli, 1967).

It has been demonstrated that a low "sludge age" (the reciprocal of the specific sludge wastage rate defined above) can cause bulking (Sladká and Zahrádka, 1970); if so, poor settling will result in further loss of sludge organisms in the effluent and a further decrease in sludge age. This could result in the washout of the organisms responsible for the bulking conditions.

From what has been said, it will be obvious that bulking sludge represents a symptomatic condition which can have several different causes, each involving a different microbial community. This is consistent with the varied, often conflicting, advice given in the literature for curing the condition. Pipes (1969) recognizes some ten different phenomena causing poor settling in sludge. The microbiological phenomena urgently need ecological and physiological studies to describe the communities involved and to elucidate the factors responsible for encouraging them. Simple identification methods, like those of Farquhar and Boyle (1971a, b), are also required, so that trained personnel in charge of sewage works can diagnose the nature of the condition and apply appropriate control measures, based upon sound ecological principles.

C. Biological Filtration

1. General Features

In the percolating filter the active biomass is retained upon the surface and pores of the filter medium, apart from that part of it removed by grazing fauna and that lost by sloughing into the effluent. Filters therefore resemble activated-sludge plants in having an inherent biomass-concentration mechanism, but much higher cell concentrations are found. Typical values for the concentration of dry suspended solids in filter slimes are of the order of 50–100 g/l, contrasting with 2–5 g/l for activated-sludge mixed liquors, although the total biomass in a filter treating a given flow may not be not very different from that in an equivalent activated-sludge plant.

Biological filters contained much greater diversity of organisms than activated-sludge plants and higher trophic levels are present (Cooke,

1959; Hawkes, 1963). However, the special features of the system, such as retention of the biomass on a solid medium, the absence of mixing and the presence of ecological successions of different members of the community, varying both with depth through the filter and with season, make it extremely difficult to model the system mathematically and to predict the effect of treatment upon wastes.

Because of the operational simplicity of biological filtration, particularly for small works, and its better tolerance of shock loadings and toxic wastes, compared with activated-sludge treatment, it was, until comparatively recently, more widely employed. Consequently, the microbial ecology of the process was explored earlier, and of recent years little information has been published, apart from studies on the removal of indicator bacteria and pathogens (including viruses), which will be considered later in thic chapter, and two recent studies upon rotating disc filters (Pretorius, 1971; Torpey *et al.*, 1971).

2. Genera of Bacteria

The bacterial flora does not appear to be markedly different from that of activated sludge, in that the dominant aerobic genera appear to be the Gram-negative rods *Zoogloea, Pseudomonas, Achromobacter, Alcaligenes* and *Flavobacterium* (Table V). James (1964) records *Zoogloea ramigera* as the most frequently isolated, but not sufficiently so to be considered dominant. The faecal indicator bacteria, coli-aerogenes, *Escherichia coli*, faecal streptococci and *Clostridium perfringens*, are universally present in filters, althought they are not indigenous members of the filter community. The genera *Nitrosomonas* and *Nitrobacter* are present, particularly in the lower layers of filters producing nitrified effluents.

3. Variations in the Bacterial Flora

Variations in the relative numbers and dominance of different microbial species occur between different levels in a biological filter and at different times of the year, besides those induced by differing operating conditions.

Because the biomass is essentially sessile, microbial successions are induced throughout the body of a filter, in response to the changes in concentrations of substrates in the percolating liquor. In the experimental filters of James (1964) and Isaac and James (1964), colony counts of aerobic bacteria growing on nutrient agar at 22° and 37°C were greatest at depths of 23–31 cm (9–12 in); with increasingly greater depths, counts of the latter bacteria fell off more rapidly. Homogenization of the filter slime appeared to affect the counts of bacteria growing at 22°C more than those growing at 37°C; from the data of his Table I (James, 1964),

TABLE V. Bacteria recorded in percolating filters[a]

Author	Bacteria	Notes
Hotchkis (1924a, b)	Oxidizing ammonium salts, nitrite, nitrate, thiosulphate Producing H_2S from protein, sulphate Albumen digesters Nitrate reducers	Population studies of bacteria in liquor from filters
Tomlinson (1941)	*Myxobacterium/Cytophaga*	Attacking moribund hyphae of *Sepedonium* spp.
Wattie (1943)	Four isolates defined by cultural reactions, not otherwise identified. All zoogloea-forming, Gram-negative rods	From experimental filter and various sewage works. Isolated from slimes by methods given in Table IV
Cooke (1959)	*Sphaerotilus*, *Zoogloea*, "many unidentified species"	Review, listing specific organisms in the community
Hutchinson and White (1964)	Thiobacilli	In filters treating coke oven and other gas industry liquors
James (1964)	*Achromobacter*[b], *Alcaligenes*[b], *Flavobacterium*[b], *Pseudomonas*[b], *Zoogloea*[b]	Together represent 90% of bacterial flora isolated from slimes developing on concrete cubes placed in filter beds
	Bacillus[b]	Commonest Gram-positive species
	Nocardia, *Chromobacterium*[b], *Sarcina*, *Streptococcus faecalis*, *Escherichia coli*[b], *Paracolobactrum*[b]	Also recorded

Harkness (1966)	Gram-negative bacteria: (a) Producing acid from glucose, motile (b) Non-motile, *Pseudomonas*, *Achromobacter*, *Flavobacterium* (c) Motile, glucose not fermented	Isolated from wire screen dosed with sewage: (a) Isolation at 5°C, predominant (b) Isolation at 20°C (c) Isolation at 30°C
Unz and Dondero (1967)	Resembling *Zoogloea*	Cells isolated by micromanipulation from washed zoogloeal masses and cultured in broth
Unz and Dondero (1970)	*Flavobacterium*, *Alcaligenes*, *Pseudomonas*	Isolation as in Unz and Dondero (1967). Strains not forming zoogloeal masses in culture
Bruce *et al*. (1970)	Zoogloeal and free-swimming bacteria, *Sphaerotilus*, fungi also present	Experimental high-rate filters
Pretorius (1971)	(a) *Sphaerotilus* (b) Zoogloeal masses	From slimes on rotating disc plant; (a) on first and second disc, accompanying thick filamentous growth, amoebae, ciliates and nematodes, (b) on subsequent discs with loop-forming fungus (*Arthrobotrys?*)
Torpey *et al*. (1971)	*Zoogloea*, *Sphaerotilus*	From slimes on rotating disc plant; present in first three of 10 discs, followed successively by ciliates, rotifers and nematodes on subsequent discs

[a] The authors' generic names and terminologies are given.
[b] Isolates from these genera produced capsular material when grown in pure culture in synthetic sewage (James, 1964).

homogenization increased counts of the former by a geometric mean factor of 18 and the latter by 5·0, compared with counts of unhomogenized material. In the surface layers of these filters, coli-aerogenes bacteria comprised some 10% of the 37°C count, but their numbers declined less rapidly with increasing depth than did those of all bacteria growing at 37°C.

The earlier studies of Hotchkiss (1924a, b) indicate, however, that the relative changes with depth of numbers of different bacteria in the liquor are much more complex. The data of her Table I (Hotchkiss, 1924b) show that, although numbers of proteolytic bacteria digesting egg albumen do not change greatly, numbers of sulphate reducers decrease, and numbers of bacteria oxidizing ammonia and nitrite increase with increasing depth. Because the bacteria involved in oxidizing ammonia and nitrite (*Nitrosomonas* spp. and *Nitrobacter* sp. respectively) are strict autotrophs and have specific requirements for these energy sources, the concentrations of ammoniacal nitrogen, nitrite and oxidized nitrogen can be used as indications of their active biomass in the filter. Nitrification has been generally found to occur in the lower levels of filters, and Harkness (1966) has considered that this may be caused rather by competition with other bacteria in the top layers than by the presence of inhibitory substances there, although with industrial sewages, the removal of inhibitors in the top layers might be important. In the alternating double filters at Minworth (Harkness, 1966), ammonia concentration fell linearly with increasing depth through the two stages, but oxidized nitrogen was not found until 92 cm (3 ft) down in the primary bed and was principally nitrite. In the second bed oxidized nitrogen increased linearly with increasing depth and the nitrite content was negligible.

There has recently been an interest in rotating disc filter units for treating small flows of sewage. With these, the biomass is supported upon discs of perforated or expanded metal, which are rotated axially so that part of each disc is submerged into a trough, through which the sewage liquor flows. The trough is divided into stages by transverse baffles to avoid short-circuiting of the flow, each stage being served by one or a group of discs. The studies of Pretorius (1971) and Torpey *et al.* (1971) demonstrate that well-defined ecological successions are formed throughout the stages of treatment. Some of their findings are summarized in Table V. The successions were particularly marked in the plants of Torpey *et al.* (1971). Their first plant (10-stage) gave rapid nitrification; the community of the first three stages was predominantly of zoogloeal bacteria and *Sphaerotilus* spp., followed by a diversified and abundant fauna of free-swimming and stalked protozoa, rotifers and nematodes; in the last four stages, activities of predators produced bare spots on the disc surfaces. The effluent was fed to a second, illuminated plant for removal of nitrate by photosynthetic algae.

The amounts of film accumulating on the filter medium fluctuate seasonally, reaching a maximum in the winter months, when low temperatures suppress the activities of the grazing fauna, and fall to a minimum around midsummer in the case of conventional works treating domestic sewage. More complex seasonal variations in film densities were recorded by Hawkes (1960) for a filter treating industrial waste. The rate of accumulation of film can be reduced by employing alternating double filtration (Tomlinson, 1946), or by reducing the frequency of dosing (Hawkes and Shepherd, 1971). In a comparison of single and alternating double filtration at Minworth (Tomlinson, 1946), the rate of film accumulation was studied by placing vessels filled with filtering medium at the surface of the filter beds. Over the period June–September, when the average sewage temperature was 17°C, film accumulated at three times the rate for September–April (sewage temperature 11°C). Similar seasonal changes in the rate of film accumulation were displayed by the double alternating filters (alternation periods $3\frac{1}{2}$, 7 and 14 days), but the maximum amounts of film accumulated were 50% greater for the single filter than for the alternating double filters. In the absence of scouring organisms, the rate of film growth at about 20°C was approximately double that in the winter, at about 11°C. In the single filter, 62% of the total dry weight of film was found in the top 30 cm (2 ft) compared with 44% for the double filter. However, in six experimental high-rate filters treating settled domestic sewage at an application rate of 6 m^3/m^3 day, there was no significant variation in film accumulation with depth and little variation with time of year or loading (Bruce, 1971; Bruce *et al.*, 1970).

D. Stabilizing and Maturation Ponds (Lagoons)

Lagooning has been used to provide a complete or major degree of treatment or as a means for providing tertiary treatment ("polishing") when effluents of high quality are demanded. The fundamental difference is one of the BOD loading rates; in the former case (stabilization ponds) loadings are high and anaerobic conditions tend to occur, particularly during the hours of darkness, when the combined respiratory activities of the algae and other forms of life and the cessation of photosynthesis combine to bring about oxygen depletion. Because oxygen for oxidation of pollutants is largely provided by the photosynthetic activities of the algal populations during daylight hours and because relatively large areas of land are required, lagoon treatment has been restricted, with notable exceptions, to regions receiving higher average solar irradiance and having higher average temperatures and lower land costs than Britain.

Relatively few microbiological studies have been made upon lagoon communities. Interest has naturally focused upon algal components of the community and factors influencing their density, as, for example, in the study by Neel and Hopkins (1956) of a lagoon in Nebraska treating raw sewage. Bacteriological studies have mainly been concerned with the effect of lagooning upon removal of faecal indicator bacteria, pathogenic bacteria and viruses (a topic to be considered in Section E of this paper). Ecological studies have largely been upon stabilization ponds, with the notable exception of the studies of the tertiary lagoons at Rye Meads, Essex (Windle Taylor, 1966).

1. Bacteria in Stabilization Ponds

Table VI lists genera and species of bacteria which have been reported in lagoons treating sewages. It is emphasized that some of these studies were restricted to particular groups of bacteria of interest, e.g. the purple photosynthetic sulphur bacteria (family Thiorhodaceae) which tend to occur during periods of overloading, in response to the production of H_2S from protein degradation under anaerobic conditions. These are strict anaerobes and are capable of an autotrophic way of life, in which they utilize H_2S as an electron donor in photosynthesis and deposit sulphur granules within the cell. Their distribution in lagoons, as in other aquatic situations, is of considerable ecological interest, as their photosynthetic pigment, bacteriochlorophyll, is activated by longer wavelengths than are the plant chlorophylls. Consequently, under stable stratified conditions—chemically and thermally—these bacteria are found in zones of natural waters below those occupied by the green and brown algae, whose photosynthetic pigments filter out the shorter active wavelengths (Stanier and Cohen-Bazire, 1957). Neel (1963) records that these bacteria were found below the algal layer in lagoons at a depth of about 1 m, at the limit of penetration by red light. In such a position they would be selectively encouraged by the factors of illumination and the combined effects in anaerobiosis and H_2S production by anaerobic bacteria in the aphotic zone. The bottom deposits and the lowest layers of stabilization ponds are characterized by anaerobic bacteria, including sulphate-reducing bacteria and methane bacteria.

2. Effects of Seasonal and Climatic Factors upon Bacterial Populations

Seasonal variations in solar irradiance and temperature considerably affect the microbial flora and the efficacy of treatment in stabilization ponds treating sewages. The mechanisms responsible for microbial variations have yet to be elucidated fully, although several workers record correlations between populations of various bacteria and water temperature or time of year. Hughes and Reuszer (1970), in a study of three farm ponds in Indiana, found maximum numbers of bacteria in late spring and early summer; significant correlation between temperature and numbers was found in one pond only, but for two ponds, numbers were positively correlated with organic carbon, the correlation with this variable being negative in the case of the third pond. In the two waste stabilization ponds at Lebanon, Ohio, studied by Horning *et al.* (1964), the mean levels of confirmed coliform bacteria, faecal coliform bacteria and faecal streptococci did not show consistent variations with season; confirmed coliform bacteria were most plentiful in the summer and winter and least in spring and autumn, whereas the other two classes of bacteria were most plentiful in the autumn and winter. Post (1970) found high negative correlation between outlet water temperature and numbers of certain faecal indicator bacteria, and bacteria growing at 20°C in a stabilization pond operating with a mean detention period of 24 h at Logan, Utah. The data indicated that some unidentified biological factors, related to incident solar radiation and air temperature, were bringing about the disappearance of these bacterial groups in the pond, one possible factor being the direct lethal effect of solar radiation, which was thought to operate similarly to the effect of solar radiation in reducing numbers of coliform bacteria in sea water (Gameson and Saxon, 1967).

The ecology of purple sulphur bacteria were studied in the primary stage of a two-stage lagoon in North Dakota, during the months of May–September 1967 and 1968 by Holm and Vennes (1970). In addition to domestic sewage, this lagoon received large quantities of potato processing wastes in the period September–May. Increases in numbers of these bacteria—which reached peak levels of 10^8 and 5×10^9/ml respectively in July 1967 and August–September 1968—were accompanied by corresponding decreases in the levels of the volatile fatty acids (principally acetate and formate), sulphide and sulphate. When sulphide could no longer be detected in the lagoon, numbers of purple sulphur bacteria declined. Isolates of the purple sulphur bacteria *Chromatium vulgaris* and *Thiocapsa floridana* were able to utilize a variety of sugars, organic acids and amino acids heterotrophically, but not formate. It was thus reasonable to relate acetate depletion in the lagoon to heterotrophic growth of the purple sulphur bacteria, but the possible relationship of formate depletion to increase in numbers of methane bacteria was not proved conclusively. Later studies by Fillipi and Vennes (1971) in the summer of 1969 showed that microbial changes could be related to changes in the level of biotin. During the periods when the anaerobic and purple sulphur bacteria were dominant, biotin levels increased from 100 ng/l (April) to

TABLE VI. Types and species of bacteria studied in lagoons treating sewages[a]

Author	Bacteria[b]	Notes[b]
Cabes *et al*. (1969)	(a) *Micrococcus*, (b) *Escherichia*, (c) *Serratia marcescens*, *P. fluorescens*, *A. faecalis*	Indoor poultry lagoon; (a) present in greatest numbers, (a, b) most active in oxidizing manure, (c) isolates also showing oxidative activities
Cooper (1963)	(d) *Thiopedia rosea*[c] (e) *Chromatium*[c]	Present in Californian lagoons treating wastes from (d) fat-rendering and (e) oil-refining
Ganapati (1963)	Purple sulphur bacteria and blue-green alga, *Merismopedia tenuissima*	In sewage lagoons at Ahmedabad, India
Gann (1966)	(f) Gram-negative rods, *Achromobacter*, *Pseudomonas*, *Flavobacterium* (g) *Bacillus*, *Streptococcus faecalis*	Sewage lagoons, loading rates 15–60 lb/acre/day. (f) Dominant, comprising 90–95 % of total viable bacteria; (g) the only common Gram-positive bacteria, but never present in significant numbers
Gann *et al*. (1968)	*Achromobacter* (65 %)[d], *Pseudomonas* (25 %)[d], *Flavobacterium* (5 %)[d]	Experimental pond (64-1) treating raw domestic sewage at loading of 34 g/m^3/day

Holm (1969)	*Thiocapsa floridana*[c], *Chromatium vinosum*[c]	Two-stage sewage lagoon also receiving potato starch wastes, Grafton, N. Dakota
Holm and Vennes (1970)	Total aerobes, coliform bacteria, methane bacteria, sulphate reducers, purple sulphur bacteria (including *C. vinosum* and *T. floridana*)	Primary lagoon of Holm (1969). Population study. Relationships between bacterial, chemical and physical changes over May–September period for two years
Neel (1963)	*Chromatium*[c], *Thiospirillum*[c], *Thiopedia*[c], unidentified rods and spirals	Photosynthetic purple sulphur bacteria in lagoons treating sewage and organic wastes, Missouri
South African Council for Scientific and Industrial Research (1964)	(h) *Bacillus* spp. (*B. cereus* var. *mycoides*, *B. licheniformis*, *B. cereus*, *B. megatherium*), *P. fluorescens*, *Micrococcus luteus*, *Sarcina lutea*, *Sarcina ureae*, unidentified chromogenic strains (i) *Azotobacter* (four species) (j) nitrogen-fixing *Pseudomonas* isolate	Study of bacterial flora of stabilization ponds. Proteolytic bacteria formed one-third of total bacteria, species of (h) were the most active protease producers. Nitrogen fixers (i) found in aerobic zone, up to 60 cm from surface, (j) found throughout depth of pond

[a] The authors' generic names and terminologies are given.
[b] Cross-references between columns shown by letters in parentheses.
[c] Purple photosynthetic sulphur bacteria (Thiorhodaceae).
[d] Percentage of total aerobic viable count.

about 10 000 ng/l (June–September) in the primary lagoon. Biotin levels rose to a similar level in the secondary lagoon. In September, biotin levels fell to near their former values, at which time the secondary lagoon entered the algal phase of growth. Pure culture studies showed that three organisms were at least partially associated with these fluctuations; the sulphur bacteria *C. vulgaris* and *T. floridana* with biotin utilization, and *Aerobacter aerogenes* with its production.

In the comprehensive study of bacterial populations in three tertiary lagoons in series at Rye Meads, Essex (Windle Taylor, 1966)—which receive fully treated activated-sludge effluent—there were no apparent seasonal changes over the period July 1962–September 1964 in the numbers of six classes of bacteria at the inlet; but of these bacteria, pronounced changes with season were noted for numbers of *Escherichia coli*, coliform bacteria and streptococci (peak counts in December–March and minimal counts in May–September) in samples taken at the outlets from the lagoons after total mean detentions of 3, 8 and 17 days. Seasonal changes at these outlets in counts of bacteria growing on nutrient agar at 37° and 22°C and those growing at 22°C upon the minimal medium of Skinner *et al.* (1952) followed the same pattern, but were not very marked. In summer months there was evidence of multiplication of saprophytic bacteria in the third lagoon and, to some extent, in the second. The colonial appearance of bacteria in these two lagoons was quite different from that of bacteria from the works effluent; chromogenic strains, with bright orange–red or pale yellow–green colonies, predominated.

IV. The Fate of Pathogenic Micro-organisms in Sewage Treatment

A. The Efficacy of Treatment Processes

This subject has been reviewed by Kabler (1959) and more recently by Pike (1971) as part of a review upon microbiology of sewage treatment, covering recent work published up to the end of 1971. Table VII lists reported reductions in various bacteria and viruses brought about by different sewage treatment processes, and is intended to be supplementary to the tables published by Kabler (1959).

In considering the effect of any treatment it is difficult to generalize because of large fluctuations in removals which occur both between and within plants. However, in broad terms it can be stated that primary settling usually has little effect in removing faecal organisms and pathogens, indeed numbers of bacteria may increase (Harkness, 1966;

TABLE VII. Percentage reductions in various micro-organisms brought about by different stages of sewage treatment

References[a]	Viable bacteria							
	22°C	37°C	*E. coli*	Coli-aerogenes	Faecal streptococci	*Salmonella*	Viruses	Remarks[b]
Primary sedimentation								
Ware and Mellon (1956)				65				Retention 9 hours
Berg (1966)							(a)	(a) Poliovirus I: retention for <3 h ineffective, 33–67% removed after 24 h
Clarke and Kabler (1964)							(b)	(b) Little or no effect
Allen and Smith (1953)			54		60			Means for seven large works
Harkness (1966)			70		44			Large works
	47	50	71	30				Norton Green, Birmingham
	260	440	72	86				Langley Mill
	66	91·8	3	13				Coleshill
Activated sludge								
Harkness (1966)	74	85	96·1	13				Treating secondary sedimentation tank effluent, Coleshill
Kelly *et al.* (1961)							94–99	Fill and draw plant, poliovirus Coxsackie B5
Clarke *et al.* (1961)							(c) 98 (d) 90	Laboratory scale, virus added: (c) Coxsackie A9, (d) poliovirus
Allen and Smith (1953)			61		84		90–98	Bio-aeration plant at large works
Clarke and Kabler (1964)								Review

Table VII. (*Contd.*)

References[a]	Viable bacteria		*E. coli*	Coli-aerogenes	Faecal streptococci	*Salmonella*	Viruses	Remarks[b]
	22°C	37°C						
Berg (1966)							(e) 97·4–99·998 (f) 79–94	Review. Author's experiments with added (e) Coxsackie A9, (f) Poliovirus I
Keller (1959)			96·0 98·6 99·7					Three works, South Africa
Pike and Carrington (1972)	53			83	92·9			Two high-rate activated sludge plants, loadings 0·49–2·4 g BOD/g day
Biological filtration								
Keller (1959)			95·9	92·7				Cydna, Johannesburg
Berg (1966)							(g)	(g) "Of little value"
Clarke and Kabler (1964)							40	"Not very effective"
Allen *et al*. (1944)		(h) 93·9 (i) 72·4		(j) 95 (i) 99·5 (k) 84 (l) 96				Single filter, loading 60 gal/yd^3/day; average temperatures: (h) 16·3°C, (i) 11·5°C; (j) December–March. Alternating double filtration, total loading 240 gal/yd^3/day; (k) December–March, (l) March–July, 16·3°C. Plate counts at 30°C
Allen and Smith (1953)			78 73–97		60 64–97·3			Seven large works, mean Five small works treating septic tank effluent
Bruce *et al*. (1970)				(m) 26–51 (n) 15–44 (o) 97·5 (p) 99·0				Six high-rate filters: means for (m) January–July, (n) August–November. Low-rate filter: (o) January–July, (p) August–November

Kelley and Sanderson (1959)							c. 40	
Tomlinson *et al*. (1962)		95·2–99·2		97·5–99·7				Averages over year for eight plants with different media
Harkness (1966)	63	98·5	98·8	96·2				Norton Green
	89	77	93·7	96·4				Langley Mill
Nupen (1970)							82	Windhoek, S. Africa; after primary settling and biological filtration with secondary settling
Lagoons								
Windle Taylor (1966) (q)	69	84·9	92·4	89	92·5	93		Lagoons at Rye Meads, Essex, receiving activated-sludge effluent; mean detention (q) 3 days, (r) 8 days, (s) 17 days
(r)	78	92·2	97·1	96·6	98·0	93		
(s)	85	95·8	99·5	99·5	99·6	93		
Horning *et al*. (1964) (t)			87·9–98·3	86–94·4	97·0–99·8			Two stabilization ponds, Lebanon, Ohio, treating raw sewage; loadings (t) 49–82 and (u) 117–194 lb BOD/acre/day respectively
(u)			80–95·2	67–94·4	87·4–98·2			
Post (1970)	39	54	88–99·5	97·7	96·2–99·0			Stabilization pond treating sewage Logan, Utah. Average retention 20 h. Viable bacteria growing at 20° and 35°C
Nupen (1970)							95	Windhoek, S. Africa; tertiary lagoons, retention time 14 days
Holm and Vennes (1970)				c. 99				Primary cell of two-stage lagoon treating domestic sewage, May–September, Grafton, North Dakota
Fitzgerald and Rohlich (1958)				98·2–99·99				Review article; 13 references to removal of bacteria

[a] References given are additional to those reviewed by Kabler (1959); removal of human enteric viruses is also reviewed by Kollins (1966).
[b] Cross-references to preceding columns shown by letters in parentheses.

Allen *et al.*, 1949). Conventional activated-sludge treatment is extremely effective at removing coli-aerogenes bacteria and enteroviruses, and removals exceeding 90% or even 99% of the micro-organisms entering have been reported. Percolating filters display similar effects upon coli-aerogenes bacteria but are ineffectual in removing virus particles (Clarke and Kabler, 1964; Berg, 1966; Malherbe and Strickland-Cholmley, 1967). Lagoon treatment may also achieve considerable removal of coli-aerogenes bacteria and viruses. The effectiveness of sewage treatments and chlorination in removing viruses has been summarized by Malherbe (1964), Clarke and Kabler (1964), Berg (1966), Kollins (1966) and Grabow (1968), and by the Committee on Environmental Quality Management of the American Society of Civil Engineers (Committee, 1970) in a consideration of the virus hazard in water. A conclusion of this committee is worth quoting:

> It is evident from the literature that virus reduction greater than 90% can be obtained by conventional activated-sludge treatment followed by chlorination to a level producing an amperometric chlorine residual of 5+ mg/l after 30 minutes contact. Information on the degree of virus destruction obtained by chlorination of effluents of primary plants, trickling filters and oxidation ponds is not clear and it must be assumed that the level of removal is relatively low.

B. Mechanisms of Removal

There is a surprising lack of quantitative information concerning the agencies responsible for removing micro-organisms in sewage treatment or the factors affecting them. In the case of percolating filters and activated-sludge treatment it is likely that the major agency is by the feeding of protozoa and larger fauna. This was demonstrated by Curds *et al.* (1968) in laboratory experiments in which six protozoa-free activated-sludge plants produced poor quality, turbid effluents. Inoculation of three of these plants with cultures of ciliated protozoa brought about a dramatic increase in effluent quality accompanied by a significant decrease in bacterial content, while the three remaining protozoa-free plants continued to deliver poor quality effluents. In the experiments of Curds and Fey (1969), the half-life of *Escherichia coli* introduced into a plant lacking protozoa was 16 h, compared with 1·8 h in a parallel plant with protozoa. This experiment also demonstrates that mortality mechanisms other than protozoan feeding are likely to be insignificant in comparison.

Bacteriophages and the predatory bacterium *Bdellovibrio bacteriovorus* can be regularly isolated from sewages and effluents. The few quantitative studies suggest, however, that neither is effective in removing bacteria during treatment. Study of *Bdellovibrio* sp. has until recently to a large degree been hampered by the lack of accurate counting procedures. Thus, the method of differential filtration through membranes of average pore diameters of 0·8, 0·65 and 0·45 μm, although commonly used for isolation, will leave only 0·1% of the original population in the final filtrate. Reasonably efficient recoveries of bdellovibrios and removal of other bacteria can be obtained by filtering through membranes of 1·2 μm average pore diameter or by differential centrifugation through a linear Ficoll gradient (Starr and Seidler, 1971).

A few hundred of phage progeny are normally released from the infected bacterial host cell after a period of about 20 min from the time of infection. Similarly, infection of host cells by *Bdellovibrio* sp. will cause a rise of about 3–5 times in counts of plaque-forming units of this organism after some 4–5 h (Varon and Shilo, 1969). Thus, if either of these parasites is active against sewage bacteria, their numbers relative to those of their hosts, or the ratio of resistant to sensitive bacteria, should increase as sewage treatment progresses. This was not so in the experiments of Ware and Mellon (1956), in which the ratio of sensitive to resistant bacteria decreased after biological filtration of crude sewage. In the experiments of Dias and Bhat (1965) a ten-fold reduction in phage count occurred after 2 h aeration of sewage with laboratory activated sludge; numbers of *Bdellovibrio* sp., attacking a variety of bacteria, were unaffected. Pretorius (1962) found no evidence that coliphage was active in reducing numbers of *Escherichia coli* in experimental sewage lagoons at Pretoria. It is clear that more quantitative work is needed to assess the possible role of *Bdellovibrio* in controlling populations of bacteria in sewage treatment and in polluted aquatic systems generally; indeed the review by Starr and Seidler (1971) noted that only two quantitative studies had been made to date upon *Bdellovibrio* sp.—by Dias and Bhat (1965) and by Klein and Casida (1967), who found that the highest titres in soils of *Bdellovibrio* sp. active against *E. coli* were from those soils known to be contaminated with sewage. The final question concerning these two obligate parasites must be whether they are indeed inactive in sewage and, if so, why and whence do they originate in the first place? One speculation may be that they are active only in situations of high host concentration, such as in faeces, where the chances of collision and host infection are much greater than they are in sewage.

The role of ciliated protozoa in controlling bacterial populations and of flagellates in consuming soluble nutrients in the activated-sludge process has been the subject of much study with experimental and theoretical dynamic modelling by Curds, and is considered by him in Chapter V and in the review of the process by Pike and Curds (1971). Amoeboid protozoa, rotifers, nematodes and annelid worms are able to ingest and digest bacterial cells; the extent to which they are able to do this has not been explored. Chang (1970) has summarized information, much resulting from his own experiments, upon the extent to which virus particles are removed or disseminated by micro-organisms and macrofauna in sewage treatment. Protozoa and rotifers will ingest, at random, virus particles with water, when their concentrations are abnormally high and these are rapidly digested. However, it is unlikely that they contribute to removal of viruses, or serve as carriers under normal conditions.

The removal of bacteria in percolating filters appears to be directly related to the bacterial content of the sewage, particularly at low bacterial contents (Allen *et al.*, 1944), to the surface area of the filtering medium (Tomlinson *et al.*, 1962) and to the BOD removal (Bruce *et al.*, 1970). The efficiency of percolating filters in removing bacteria from sewage appears to display short-term and longer-term changes; the former are often considerable, and where the latter can be discerned—as for example in the data of Allen *et al.* (1944, 1949) and Bruce *et al.* (1970)—they are not consistently related to the seasons of the year. The data of Allen *et al.* (1944), summarized in Table VII, show clearly that the efficiencies of both the single and double alternating filters were higher over the spring and summer period than in the winter.

Bacterial removals in oxidation and stabilization ponds have, in the past, been attributed to a variety of agencies, but none has been explored extensively. An early review by Fitzgerald and Rohlich (1958) cites such mechanisms as production of antibacterial toxins by algae, sedimentation, competition, and the effect of photosynthesis in raising the pH of the water as high as 9·0–9·5 during the daytime by assimilation of CO_2. From what has been said earlier in Section D of this chapter, it is quite apparent that removal efficiencies in lagoons are affected markedly by season or related factors, such as water temperature. Post (1970) suggested that the effect of water temperature might be attributable to lethal effect of solar radiation, and it is in this connection that work at WPRL, upon the lethal effect of daylight and simulated daylight on coli-aerogenes bacteria in mixtures of sea or fresh water and sewage, may be relevant (Gameson and Saxon, 1967; Gameson *et al.*, 1968; Ministry of Technology, 1969, 1970). Radiation lethal to coli-aerogenes bacteria can exert measurable effects in the sea to some 5 m depth and is apparently of the long-wavelength ultraviolet spectrum present in daylight (=300–400 nm). Death follows the usual exponential, first-order law of decay under constant irradiance, and the rates of death in the light are considerably greater than in the dark. In experiments with mixtures of sewage and natural waters from the East Anglian coast, estuaries and eutrophic rivers illuminated under simulated daylight ($\lambda = 360$–700 nm) at 20°C, it was shown that the mortality rate was inversely proportional to salinity—being in river water about one-third its value in sea water—and was not significantly affected by season of the year and content of chlorophyll pigments. The mortality rate was only slightly affected by water temperature over the range 8–39°C.

The removal of viruses in the activated-sludge process can be explained by adsorption onto the floc surfaces (Clarke *et al.*, 1961; Lund, 1970). Aeration of sewage itself does not bring about removal of viruses and bacteriophage; the presence of activated-sludge solids is required (Kelly *et al.*, 1961). Chang (1970), in his survey upon interactions between viruses and higher microbial forms, concluded that the picornaviruses typified by poliovirus, being resistant to external proteolysis, would be unaffected by activities of proteolytic bacteria; bacterial growth might increase survival by lowering the oxygen tension, whereas active algal growth might shorten their survival if they caused sufficient increase in pH and dissolved oxygen.

References

AARONSON, S. (1970). "Experimental Microbial Ecology" Academic Press, New York and London.

ADAMSE, A. D. (1968). *Wat. Res.* **2,** 665–671.

ADAMSE, A. D. (1970). *Wat. Res.* **4,** 797–803.

AINSWORTH, G. C. and SNEATH, P. H. A. (Eds) (1962). "Microbial Classification." 12th Symp. Soc. gen. Microbiol. Cambridge University Press, Cambridge, England.

ALLEN, L. A. (1944). *J. Hyg., Camb.* **43,** 424–431.

ALLEN, L. A. and SMITH, H. M. (1953). *Wat. Sanit. Engr* **4,** 6–10.

ALLEN, L. A., TOMLINSON, T. G. and NORTON, I. L. (1944). *J. Proc. Inst. Sew. Purif.* 115–132.

ALLEN, L. A., BROOKS, E. and WILLIAMS, I. L. (1949). *J. Hyg., Camb.* **47,** 303–319.

ANDERSON, R. E. (1968). "Isolation and Generic Identification of the Bacteria from Activated Sludge Flocs, with Studies of Floc Formation". Thesis, University of Wisconsin, U.S.A. Abstract in *Wat. Pollut. Abstr.* **42,** 263.

ANGELBECK, D. I. and KIRSCH, E. J. (1969). *Appl. Microbiol.* **17,** 435–440.

ANNISON, E. F. and LEWIS, D. (1959). "Metabolism in the Rumen." Methuen, London.

ANON (1970). *Proc. Biochem.* **5**(3), 35.

ARDERN, E. and LOCKETT, W. T. (1914). *J. Soc. Chem. Ind.* **33,** No. 10. Reprinted (1954) in *J. Proc. Inst. Sew. Purif.* 175–193.

AUSTIN, B. L. and FORSTER, C. F. (1969). *Wat. Waste Treat. J.* **12,** 208–210.
BABIUK, L. A. and PAUL, E. A. (1970). *Can. J. Microbiol.* **16,** 57–62.
BAUMANN, P. (1968). *J. Bact.* **96,** 39–42.
BENEDICT, R. G. and CARLSON, D. A. (1971). *Wat. Res.* **5,** 1023–1030.
BERG, G. (1966). *Hlth Lab. Sci.* **1,** 90–100.
BERG, G. (Ed.) (1967). "Transmission of Viruses by the Water Route." Interscience, New York.
BERG, G. (1971). *J. sanit. Engng Div., Am. Soc. civ. Engrs* **97,** SA6, 867–882.
BISSET, K. A. (1955). "The Cytology and Life History of Bacteria" (2nd Edn). E. and S. Livingstone, Edinburgh.
BISSET, K. A. and MOORE, F. W. (1963). "Bacteria" (3rd Edn). E. and S. Livingstone, Edinburgh.
BOON, A. G. and BURGESS, D. (1972). *Wat. Pollut. Control,* **71,** 493–522.
BOTT, T. L. and BROCK, T. D. (1970). *Limnol. Oceanogr.* **15,** 333–342.
BREED, R. S., MURRAY, E. G. D. and SMITH, N. R. (1957). "Bergey's Manual of Determinative Bacteriology" (7th Edn). Williams and Wilkins, Baltimore.
BREZONIK, P. L. and PATTERSON, J. W. (1971). *J. Sanit. Engng Div., Am. Soc. civ. Engrs* **97,** SA6, 813–824.
BROCK, T. D. (1971). *Bact. Rev.* **35,** 39–58.
BRUCE, A. M. (1971). In *Proc. 5th int. Conf. Wat. Pollut. Res., San Francisco* 1970, Vol. 1, pp. II-14/1–8. Pergamon Press. Oxford.
BRUCE, A. M., MERKENS, J. C. and MACMILLAN, S. C. (1970). *Inst. publ. Hlth Engrs J.* **69,** 178–207.
BUTTERFIELD, C. T. (1935). *Publ. Hlth Rept., Wash.* **50,** 671–684.
CABES, L. J., COLMER, A. R., BARR, H. T. and TOWER, B. A. (1969). *Poultry Sci.* **48,** No. 1, 54–63. In *Wat. Pollut. Abstr.* (1970) **43,** No. 1200.
CAMPBELL, L. A. (1972). *Wat. Pollut. Control* **110,** 14–17.
CASIDA, L. A. (1968). *Science* **159,** 199–200.
CHANG, S. L. (1970). *J. sanit. Engng Div., Am. Soc. civ. Engrs* **96,** SA1, 151–161.
CLARKE, N. A. and KABLER, P. W. (1964). *Hlth Lab. Sci.* **1,** 44–50.
CLARKE, N. A., STEVENSON, R. E., CHANG, S. L. and KABLER, P. W. (1961). *Am. J. publ. Hlth* **51,** 1118–1129.
COACKLEY, P. (1969). *Process Biochem.* **4**(10), 27–29, 37.
COLLINS, V. G. (1963). *Proc. Soc. Wat. Treat. Examn* **12,** 40–73.
COMMITTEE (1970). *J. sanit. Engng Div., Am. Soc. civ. Engrs* **96,** SA1, 111–150.
COOKE, W. B. (1959). *Ecology* **40,** 273–291.
COOKE, W. B. and LUDZACK, F. J. (1958). *Sew. ind. Wastes* **30,** 1490–1495.

Cooper, R. C. (1963). *Devs. ind. Microbiol.* **4,** 95–103. In *Wat. Pollut. Abstr.* (1965) **38,** No. 842.
Cowan, S. T. and Steel, K. J. (1966). "Manual for the Identification of Medical Bacteria." Cambridge University Press, Cambridge, England.
Crabtree, K., McCoy, E., Boyle, W. C. and Rohlich, G. A. (1965). *Appl. Microbiol.* **13,** 218–226.
Cruickshank, E., Duguid, J. P. and Swain, R. H. A. (Eds) (1965). "Medical Microbiology" (11th Edn). E. and S. Livingstone, Edinburgh.
Culbertson, C. G. (1971). *Ann. Rev. Microbiol.* **25,** 231–254.
Curds, C. R. (1971). *Wat. Res.* **5,** 1049–1066.
Curds, C. R. and Cockburn, A. (1971). *J. gen. Microbiol.* **66,** 95–108.
Curds, C. R. and Fey, G. J. (1969). *Wat. Res.* **3,** 853–867.
Curds, C. R., Cockburn, A. and Vandyke, J. M. (1968). *Wat. Pollut. Control* **67,** 312–329.
Curtis, E. J. C. (1969). *Wat. Res.* **3,** 289–311.
Curtis, E. J. C. and Curds, C. R. (1971). *Wat. Res.* **5,** 1147–1159.
Cyrus, Z. and Sladká, A. (1970). *Hydrobiologia* **35,** 383–396.
Dennis, J. M. (1959). *J. Am. Wat. Wks Assn* **51,** 1288–1298.
Department of the Environment (1971). "Water Pollution Research 1970." H.M. Stationery Office, London.
Department of the Environment (1972). "Water Pollution Research, 1971." H.M. Stationery Office, London.
Department of the Environment (1973). "Water Pollution Research, 1972." H.M. Stationery Office, London.
Dias, F. F. and Bhat, J. V. (1964). *Appl. Microbiol.* **12,** 412–417.
Dias, F. F. and Bhat, J. V. (1965). *Appl. Microbiol.* **13,** 257–261.
Dias, F. F. Dondero, N. C. and Finstein, M. S. (1968). *Appl. Microbiol.* **16,** 1191–1199.
Dondero, N. C. (1963). *J. Bact.* **85,** 1171–1173.
Downing, A. L., Painter, H. A. and Knowles, G. (1964). *J. Proc. Inst. Sew. Purif.* 130–158.
Elliot, L. P. and Rowe, D. R. (1971). *Wat. Sew. Wks* **118,** 260–261.
El-Shazly, K. and Hungate, R. E. (1965). *Appl. Microbiol.* **13,** 62–69.
Farquhar, G. J. and Boyle, W. C. (1971a). *J. Wat. Pollut. Control Fed.* **43,** 604–622.
Farquhar, G. J. and Boyle, W. C. (1971b). *J. Wat. Pollut. Control Fed.* **43,** 779–798.
Farquhar, G. J. and Boyle, W. C. (1972). *J. Wat. Pollut. Control Fed.* **44,** 14–24.
Fillipi, G. M. and Vennes, J. W. (1971). *Appl. Microbiol.* **22,** 49–54.
Fitzgerald, G. P. and Rohlich, G. A. (1958). *Sew. ind. Wastes* **30,** 1213–1224.
Forster, C. F. (1971). *Wat. Res.* **5,** 861–870.
Gameson, A. L. H. and Saxon, J. R. (1967). *Wat. Res.* **1,** 279–295.

GAMESON, A. L. H., BUFTON, A. W. J. and GOULD, D. J. (1967). *Wat. Pollut. Control* **66,** 501–523.

GAMESON, A. L. H., MUNRO, D. and PIKE, E. B. (1970). *In* "Water Pollution Control in Coastal Areas", pp. 34–54. Institute of Water Pollution Control, London.

GAMESON, A. L. H., PIKE, E. B. and MUNRO, D. (1968). *Chemy Ind.* 1582–1589.

GANAPATI, S. V. (1963). *In* Proc. Symp. "Waste Treatment by Oxidation Ponds", October 29–30, 1963. (Discussion of paper by Neel, J. K.). Central Public Health Engineering Research Unit, Nagpur. In *Wat. Pollut. Abstr.* (1966) **39,** No. 272.

GANN, J. D. (1966). "Studies on the Aerobic Bacteriology of Waste Stabilization Pounds." Thesis, University of Oklahoma, U.S.A. 88 pp. *Diss. Abstr.* (1967) **27B,** 1965. In *Wat. Pollut. Abstr.* (1967) **40,** No. 1485.

GANN, J. D., COLLIER, R. E. and LAWRENCE, C. H. (1968). *J. Wat. Pollut. Control Fed.* **40,** 185–191.

GAYFORD, C. G. and RICHARDS, J. P. (1970). *J. appl. Bact.* **33,** 342–350.

GELDREICH, E. E., BORDNER, R. H., HUFF, C. B., CLARK, H. F. and KABLER, P. W. (1962). *J. Wat. Pollut. Control Fed.* **34,** 295–301.

GENETELLI, E. J. (1967). *J. Wat. Pollut. Control Fed.* **39,** R32–R44.

GIBBS, B. M. and SHAPTON, D. A. (Eds) (1968). "Identification Methods for Microbiologists", Part B. Soc. Appl. Bact. Techn. Ser. No. 2. Academic Press, London and New York.

GIBBS, B. M. and SKINNER, F. A. (Eds) (1966). "Identification Methods for Microbiologists", Part A. Soc. Appl. Bact. Techn. Ser. No. 1. Academic Press, London and New York.

GILS, H. W. VAN (1964). "Bacteriology of Activated Sludge." IG-TNO Report No. 32. Research Institute for Public Health Engineering, The Hague, Holland.

GOODFELLOW, M. and GRAY, T. R. C. (1966). *In* "Identification Methods for Microbiologists", Part A (Eds B. M. Gibbs and F. A. Skinner), pp. 117–123. Academic Press, London and New York.

GRABOW, W. O. K. (1968). *Wat. Res.* **2,** 675–701.

HALE, C. M. F. (1953). *Lab. Practice* **2,** 115–116.

HARDT, F. W., CLESCERI, L. S., NEMEROW, N. L., and WASHINGTON, D. R. (1970). *J. Wat. Pollut. Control Fed.* **42,** 2135–2148.

HARKNESS, N. (1966). *J. Proc. Inst. Sew. Purif.*, 542–557.

HARRIS, R. F. and SOMMERS, L. E. (1968). *Appl. Microbiol.* **16,** 330–334.

HATTINGH, W. H. J. (1963). *Wat. Waste Treat.* **9,** 476–480.

HAWKES, H. A. (1960). *In* "Waste Treatment" (Ed. P. C. G. Isaac), pp. 52–98. Pergamon Press, Oxford.

HAWKES, H. A. (1963). "The Ecology of Waste Water Treatment." Pergamon Press, Oxford.

HAWKES, H. A. (1971). *In* "Microbial Aspects of Pollution". Soc. Appl.

Bact. Symp. Ser. No. 1 (Eds G. Sykes and F. A. Skinner), pp. 149–179. Academic Press, London and New York.

HAWKES, H. A. and SHEPHERD, M. R. N. (1971). In *Proc. 5th int. Conf. Wat. Pollut. Res., San Francisco 1970*, Vol. 1, pp. II-11/1–8.

HENDRIE, M. S. and SHEWAN, J. M. (1966). *In* "Identification Methods for Microbiologists", Part A (Eds B. M. Gibbs and D. A. Shapton), pp. 1–7. Academic Press, London and New York.

HERBERT, D. (1961). *In* "Continuous Culture of Micro-organisms", Monograph No. 12, pp. 21–53. Society of Chemical Industry, London.

HEUKELEKIAN, H. and LITTMAN, M. L. (1939). *Sew. Wks J.* **11,** 752–763.

HILL, I. R. (1970). *In* "Automation, Mechanization and Data Handling in Microbiology" (Eds A. Baillie and R. J. Gilbert), pp. 175–189. Academic Press, London and New York.

HOLM, H. W. (1969). "An Ecological Study of an Overloaded Oxidation Lagoon Containing High Populations of Purple Sulphur Bacteria." Thesis, University of North Dakota, U.S.A. 191 pp. *Diss. Abstr.* (1969) **30** B, 1807–1808. In *Wat. Pollut. Abstr.* (1970) **43,** No. 1567.

HOLM, H. W. and VENNES, J. W. (1970). *Appl. Microbiol.* **19,** 988–996.

HOLM-HANSEN, O. and BOOTH, C. R. (1966). *Limnol. Oceanogr.* **11,** 510–519.

HORNING, W. B., PORGES, R., CLARKE, H. F. and COOKE, W. B. (1964). "Waste Stabilization Pond Study, Lebanon, Ohio." Public Health Service Publication No. 999-WP-16, U.S. Department of Health, Education and Welfare, Public Health Service, Cincinnati, Ohio, U.S.A.

HOTCHKISS, M. (1924a). *J. Bact.* **9,** 437–454.

HOTCHKISS, M. (1924b). *J. Bact.* **9,** 455–461.

HOUTE, J. VAN and GIBBONS, R. J. (1966). *Antonie van Leeuwenhoek* **32,** 212–222.

HUGHES, L. B. and REUSZER, H. W. (1970). *Indiana Acad. Sci.* **79,** 423–431. In *Wat. Pollut. Abstr.* (1971) **44,** No. 1758.

HUTCHINSON, M. and WHITE, D. (1964). *J. appl. Bact.* **27,** 244–251.

INTERNATIONAL CODE OF NOMENCLATURE OF BACTERIA AND VIRUSES (1959). 156 pp. Iowa State College Press, Ames, Iowa, U.S.A.

INTERNATIONAL CODE OF NOMENCLATURE OF BACTERIA (1966). *Int. J. syst. Bact.* **16,** 459–490.

IRGENS, R. L. (1971). "DNA Concentration as an Estimate of Sludge Biomass." Water Pollution Control Research Series No. 17070DH002/71, Water Quality Office, Environmental Protection Agency. U.S. Government Printing Office, Washington, U.S.A.

ISAAC, P. C. G. and JAMES, A. (1964). *Verh. int. Ver. Limnol.* **15,** 620–630.

JAMES, A. (1964). *J. appl. Bact.* **27,** 197–207.

JANNASCH, H. W. (1958). *J. gen. Microbiol.* **18,** 609–620.

JONES, G. L. and CARRINGTON, E. G. (1972). *J. appl. Bact.* **35,** 395–405.

Jasewicz, L. and Porges, N. (1956). *Sew. ind. Wastes* **28,** 1130–1136.
Kabler, P. W. (1959). *Sew ind. Wastes* **31,** 1373–1382.
Keller, P. (1959). *J. Hyg., Camb.* **57,** 410–426.
Kelly, S. and Sanderson, W. W. (1959). *Sew ind. Wastes* **31,** 683–689.
Kelly, S., Sanderson, W. W. and Neidl, C. (1961). *J. Wat. Pollut. Control Fed.* **33,** 1056–1062.
Klein, D. A. and Casida, L. E. (1967). *Can. J. Microbiol.* **13,** 1235–1241.
Kollins, S. A. (1966). *Adv. appl. Microbiol.* **8,** 145–193.
Lechevalier, H. (1964). *In* "Principles and Applications in Aquatic Microbiology" (Eds H. Heukelekian and N. C. Dondero), pp. 230–253. John Wiley, New York.
Lighthart, B. and Oglesby, R. T. (1969). *J. Wat. Pollut. Control Fed.* **41,** R267–R281.
Linton, K. B., Lee, P. A., Richmond, M. H., Gillespie, W. A., Rowland, A. J. and Baker, V. N. (1972). *J. Hyg., Camb.* **70,** 99–104.
Lochhead, A. G. and Chase, F. E. (1943). *Soil Sci.* **55,** 185–195.
Lockett, W. T. (1954a) *J. Proc. Inst. Sew. Purif.*, 189–193.
Lockett, W. T. (1954b) *J. Proc. Inst. Sew. Purif.*, 194–210.
Loveless, J. E. and Painter, H. A. (1968). *J. gen. Microbiol.* **52,** 1–14.
Lund, E. (1970). In *Proc. 5th int. Conf. Wat. Pollut. Res, San Francisco 1970,* Vol. 1, pp. II-11/1–8.
Malherbe, H. H. (1964). *J. Proc. Inst. Sew. Purif.* pp. 210–214.
Malherbe, H. H. and Strickland-Cholmley, M. (1967). *In* "Transmission of Viruses by the Water Route" (Ed. G. Berg), pp. 379–387). Interscience, New York.
Marshall, K. C., Stout, R. and Mitchell, R. (1971). *J. gen. Microbiol.* **68,** 337–348.
Mata, L. J., Carrillo, C. and Villatoro, E. (1969). *Appl. Microbiol.* **17,** 596–602.
Maynard Smith, J. (1969). *In* "Microbial Growth", 19th Symp. Soc. Gen. Microbiol. (Eds P. Meadow and S. J. Pirt), pp. 1–13. Cambridge University Press, Cambridge, England.
McCoy, J. H. (1971). *In* "Microbial Aspects of Pollution". Soc. Appl. Bact. Symp. Ser. No. 1 (Eds G. Sykes and F. A. Skinner), pp. 33–50. Academic Press, London and New York.
McKinney, R. E. (1952). *Sew ind. Wastes* **24,** 280–287.
McKinney, R. E. (1956). *In* "Biological Treatment of Sewage and Industrial Wastes" (Eds J. McCabe and W. W. Eckenfelder), Vol. 1, pp. 88–100. Reinhold Publishing Corporation, New York.
McKinney, R. E. and Horwood, M. P. (1952). *Sew. ind. Wastes* **24,** 117–123.
McKinney, R. E. and Weichlein, R. G. (1953). *Appl. Microbiol.* **1,** 259–261.
Ministry of Technology (1968). "Water Pollution Research, 1967." H.M. Stationery Office, London.

MINISTRY OF TECHNOLOGY (1969). "Water Pollution Research, 1968." H.M. Stationery Office, London.
MINISTRY OF TECHNOLOGY (1970). "Water Pollution Research 1969." H.M. Stationery Office, London.
MOORE, B. (1970a). *Rev. int. Oceanogr. med.* **18–19,** 192–223.
MOORE, B. (1970b). *In* "Water Pollution Control in Coastal Areas", pp. 22–33. Institute of Water Pollution Control, London.
MOORE, B. (1971). *In* "Microbial Aspects of Pollution." Soc. Appl. Bact. Symp. Ser. No. 1 (Eds G. Sykes and F. A. Skinner), pp. 11–32. Academic Press, London and New York.
MOSLEY, J. W. (1967). *In* "Transmission of Viruses by the Water Route" (Ed. G. Berg), pp. 5–23. Interscience Publishers, New York.
MÜLDER, E. G. (1964a). *In* "Principles and Applications in Aquatic Microbiology" (Eds H. Heukelekian and N. C. Dondero), pp. 254–279. John Wiley and Son, New York.
MÜLDER, E. G. (1964b). *J. appl. Bact.* **27,** 151–173.
MÜLDER, E. G., ANTHEUNISSE, J. and CROMBACH, W. H. J. (1971). *In* "Microbial Aspects of Pollution." Soc. Appl. Bact. Symp. Ser. No. 1 (Eds G. Sykes and F. A. Skinner), pp. 71–89. Academic Press, London and New York.
NEEL, J. K. (1963). *In* Proc. Symp. "Waste Treatment by Oxidation Ponds", October 29–30, 1963. Central Public Health Research Institute, Nagpur. In *Wat. Pollut. Abstr.* (1966) **39,** No. 272.
NEEL, J. K. and HOPKINS, G. J. (1956). *Sew. Ind. Wastes* **28,** 1326–1356.
NEWTON, B. A. (1972). *In* "Microbial Pathogenicity in Man and Animals" (Eds H. Smith and J. H. Pearce), 22nd Symp. Soc. Gen. Microbiol., pp. 269–301. Cambridge University Press, Cambridge, England.
NORROS, J. R. and RIBBONS, D. A. (Eds) (1969) "Methods in Microbiology", Vol 3B. Academic Press, London and New York.
NUPEN, E. M. (1970). *Wat. Res.* **4,** 661–672.
PAINTER, H. A. and HOPWOOD, A. P. (1967). *Rep. Progr. appl. Chem.* **52,** 491–504.
PAINTER, H. A. and HOPWOOD, A. P. (1969). *Rep. Progr. appl. Chem.* **54,** 405–417.
PAINTER, H. A., DENTON, R. S. and QUARMBY, C. (1968). *Wat. Res.* **2,** 427–447.
PASVEER, A. (1969). *J. Wat. Pollut. Control Fed.* **41,** 1340–1352.
PATTERSON, J. W., BREZONIK, P. L. and PUTNAM, H. D. (1969). *Proc. 24th ind. Waste Conf., Univ. Purdue 1969. Engng Extn Ser.* **135,** 127–154.
PATTERSON, J. W., BREZONIK, P. L., and PUTNAM, H. D. (1970). *Envir. Sci. Technol.* **4,** 569–575.
PETER, G. and WUHRMANN, K. (1971). *Proc. 5th int. Conf. Wat. Pollut. Res., San Francisco 1970*, Vol 1, pp. II-1/1–8. Pergamon Press, Oxford.
PHAUP, J. D. (1968). *Wat. Res.* **2,** 597–614.
PIKE, E. B. (1971). *Rep. Progr. appl. Chem.* **56,** 582–598.

PIKE, E. B. and CARRINGTON, E. G. (1972). *Wat. Pollut. Control* **71,** 583–605.

PIKE, E. B. and CURDS, C. R. (1971). *In* "Microbial Aspects of Pollution." Soc. Appl. Bact. Symp. Ser. No, 1 (Eds G. Sykes and F. A. Skinner), pp. 123–147. Academic Press, London and New York.

PIKE, E. B. and GAMESON, A. L. H. (1970). *Wat. Pollut. Control* **69,** 355–382.

PIKE, E. B., CARRINGTON, E. G. and ASHBURNER, P. A. (1972). *J. appl. Bact.* **35,** 309–321.

PIPES, W. O. (1965). *Proc. 20th ind. Waste Conf., Purdue Univ. Engng Extn Ser.* No. 118, 647–656.

PIPES, W. O. (1966). *Adv. appl. Microbiol.* **8,** 77–103.

PIPES, W. O. (1967). "Ecology of *Sphaerotilus* in Activated Sludge." 3rd Ann. Rept., Dept. Civ. Engng, Northwestern University, Evanston, Illinois, U.S.A.

PIPES, W. O. (1968). "An Atlas of Activated Sludge Types." Dept. Civ. Engng, Northwestern University, Evanston, Illinois, U.S.A.

PIPES, W. O. (1969). *J. Wat. Pollut. Control Fed.* **41,** 714–724.

PIPES, W. O. and JENKINS, D. (1965). *Int. J. Air Wat. Pollut.* **9,** 495–500.

PIPES, W. O. and JONES, P. H. (1963). *Biotechnol. Bioengng* **5,** 287–307.

PIRT, S. J. (1965). *Proc. R. Soc.* B **163,** 224–231.

PIRT, S. J. and KUROWSKI, W. M. (1970). *J. gen. Microbiol.* **63,** 357–366.

POST, F. J. (1970). *Wat. Res.* **4,** 341–351.

POSTGATE, J. R. (1969). *In* "Methods in Microbiology", Vol. 1 (Eds J. R. Norrris and D. W. Ribbons), pp. 611–628. Academic Press, London and New York.

POSTGATE, J. R. and HUNTER, J. R. (1963). *Nature, Lond.* **198,** 273.

PRAKASAM, T. B. S. and DONDERO, N. C. (1967a). *Appl. Microbiol.* **15,** 461–467.

PRAKASAM, T. B. S. and DONDERO, N. C. (1967b). *Appl. Microbiol.* **15,** 1122–1127.

PRAKASAM, T. B. S. and DONDERO, N. C. (1967c). *Appl. Microbiol.* **15,** 1128–1137.

PRAKASAM, T. B. S. and DONDERO, N. C. (1970). *Appl. Microbiol.* **19,** 671–680.

PRETORIUS, W. A. (1962). *J. Hyg., Camb.* **60,** 279–281.

PRETORIUS, W. A. (1971). *Wat. Res.* **5,** 1141–1146.

REGNIER, A. P. and PARK, R. W. A. (1972). *Nature, Lond.* **239,** 408–410.

REPORT (1969). "The Bacteriological Examination of Water Supplies" (4th Edn). *Rep. publ. Hlth med. Subj. Lond.* No. 71. Her Majesty's Stationery Office, London.

RIVIÈRE, J. W. M. LA (1972). *In* "Water Pollution Microbiology" (Ed. R. Mitchell), pp. 365–388. Wiley-Interscience, New York.

ROTMAN, B. and PAPERMASTER, B. W. (1966). *Proc. N.Y. Acad. Sci.* **55,** 134–141.

SCARPINO, P. V. (1971). *In* "Water and Water Pollution Handbook",

Vol. 2 (Ed. L. L. Ciaccio), pp. 639–761. Marcell Dekker, New York.
SCHELTINGA, H. M. J. (1969). *Wat. Pollut. Control* **68,** 403–413.
SCHOFIELD, T. (1971). *Wat. Pollut. Control* **70,** 32–47.
SHAPTON, D. A. and GOULD, G. W. (eds.) (1969). "Isolation Methods for Microbiologists." Academic Press, London.
SHARPE, A. N., WOODROW, M. N. and JACKSON, A. K. (1970). *J. appl. Bact.* **33,** 758–767.
SHUVAL, H. I. and KATZNELSON, E. (1972). *In* "Water Pollution Microbiology" (Ed. R. Mitchell), pp. 347–361. Wiley-Interscience, New York.
SKERMAN, V. B. D. (1967). "A Guide to the Identification of the Genera of Bacteria" (2nd Edn). Williams and Wilkins, Baltimore, Maryland, U.S.A.
SKINNER, F. A., JONES, P. C. T. and MOLLISON, J. E. (1952). *J. gen. Microbiol.* **6,** 261–271.
SLADKÁ, A. and ZAHRÁDKA, V. (1970). "Morphology of Activated Sludge." Technical Paper No. 126, Water Research Institute, Prague-Podbaba.
SNEATH, P. H. A. (1957a). *J. gen. Microbiol.* **17,** 184–200.
SNEATH, P. H. A. (1957b). *J. gen. Microbiol.* **17,** 201–226.
SNEATH, P. H. A. (1962). *In* "Microbial Classification" (Eds G. C. Ainsworth and P. H. A. Sneath), 12th Symp. Soc. Gen. Microbiol., pp. 289–332. Cambridge University Press, Cambridge, England.
SOKAL, R. R. and SNEATH, P. H. A. (1963). "Principles of Numerical Taxonomy." Freeman, San Francisco.
SOUTH AFRICAN COUNCIL FOR SCIENTIFIC AND INDUSTRIAL RESEARCH (1964). National Institute for Water Research, Director's Report for 1963. C.S.I.R. Spec. Rep. No. WAT29. 92 pp.
STANIER, R. Y. and COHEN-BAZIRE, G. (1957). *In* "Microbial Ecology" (Eds R. E. O. Williams and C. C. Spicer), 7th Symp. Soc. gen. Microbiol., pp. 56–89. Cambridge University Press, Cambridge, England.
STARR, M. P. and HUANG, J. C.-C. (1972). *Adv. Microbial Physiol.* **8,** 215–261.
STARR, M. P. and SEIDLER, R. J. (1971). *A. Rev. Microbiol.* **25,** 649–678.
STRANGE, R. E. (1972). *Adv. Microb. Physiol.* **8,** 105–141.
STUMM-ZOLLINGER, E. (1968). *J. Wat. Pollut. Control Fed.* **40,** R213–R229.
SUNDMAN, V. (1970). *Can. J. Microbiol.* **16,** 455–464.
SUNDMAN, V. and GYLLENBERG, H. G. (1967). Suomal. Tiedeakat. Toim., Sarja IV, No. 112, 1967.
TAUB, F. B. (1971). *In* "The Structure and Function of Freshwater Microbial Communities" (Ed. J. Cairns). Res. Div. Monograph No. 3. Virginia Polytechnic Institute and State University, Blacksburg, Virginia, U.S.A.
TEMPEST, D. W., HERBERT, D. and PHIPPS, P. J. (1967). *In* "Microbial

Physiology and Continuous Culture" (Eds R. E. Strange and D. W. Tempest), pp. 240–254. H.M. Stationery Office, London.
TENNEY, M. W. and STUMM, W. (1965). *J. Wat. Pollut. Control Fed.* **37,** 1370–1388.
TEZUKA, Y. (1967). *Appl. Microbiol.* **15,** 1256.
TEZUKA, Y. (1969). *Appl. Microbiol.* **17,** 222–231.
THIEL, P. G., TOERIEN, D. F., HATTINGH, W. H. J., KOTZE, J. P. and SIEBERT, M. L. (1968). *Wat. Res.* **2,** 391–408.
TOERIEN, D. F. (1970a). *Wat. Res.* **4,** 285–303.
TOERIEN, D. F. (1970b). *Wat. Res.* **4,** 305–314.
TOERIEN, D. F., HATTINGH, W. H. J., KOTZE, J. P., THIEL, P. G. and SIEBERT, M. L. (1969). *Wat. Res.* **3,** 129–140.
TOMCSIK, J. and GUEX-HOLZER, S. (1954a). *J. gen. Microbiol.* **10,** 97–109.
TOMCSIK, J. and GUEX-HOLZER, S. (1954b). *J. gen. Microbiol.* **10,** 317–324.
TOMLINSON, T. G. (1941). *J. Proc. Inst. Sew. Purif.*, 39–76.
TOMLINSON, T. G. (1946). *J. Proc. Inst. Sew. Purif.*, 168–183.
TOMLINSON, T. G., LOVELESS, J. E. and SEAR, L. G. (1962). *J. Hyg., Camb.* **60,** 365–377.
TOMLINSON, T. G., BOON, A. G. and TROTMAN, C. N. A. (1966). *J. appl. Bact.* **29,** 266–291.
TORPEY, W. N., HEUKELEKIAN, H., KAPLOVSKY, A. J. and EPSTEIN, R. (1971). *J. Wat. Pollut. Control Fed.* **43,** 2181–2188.
UNZ, R. F. (1971). *Int. J. syst. Bact.* **21,** 91–99.
UNZ, R. R. and DONDERO, N. C. (1967). *Can. J. Microbiol.* **13,** 1671–1682.
UNZ, R. R. and DONDERO, N. C. (1970). *Wat. Res.* **4,** 575–579.
UNZ, R. R. and FARRAH, S. R. (1972). *Appl. Microbiol.* **23,** 524–530.
VARON, M. and SHILO, M. (1969). *J. Bact.* **99,** 136–141.
WALLEN, L. L. and DAVIS, E. N. (1972). *Envir. Sci. Technol.* **6,** 161–164.
WARE, G. C. (1961). *In* "Continuous Culture of Micro-organisms", Monograph No. 12, pp. 165–174. Society of Chemical Industry, London.
WARE, G. C. and MELLON, M. A. (1956). *J. Hyg., Camb.* **54,** 99–101.
WATT, P. R., JEFFERIES, L. and PRICE, S. A. (1966). *In* "Identification Methods for Microbiologists", Part A (Eds B. M. Gibbs and F. A. Skinner), pp. 125–129. Academic Press, London and New York.
WATTIE, E. (1943). *Sew Wks J.* **15,** 476–490.
WILLIAMS, A. R., STAFFORD, D. A., CALLELY, A. G. and HUGHES, D. E. (1970). *J. appl. Bact.* **33,** 656–663.
WILLIAMS, A. R., FORSTER, C. F. and HUGHES, D. E. (1971). *Effl. Wat. Treat. J.* **11,** 83–86.
WILLIAMS SMITH, H. (1970). *Nature, Lond.* **228,** 1286–1288.
WILLIAMS SMITH, H. (1971). *Nature, Lond.* **234,** 155–156.
WILSON, G. S. and MILES, A. A. (1975). "Topley and Wilson's 'Principles of Bacteriology and Immunity' " (6th Edn). Edward Arnold, London.

WINDLE TAYLOR, E. (1966). "41st Report on the Results of the Bacteriological Chemical and Biological Examination of the London Waters for the Years 1963–1964." Metropolitan Water Board, London.
WINTER, F. H., YORK, G. K. and EL-NAKHAL, H. (1971). *Appl. Microbiol.* **22,** 89–92.
WISHART, D. (1969). "CLUSTAN IA User Manual". Obtainable with program on magnetic tape from Computing Laboratory, University of St. Andrews, North Haugh, St. Andrews, Fife, Scotland.
WOOLDRIDGE, W. R. and STANDFAST, A. F. B. (1933). *Biochem. J.* **27,** 183–192.
WOOLDRIDGE, W. R. and STANDFAST, A. F. B. (1936). *Biochem. J.* **30,** 1542–1553.

2

Anaerobic Bacteria

R. F. Crowther

Severn–Trent Water Authority
Upper Tame Division
Tame Valley House
156–170 Newhall St
Birmingham B3 1SE
England

N. Harkness

Severn–Trent Water Authority
Abelson House
2297 Coventry Rd
Sheldon
Birmingham B26 3PR
England

I. Introduction

Since the discovery of anaerobic life, there have been attempts to divide micro-organisms into groups based on their responses to free oxygen. For the purposes of this chapter on the anaerobic micro-organisms in used-water treatment, anaerobic organisms can be defined as of two types: obligate anaerobes, or organisms that cannot grow in the presence of oxygen, and facultative anaerobes, which are organisms able to grow with and without oxygen present. These latter organisms bridge the gap between anaerobes and aerobes. Aerobes are organisms only capable of growth when free oxygen is present. Another term which might occasionally appear is micro-aerophile. Micro-aerophiles are organisms which prefer lower oxygen levels. These organisms could be regarded as bridging the gap between obligate and facultative anaerobes. It can be clearly seen that these terms are vague: there are no absolute limits by which each term can be defined, and there exists a continuous spectrum of organisms with varying responses to oxygen.

Microbial activity may be beneficial or undesirable depending upon circumstances. With anaerobic micro-organisms both possibilities occur in used-water treatment. An example of the beneficial action of anaerobic micro-organisms is in the process of sludge digestion where their activity converts the unstable sludges to a more stable form. Amongst the undesirable effects of anaerobic activity are the production of smells and denitrification in the settlement stages of the activated-sludge process. Sludge digestion is the major anaerobic biological process of used-water treatment but the role of anaerobes in other aspects of used-water treatment cannot be ignored, and will be discussed accordingly.

II. Occurrence of Anaerobic Bacteria in Aerobic Processes

A. Anaerobes in Used Water

The anaerobic micro-organisms in used water may have been derived from the original waste or may have been initially present in small numbers and selectively encouraged to grow by the conditions in the various stages and processes. Possible sources of anaerobic micro-organisms in used waters are land drainage and stormwater, faecal matter and biological processes in industry. In sewage, faecal bacteria will generally outnumber other micro-organisms present. Although the characteristic bacteria that are used to indicate faecal pollution are the coliforms, *Escherichia coli* I, enterococci and *Clostridium welchii* (an obligate anaerobe) these organisms may not be the most important faecal bacteria numerically. For example, van Houte and Gibbons (1966) found that *Bacteroides* sp. numbered 10^{10}/g of faeces, compared with 10^6–10^8/g for coliforms, streptococci and lactobacilli, and that these obligate anaerobes outnumbered the facultative bacteria by a factor of 40. In contrast, Post *et al.* (1967) found that the counts of *Bacteroides* in faeces and raw sewage did not exceed the coliform numbers, averaging only 20% of the coliform population in faeces and 0·5% in raw sewage. The low numbers of *Bacteroides* sp. counted were unexplained. van Houte and Gibbons did not appear to use more stringent anaerobic techniques, and all their isolates were identified mainly as *Bacteroides* sp.

Counting and classifying anaerobic bacteria is a difficult exercise and, in fact, no really satisfactory figures are available for the total anaerobic counts of sewage. Presumably these are very high, since the coliform counts number 10^6–10^7/ml and represent only a small proportion of the total.

The aerobes and facultative anaerobes in sewage by plate count methods number of the order of 10^7–10^8/ml. In the presence of an easily degradable substrate the dissolved oxygen concentration of the sewage is rapidly lowered. The result of aerobic activity, in the absence of replaceable oxygen, is therefore to provide an anaerobic environment.

Anaerobic activity is indicated usually by offensive smells, particularly that of hydrogen sulphide. However, other products of anaerobic growth such as amines and skatole also smell offensively. Release of these gases to the atmosphere is highly objectionable, and without proper ventilation, their presence in the sewer could be fatal to maintenance men. In addition, the presence of sulphide can be harmful to the sewer fabric, being converted by *Thiobacillus* sp. oxidatively to sulphate, and in particular sulphuric acid, which attacks the concrete of the sewer wall (Postgate, 1959). Organisms producing sulphide can do so in two ways; either they break down protein into amino acids and thereby degrade the sulphur-containing amino acids, cysteine, cystine and methionine, or they reduce the sulphate present in the sewage. The ability to produce sulphide from protein is quite common amongst bacteria, for instance, *Proteus*, *Bacteroides* sp. and some *Clostridium* spp. can produce sulphide from protein. These organisms can all grow anaerobically, and *Bacteroides* sp. are obligate anaerobes. It is thought, however, that most of the sulphide produced from sewage is by the reduction of sulphates (Heukelekian, 1948). This is brought about by the anaerobic sulphate-splitting bacteria *Desulfovibrio desulfuricans*. Heukelekian (1948) found 60–600 sulphate-splitting bacteria/ml of sewage. In experiments on stored sewage, he found an increase in the proportions of sulphate-splitting bacteria after 14 days, and the numbers of organisms rose from about 100/ml to nearly 100 000/ml. This coincided with the maximum production of sulphide. During the first seven days there was little increase in the numbers of sulphate-splitting bacteria, or in sulphide production. This lag

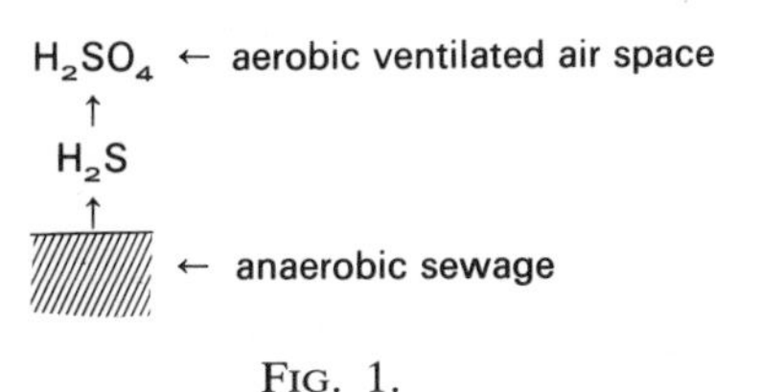

FIG. 1.

in their growth was thought to be due to the initially high oxygen concentration which did not permit the growth of the anaerobic organisms until the oxygen tension had been lowered by aerobic activity. Toerien *et al.* (1968) found $3\text{–}5 \times 10^4$/ml in raw sewage sludge, indicating much lower figures in sewage. Similar figures were obtained by Mara and Williams (1970) in counts on raw sewage sludge using various methods.

The biochemistry of sulphate reducing bacteria has been examined in detail by Postgate (1959).

B. Anaerobes in Primary Sedimentation Tanks

The retention times provided by sedimentation tanks now vary considerably in different countries. The United States use fairly short retention times of 2–4 h, whereas in Britain and parts of Europe, longer retention times of 6–8 h are used. It is possible that during sedimentation tank treatment, an increase in the numbers of aerobic and facultative bacteria may take place (Harkness, 1966). This would further reduce the dissolved oxygen concentration. However, it is unlikely that significant anaerobic action takes place during sedimentation tank treatment. Sulphide production is not usually encountered, possibly because of the time lag shown by Heukelekian (1948). Some anaerobic fermentation of the settled solids in the sedimentations tanks may occur, however, causing the sludge to rise if this is not removed frequently enough.

C. Anaerobes in Percolating Filters and Activated Sludge

These two processes are basically aerobic; consequently, any obligate anaerobic micro-organisms occurring will probably be due to chance, and will not be active. Only when anaerobic conditions prevail will anaerobic micro-organisms become dominant. In filters this will be when the spaces between the stones have become blocked with film growth, due either to small filter media, gross overloading or toxicity to the filter fauna. Anaerobic activity may also play some part in the inner layers of the filter film. In activated-sludge plants anaerobic conditions may occur, due to poor design or insufficient aeration, which leads to an inability to supply sufficient air to meet the demand of the growing and respiring sludges. Wherever this occurs, the result is the production of smells and poorly settling sludges.

Facultative anaerobes might be involved in the aerobic purification processes in an aerobic capacity, but there does not appear to be much evidence for this. In the activated-sludge process, Lighthart and Oglesby

(1969) found that none of the dominant organisms isolated from the biomass of an activated-sludge plant was able to grow anaerobically at 20°C. Other workers have not included anaerobic growth as a test in their investigations of activated-sludge and filter-bed bacteria, but from the identification of organisms isolated, it would seem that the organisms predominating are aerobic, and that anaerobic organisms present have been introduced from the settled sewage or have occurred as a result of malfunction of the plant.

D. Anaerobes in Final Sedimentation Tanks

Due to their highly aerobic character, the solids settling out in humus tanks, and particularly those in the activated-sludge separating tanks, are usually retained for as short a time as possible. This is to minimize the development of anaerobic conditions which may result in the sludge rising from the bottom of the tank or in the incorporation of unwanted facultative anaerobes into the activated sludges. One of the reasons for rising sludge is the conversion of nitrate, formed by nitrification during treatment, to gaseous nitrogen by the process of denitrification. The attachment of the nitrogen bubbles to the sludge and humus particles, and high rates of denitrification, may cause the mass of sludge to rise. This should not be confused with activated-sludge bulking, which is the result of other factors and not necessarily the dissolved oxygen concentration (Pipes, 1967). Painter (1970) reviewed the metabolism of inorganic nitrogen by micro-organisms, and it would appear that the conversion of nitrate to nitrogen via nitrite only occurs at low oxygen concentrations. Organisms capable of this activity are the facultative anaerobes *Pseudomonas* sp., *Achromobacterium* sp., *Spirillum* sp., *Micrococcus* sp. and *Xanthomonas* sp. Although there is some evidence for aerobic denitrification, no confirmation has been possible, due to the problems of obtaining reliable dissolved oxygen concentrations. Kiff (1972) showed denitrification to be severely reduced by dissolved oxygen levels as low as 2% saturation.

III. Occurrence and Role of Anaerobes in Anaerobic Sludge Digesters

Most previous references to anaerobic micro-organisms have been with regard to their being a nuisance in the operating of a treatment plant. In sludge digestion, the activities of anaerobic micro-organisms are used to advantage in converting the highly putrescible raw sludge, withdrawn

TABLE I. Chemical analysis of the settled fraction of sewage. After Balmat from Heukelekian (1957)

Protein	25·1%
Cellulose	14·2%
Hemi-cellulose	3·1%
Lignin	6·1%
Ether soluble matter (fats and grease)	19·0%

from sedimentation tanks, to a stable and disposable product which does not give rise to offensive smells and does not encourage rodents or insects. McCarty (1964) lists the advantages of treating sludge anaerobically; these are a high degree of stabilization, low production of extra biological sludge, low nutrient requirements, no oxygen requirements, and the evolution of the valuable end-product methane. To discuss the activity of anaerobic organisms in digestion, it is important to realize what substrates for growth are present in the raw sludge. Heukelekian (1957) lists chemical analyses of raw and digested sludges, and these are given in Table I.

Although variations between sludges from different sources do occur, these analyses and others which give similar results, show that the major constituents of raw sludge are protein, fats, and polysaccharides. The stabilization of these compounds, involving three stages, results in their conversion to the end-products of metabolism, mainly methane, carbon dioxide, water and new bacterial cells. This three-stage process is sometimes regarded as a two-stage process due to the complex interrelations of the first two stages. As shown diagrammatically in Fig. 2, the process consists of the three stages hydrolysis, acid formation, and methane formation (methanogenesis). Various terms have been used to describe these stages; Toerien and Hattingh (1969) regarded the process as a two-stage one and termed the stages as non-methanogenic, and methanogenic. They used these terms to avoid any erroneous interpretation of the names of the stages, in particular the acid-forming stage which not only produces acid, but also smaller quantities of alcohols and ketones. However, there is generally little disagreement concerning the biochemical activities involved.

A. Anaerobic Hydrolytic Bacteria

As shown in Fig. 2, there are three major macromolecules found in raw sludge: proteins, fats and polysaccharides. In the digestion of sludge these

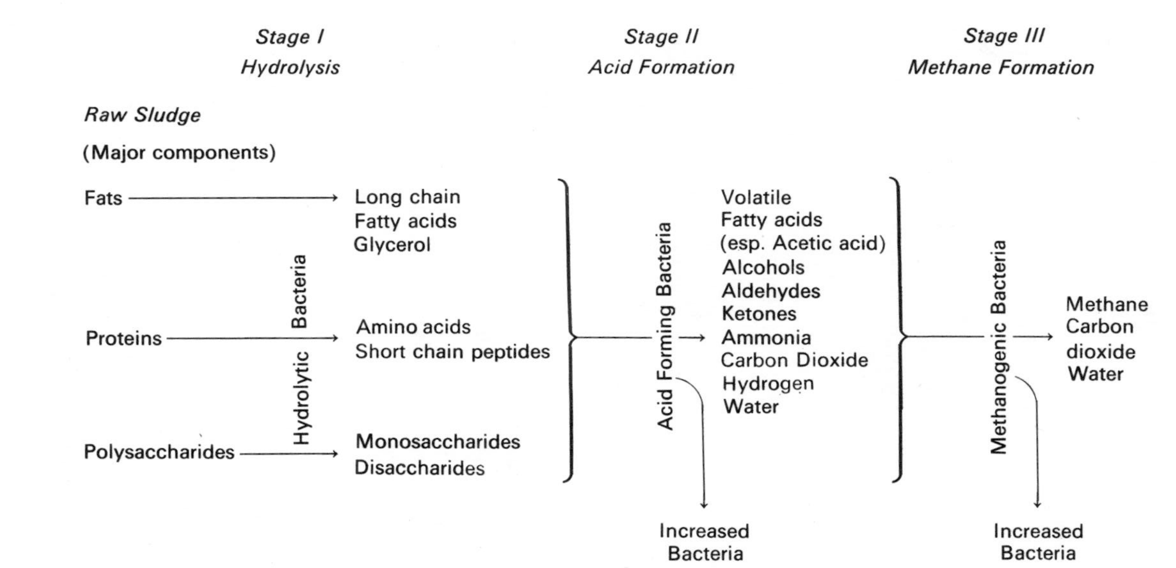

FIG. 2. Three stages of sludge digestion. (Some hydrolytic bacteria are known to be able to carry out stages I and II.)

are converted by hydrolysis or liquefaction into their respective basic components, proteins being converted to amino acids, fats to glycerol and long-chain fatty acids, and polysaccharides into monosaccharides or disaccharides.

1. *Hydrolysis of Protein (Proteolysis)*

Harkness (1966) examined digesting sewage sludges for proteolytic bacteria. He found 7×10^4/ml and $2{\cdot}5 \times 10^4$/ml of proteolytic bacteria. In the first case, this corresponded to $3{\cdot}5 \times 10^6$/g of solids. It was also shown that most of the organisms isolated were spore-formers. Siebert and Toerien (1969) examined the proteolytic bacteria present in anaerobic sludge using more stringent anaerobic techniques and found numbers of 65×10^6/ml. Of these 65% were spore-formers, including seven *Clostridium* spp. Cocci comprised 21% of the isolates, and the remainder were non-sporing rods and bifid-like bacteria. The *Clostridium* spp. showed more active proteolysis than the other bacteria isolated.

Toerien (1970a), in listing the properties of the major isolates from a digester, showed that ability to liquefy gelatin was only possessed by 4% of the cultures and to hydrolyse casein by only 7% of the cultures. Hobson and Shaw (1971) studied the activity of anaerobic bacteria in digesters with particular reference to digestion of piggery waste. Amongst the proteolytic anaerobic bacteria which they isolated were *Clostridium* sp. and organisms provisionally identified as *Bacteroides* sp., although the numbers and relative importance of the various isolates were not investigated.

The aerobic and facultative proteolytic bacteria do not appear to be as significant as anaerobic proteolytic bacteria. Kotze *et al.* (1968) gave aerobic proteolytic counts of $1–80 \times 10^5$/g of volatile suspended solids (VSS). This corresponded to $1{\cdot}2–150 \times 10^3$/ml of digesting sludge. Of the aerobes and the facultative anaerobes, the bacilli were seen to be the major proteolytic organisms (Toerien, 1967b).

Protein hydrolysis involves the action of the extracellular proteases, enzymes which convert protein into the smaller sub-units, polypeptides, oligopeptides and amino acids. The smaller units are then able to be broken down intracellularly, since they are able to pass through the cell wall of many bacteria. The ability to utilize the protein sub-units is possessed by a large number of bacteria, whereas protease activity is restricted to a relatively small number of types. Hattingh *et al.* (1967) showed that the activity of the proteolytic enzymes in digesting sludge was sufficient to hydrolyse more than 50 times the daily input of protein. Thus, even with such relatively low numbers of proteolytic bacteria compared with the total bacterial count, there are more than enough to hydrolyse the protein content of raw sewage sludge.

2. Hydrolysis of Lipids (Lipolysis)

Enhanced gas production by the addition of vegetable oils to anaerobic digesters has been recognized for years. Toerien (1967b) has studied the organisms isolated from digested sludge enriched with sunflower oil. Microscopic examination of the enrichment indicated the presence of vibrios with some spirillae and rods, but these organisms were apparently not isolated in subsequent studies. *Micrococcus* sp., *Bacillus* sp., *Streptomyces* sp., *Alcaligenes* sp. and *Pseudomonas* sp., all isolated from the enrichment, were thought to have been associated with the breakdown products of the oil, rather than being primarily responsible for lipolysis. All these organisms are aerobic or facultatively anaerobic, and were not considered to be a major part of the flora in the enrichment. It was considered that the vibrios indicated by microscopic examination of the enrichment were probably obligate anaerobes and were in fact the major lipolytic organisms. Kotze *et al.* (1968) subsequently investigated the number of lipolytic organisms in a digester operated with raw sewage sludge. They found 32×10^5/g VSS. corresponding to a count of 7×10^4/ml. These counts were of aerobic or facultative anaerobic organisms and no counts of obligate anaerobes were made. Hobson and Shaw (1971) did not carry out lipolytic bacterial counts on their piggery waste sludge, although they suggested such organisms could be important, particularly in domestic anaerobic digesters.

The reduction in amounts of ether-soluble matter during digestion, for example, from some 19·7% of total solids in raw sludge to only 4·6% of primary digested sludge in a domestic treatment plant recently examined by the present authors, shows that lipolytic organisms do in fact play a very large role in the stabilization of raw sludge. For this reason, it is surprising that so little is known about these organisms.

3. Hydrolysis of Polysaccharides

Hungate (1950) isolated anaerobic mesophilic cellulolytic bacteria from sewage sludge. No isolates were made from raw sludge, but organisms were isolated from digesting sludge in a two-stage process. More organisms (2600/ml) were isolated from the bottom of the secondary tank than from the bottom of the primary tank (880/ml). It was concluded that the organisms, concerned with the decomposition of cellulose, developed during the process. Three isolates were made from the digesting sludge which appeared to resemble cellulolytic bacteria from other sources. These were gram-negative rods, variable in size and form, particularly when growing on cellulose. Motility, but not spore formation, was observed. Maki (1954) studied the cellulolytic bacteria in a municipal sewage plant and found members ranging from 16 000/ml to 970 000/ml. These were much higher than those obtained by Hungate (1950). Maki

(1954) did not find cellulolytic organisms in crude sludge, but in contrast to the observations of Hungate, he found fewer organisms in the secondary tank than in the primary tank. Ten isolates were obtained and these appeared to fall into two groups, one being less active than the other. The more active organisms were motile, gram-negative, and variable in size. One active cellulolytic organism appeared to contain spores, although these were no longer clearly observable after prolonged subculture. The ability of the organisms to remain viable after pasteurization, however, suggested some resistant forms were still present. The organisms of Maki and Hungate (Table II) do not appear to be the same from analyses of their fermentation products.

TABLE II. Fermentation products of cellulolytic organisms found by Hungate (1950) and Maki (1954)

	Hungate (1950)		Maki (1954)		
	Strain B	Strain D	Strain E	Strain 4b	Strain 13a
Cellulose used	204·8 mg	177·9 mg	58·9 mg	64·6 mg	89·9 mg
Fermentation products					
H_2	0·62	0·48	0·15	0·07	0·20
CO_2	0·45	0·26	0·63	0·31	0·74
Ethanol	0·34	0·61	—	0·15	0·54
Formic acid	—	None	None	None	0·43
Acetic acid	0·51	0·42	0·16	0·50	0·04
Lactic acid	None	None	0·06	None	0·03
Succinic acid	None	0·20			
Glucose remaining	90·0 mg	None			

Concentration of fermentation products expressed as mmol

Hobson and Shaw (1971) also counted the anaerobic cellulolytic bacteria and found numbers of 10^4–10^5/ml. One isolate was a Gram-positive, curved rod often occurring in short chains which produced mainly propionic acid with traces of formic, succinic and acetic acids. The organism appeared to maintain its cellulolytic activity during stock culturing, whilst all the other strains lost this ability. The remainder of the isolates were Gram-negative cocco-bacilli of various morphologies.

Hobson and Shaw (1971) found more than 4×10^4/ml of organisms able to hydrolyse hemi-cellulose (Xylan). Of ten isolates, nine were identified as *Bacteroides ruminicola*, a Gram-negative cocco-bacillus. None were shown to be cellulolytic. The ability to hydrolyse starch was found to be the most common hydrolytic activity amongst the various isolates, and most of the organisms provisionally characterized as *Bacteroides* sp.

showed amylolytic activity. Other Gram-negative, pleomorphic, cocco-bacilli were also shown to be amylolytic, but were not identified. Toerien (1967b) examined the facultative anaerobic and aerobic organisms able to break down cellulose and starch. He found *Bacillus* sp., *Alcaligenes* sp., *Proteus* sp. and *Pseudomonas* sp. in the cellulose enrichment, and *Micrococcus* sp., *Bacillus* sp. and *Pseudomonas* sp. in the starch enrichment. However, in his conclusions he points out that all the organisms isolated organism have been part of a secondary flora utilizing the end-products of the primary hydrolysis, and that obligate anaerobic organisms are primarily responsible for hydrolysis. Later, he was able to show that some 19% of his anaerobic isolates hydrolysed starch (Toerien, 1970a). There is a general consensus, therefore, that anaerobes, although difficult to isolate, are primarily responsible for the hydrolysis of macromolecules in digesting sludge.

B. Acid-forming Bacteria

It is often difficult to distinguish the acid-forming organisms from the hydrolytic organisms, since many are able to carry out both processes (Fig. 2). It is therefore understandable why authors prefer to link the two processes together. The hydrolysis of the macromolecules gives rise to the formation of simpler compounds readily utilized by a greater number of micro-organisms. As indicated in the previous section, the number of hydrolytic organisms appear to be only a small proportion of the very large number of bacteria present in digesting sludge, which suggests that much of the flora of digesting sludge is responsible for this secondary process.

Total counts for the bacterial flora were attempted quite early during the development of digestion as a means of sludge treatment. However, counts of bacteria in anaerobic digesters must be critically examined on at least three points. Firstly, in trying to assess the number of anaerobic organisms present, the techniques employed should be adequate for the development of strict anaerobes; secondly, the correct growth factors should be present in the isolation medium; thirdly, the organisms which occur in clumps, chains or aggregates around particles should be adequately dispersed. Because of these factors, counts will almost always be low. The maintenance of strict anaerobiosis in counting techniques appears to have been seriously considered only in more recent work, e.g. Toerien *et al.* (1967), Mah and Sussmann (1968) and Kirsh (1969). For this reason we intend to consider only the more recent counts, and the earlier work, e.g. O'Shaughnessy (1914), Gaub (1924), Hotchkiss (1924) and Ruchhoft *et al.* (1930), will not be considered, since anaerobic techniques were not generally used. The paper of Toerien and Hattingh (1969) on the microbiology of digestion, however, covers the results of these earlier workers.

Until the work of Toerien *et al.* (1967) it had been generally thought that the majority of bacteria in the acid-forming phase were facultative anaerobes. They showed that this was not so, and that obligate anaerobic micro-organisms occurred in far greater numbers than the aerobic and facultative anaerobic bacteria. Ratios of obligate to facultative or aerobic organisms of 1:100–200 were recorded. This was confirmed by Mah and Sussmann (1968) who reported numbers of obligate anaerobes of 10 to 100 times more than aerobic bacteria. Kirsch (1969) found the predominant bacterial flora consisted of Gram-negative, non spore-forming, obligate anaerobic bacilli. Furthermore he showed that the selection of a suitable medium for counting these organisms was important, as well as extended incubation times. Mah and Sussmann (1968) also showed that extended periods of incubation were significant. The effects of the medium on aerobic and anaerobic counts were taken into account, so that their ratios of 10 to 100 times the number of anaerobes as compared with aerobes is probably the most reliable data currently available. Toerien *et al.* (1967) did not use comparable media nor did they use similar times of incubation, which may perhaps account for their higher ratios. In terms of numbers, Toerien *et al.* (1967) found $31–1500 \times 10^7$/ml of obligate, anaerobic, acid-forming bacteria. Mah and Sussmann (1968) found 10^8–10^9/ml, and Kirsch (1969) found 52×10^7/ml. Hobson and Shaw (1971), examining a piggery waste sludge, found $2{\cdot}4 \times 10^7$/ml of anaerobic bacteria in their domestic sludge inoculum. Thiel *et al.* (1968) showed anaerobic counts for $89–550 \times 10^7$/ml. Microscopic counts showed far higher numbers of bacteria, but these may include methanogenic organisms not recovered by the techniques used for viable counts. Separation of bacteria from the particles, and the dispersion of the chains and micro-colonies so clearly visible under the microscope, remain very difficult problems.

However, in correlation-coefficient studies, Toerien *et al.* (1967) found highly significant correlations between anaerobic counts and other digester parameters, particularly DNA and VSS, although the numbers of aerobic and facultative anaerobic bacteria did not significantly correlate with any of these parameters. In contrast, Thiel *et al.* (1968) found that the number of aerobes did correlate with many parameters. Bearing in mind the work on relative numbers of aerobes to anaerobes, it would seem that the aerobes play only a minor part in the total role of bacteria in digestion.

The taxonomy of the obligate anaerobic bacteria is far from precise and the present confusion surrounding this subject is reflected in the comments and work of the people involved in anaerobic digestion studies. Standard references for classification and identification, for instance Bergey's Manual (Breed *et al.*, 1957), Prevot and Fredette (1966) and Skerman (1967), fail to be of much help in identification. Toerien (1970a) attempted to use such means of identification, but found them to be most unsatisfactory. He describes 92 numerically important bacterial cultures isolated from an anaerobic digester. All the isolates were micro-aerophiles or obligate anaerobes. The identity of the isolates was reported at generic level with the majority of the organisms classified into genera. Of the organisms tested, 50 were of the *Corynebacterium, Lactobacillus, Ramibacterium, Actinomyces* or *Bifidobacterium* genera, and the rest were of the genera *Ramibacterium* (1), *Eubacterium* (6), *Lactobacillus* (8), *Clostridium* (7), *Bacteroides* (8), *Sphaerophorus* or *Fusobacterium* (4), *Vibrio* or *Spirillum* (1), *Peptococcus* (5), *Veillonella* (1), and an unknown crescent-shaped, Gram-negative bacterium (1). Kirsch (1969) found that the predominant bacterial flora consisted of Gram-negative, non spore-forming, obligately anaerobic bacilli. From 72 isolates, 3% were Gram-positive spore-formers, and 25% were motile, Gram-negative, and non spore-forming. Of the motile bacilli, all fermented carbohydrates to produce gas containing measurable amounts of hydrogen. These organisms also produced acetic, propionic and butyric acids in sterile sewage sludge, but produced only acetic acid from fermentable carbohydrates. All possessed single or multiple, lateral or sub-terminal flagella. The non-motile organisms failed to produce hydrogen from carbohydrates and did not produce butyric acid from sterile sewage sludge. From these data, 19 of the original 72 isolates were placed in a common group. All these organisms showed enhanced response to digester fluid enrichment. No further identification was attempted. Hobson and Shaw (1971) also avoided any formal identification of their isolates with the exception of those possessing hydrolytic activity.

The lack of characterization of organisms in anaerobic digesters due to the difficulties involved has given rise to the use of other taxonomic procedures. Toerien (1970b) examined organisms previously isolated and classified (Toerien, 1970a). Using a computer he presented a hierarchical Adansonian classification (agglomerative–polythetic) of the isolates based on similarity coefficients using 82 characteristics. By clustering those organisms showing close correlations, he found eight groups, or taxa, of organisms with similar physiological characteristics. The analysis was successful in that it demonstrated the presence of distinct groups of bacteria amongst the numbers tested. The metabolic end-product pattern of the groups were also shown to be homologous. Toerien (1970c) then attempted another computer analysis, this time using the principle component method. This is an analysis whereby each characteristic of the isolates is compared in order to determine any relationships between them. It was found that this method was successful since interrelationships between characteristics were revealed which could be interpreted ecologically. These interrelationships showed marked similarity to those determined from the agglomerative–polythetic classification.

The facultative anaerobes of digesters receiving raw sewage sludge were examined by Toerien (1967a). The major organisms were *Bacillus*, *Micrococcus* and *Pseudomonas* sp. Coliforms were present only in small numbers. The persistence of these organisms despite expected washout, suggested that they do play a role, though probably a minor one.

The biochemical activities of these facultative organisms will be extremely varied. Of the hydrolytic products from the first stage, the monosaccharides and disaccharides, long-chain fatty acids, glycerol, amino acids and short-chain peptides, provide the main carbon sources for growth. The major detectable end-products from the activity of these organisms are saturated fatty acids, carbon dioxide and ammonia. Other products such as alcohols, aldehydes and ketones occur in minute quantities, and the significance of these substances remains unknown. Toerien and Hattingh (1969) suggested that a high turnover rate of alcohols and ketones could account for the low level of these metabolites. The rapid production of acids induced by the addition of alcohols to digesting sludge would support this view. The products of cellulose breakdown have been shown by Hungate (1950) and Maki (1954) to include ethanol. It is possible, therefore, that alcohols may be more important intermediate products in digestion than their detectable concentrations would support. Hydrogen is another product known to be produced by the various anaerobic bacteria, although not detectable in digester gases. Hungate (1950) and Maki (1954) show hydrogen production from pure cultures of cellulolytic bacteria on cellulose. Kirsch (1969) showed that 20–30% of the isolates could ferment carbohydrates to yield hydrogen as well as butyric, acetic and propionic acids from sterile sewage sludge. The absence of hydrogen from normal digester gases suggests its rapid utilization by other bacteria, and it is probable that methane-forming bacteria utilize hydrogen as quickly as it is produced. The "S" organism, discovered by Bryant *et al.* (1967), growing symbiotically in cultures of *Methanobacillus omelianskii*, appears to be an organism providing hydrogen and ethanol. The methanogenic partner utilized the hydrogen produced at sufficient rate to maintain the activity of the "S" organism, which became inhibited by excessive hydrogen. Thiel (1969) showed that by selectively inhibiting methanogenesis with methane analogues, such as chloroform, hydrogen was able to accumulate, suggesting the important role of hydrogen as a substrate for methanogenesis in digestion. The ability of the anaerobic methanogenic bacteria to grow on carbon dioxide and hydrogen will be discussed in the next section.

Toerien (1970a) lists the fatty acid end-products of his anaerobic isolates. He reports acetic, propionic and lactic acids as being the most frequently produced acid end-products, with 80, 62 and 64 isolates out of 92 producing these acids respectively. Butyric acid was produced by only nine isolates. No analyses of gases produced were given.

The breakdown of lipid appears to be by β-oxidation of the long-chain fatty acid. The glycerol portion of the lipid is probably degraded to pyruvic acid and hence to the volatile fatty acids or alcohols.

The degradation of proteins to polypeptides and amino acids by protease activity is carried out by the proteolytic organisms, as mentioned earlier. The resulting amino acids can be degraded by many mechanisms, depending on which organism is involved. Sokatch (1969) lists some of the proteases and peptidases produced from various organisms. With such large numbers of amino acids it is not intended to examine the breakdown of each amino acid individually. Toerien and Kotze (1970) further examined the intermediary metabolism of the non-methanogenic bacteria with reference to amino acid metabolism. This involved examining and assaying for a number of enzymes capable of converting amino acids. All the organisms tested were from previous isolations (Toerien, 1970a, b, c), being representatives of each group resolved by the computerized analyses. The above workers found that the metabolism of amino acids followed the pathways already known.

The breakdown of carbohydrates was also examined by Toerien and Kotze (1970). They examined the same isolates for the presence of a number of enzymes associated with the carbohydrate catabolism and found that glycolytic enzymes, as well as some of the citric acid cycle enzymes, occurred in all isolates examined. The fructose-6-phosphate phosphoketolase pathway was shown to be used by the bifid-like isolates.

The work of Hattingh *et al.* (1967), Thiel *et al.* (1968) and Kotze *et al.* (1968) demonstrated the presence of a large number of enzymes in anaerobic digesters, and provided a complementary approach to the study of the activity of laboratory isolates.

It can be seen that our knowledge concerning the micro-organisms and biochemical activities occurring in the hydrolytic and acid-forming stages of anaerobic digestion is still somewhat limited. The classical methods of bacteriology, involving identification and classification, fail to provide satisfactory concepts of the ecology of sludge digestion. Computerized bacteriology seems likely to aid the bacteriologist greatly, not only in the study of anaerobic digestion, but also in other ecological problems where a wide variety of micro-organisms apparently exist harmoniously.

C. Anaerobic Methanogenic Bacteria

These bacteria are responsible for the conversion of the end-products of the acid-forming phase into the gaseous hydrocarbon, methane. Wolfe (1971) pointed out what an ideal end-product methane is, being poorly soluble, inert under anaerobic conditions, non-toxic, and able to escape from the anaerobic environment. The bacterial formation of methane is not peculiar to anaerobic digesters. It also occurs in other environments, for instance in the black mud of lakes and swamps, and in the digestive tracts of animals, particularly in the rumen of herbivorous animals and in the caecum of non-ruminant animals. It is believed that the micro-organisms responsible for methanogenesis in each of these different environments are similar; the study of the methanogenic bacteria has involved examination of isolates from each of these environments, but in spite of several notable studies (Barker, 1956; Wolfe, 1971; Kirsch and Sykes, 1971) our knowledge of these bacteria is still incomplete. A particular problem in the study of these bacteria appears to be the difficulty of obtaining the organisms in pure culture (Barker, 1956). Because of their extreme obligate anaerobic nature there have been almost insurmountable problems in this respect; a good example of this is illustrated by the organism *Methanobacillus omelianskii* isolated by Barker (1940). Although for many years considered to be a characterized pure culture, providing methane from ethanol, Bryant *et al.* (1967) showed that when examined in detail the culture consisted of two different organisms growing symbiotically.

The methanogenic bacteria believed isolated as pure cultures from digestion tanks are shown in Table III. Of the methanogenic bacteria that have not been obtained in pure culture, Barker (1956) lists four of these, being *Methanobacterium suboxydans*, *Methanobacterium söhngenii*, *Methanosarcina methanica* and *Methanococcus mazei*.

Considering the organisms from Table III, *Methanobacterium ruminantium* was the name given by Smith and Hungate (1958) to a non-motile, non-sporing, Gram-positive, encapsulated rod with rounded ends. It was originally isolated from the bovine rumen. The organism is able to grow with formate, or carbon dioxide and hydrogen as substrates.

Methanobacterium strain MoH was isolated from a culture of *Methanobacillus omelianskii* (Bryant *et al.*, 1967). The culture of *M. omelianskii* under phase-contrast microscopy appeared to show two distinct morphological types. These were (1) an irregularly curved slender

TABLE III. Methanogenic organisms isolated as pure culture from digesting sludge (Wolfe, 1971)

Organisms	Source	Morphology	Gram reaction	Substrates
Methanobacterium ruminatium	rumen and sludge	coccus to short rod in chains	positive	H_2+CO_2, formate
Methanobacterium strain MoH	*Methanobacillus omelianskii*	irregularly curved rod	variable	H_2+CO_2
Methanobacterium formicicum	mud and sludge	irregularly curved rod	variable	H_2+CO_2, formate
Methanosarcina barkerii	mud and sludge	sarcina	positive	H_2+CO_2, methanol, acetate
Methanospirillum sp.	sludge	spirillum	positive	H_2+CO_2, formate
Methanococcus sp.	sludge	coccus	positive	H_2+CO_2, formate

rod with rounded ends, occurring singularly or in short chains, which was the predominant type, and (2) a curved rod, wider and shorter, and with less rounded ends. When grown on a medium in a hydrogen–carbon dioxide atmosphere, it was found that *M. omelianskii* rapidly lost its ability to grow on ethanol. In an ethanol–hydrogen–carbon dioxide medium only one type of colony was present, that of the slender rod organism. This was found to produce methane from hydrogen and carbon dioxide, but was unable to produce methane from ethanol and carbon dioxide. By using an ethanol agar with a nitrogen–carbon dioxide atmosphere, only one colony type was isolated, being morphologically similar to the shorter rod observed in the original culture. This shorter rod failed to produce methane, but produced small amounts of hydrogen and acetate. Ethanol needed to be present before growth was possible. A mixture of the two cultures produced excellent growth with the production of methane from ethanol in a carbon dioxide atmosphere. *Methanobacterium ruminantium* mixed with the non-methanogenic organism also showed the ability to produce methane from ethanol. This evidence would seem to be quite conclusive in showing that *Methanobacillus omelianskii* is, in fact, two symbiotic organisms. Further evidence of this is given by DNA analysis, which showed two forms of DNA in the original *M. omelianskii* culture but only one in the resolved methanogenic isolate. This organism is now called *Methanobacterium* strain MoH. The non-methanogenic partner is termed the "S" organism. These "S" organisms grow symbiotically on ethanol because their ability to ferment the ethanol to hydrogen and acetate. *Methanobacterium* strain MoH is able to use the hydrogen so produced to reduce carbon dioxide to methane. The removal of the hydrogen is important because the "S" organism is inhibited by the hydrogen it produces.

Methanobacterium formicicum was first isolated by Schnellen in 1947 (Barker, 1956). It is able to grow on formate, and carbon dioxide and hydrogen. *M. formicicum* was also isolated by Mylroie and Hungate (1954) from digesting sludge; they found that it was the most numerous methanogenic bacteria occurring in the sludges producing methane that were tested. Although not ruling out the role played by other methanogenic bacteria, it was concluded that *M. formicicum* plays the major role in sludge fermentation. Its effect when inoculated into sludge is to produce rapid conversion of formate to methane and carbon dioxide, which would confirm this finding. Smith (1961) also found an organism resembling *M. formicicum* in a domestic sewage sludge, although *M. ruminantium* appeared to have been more numerous in that case. Smith (1965) examined the substrate utilization of *M. formicicum*, and showed that acetate, glucose and methanol were not utilized, whereas hydrogen and formate gave enhanced production of methane. Later, Smith (1966) found *M. formicicum* in sludge in numbers greater than 1×10^7/ml.

Methanosarcina barkerii was also isolated by Schnellen in 1947 (Barker, 1956). This is a sarcina which is able to produce methane from methanol, acetate, and hydrogen and carbon dioxide. Kluyver and Schnellen (1947) showed that *M. barkerii* also acts on carbon monoxide in the absence of hydrogen, unlike *Methanobacillus omelianskii*, which was unable to utilize carbon monoxide. Smith (1965) confirmed that *M. barkerii* was able to grow on hydrogen, methanol and acetate, whereas glucose and formate were not utilized, and later (Smith, 1966) found it to be present in sludge in numbers exceeding 1×10^6/ml, although he did not consider it to be amongst the most predominant of methanogenic bacteria. Wolfe (1971) pointed out that *M. barkerii* is noticeably different from the other methanogenic bacteria, in that hydrogen and carbon dioxide or formate are not the preferred substrates for growth.

The organism *Methanospirillum* sp. listed by Wolfe (1971) does not appear to have been isolated by any other worker and information on the physiology and biochemistry of this organism is limited. Wolfe (1971) describes it as being a Gram-positive spirillum, able to produce methane from hydrogen and carbon dioxide, and formate. No other information about this organism is given, although photographs showing the flagella and spiral form of the organism are given in the paper. Similarly the identity of the *Methanococcus* sp. also remains a mystery, although Wolfe listed it to be a Gram-positive coccus producing methane from hydrogen and carbon dioxide, and formate. It could be the same *Methanococcus* sp. as that isolated by Smith (1966), who found a new species of

Methanococcus in sludge in numbers exceeding 1×10^6/ml. That organism was also able to oxidize molecular hydrogen, and formic acid. Another methane bacterium, also isolated by Smith (1966), was one from a propionic acid enrichment, and was found in sludge in numbers greater than 1×10^7/ml. Neither of these latter organisms was examined for the ability to utilize substrates other than hydrogen.

The following four organisms were described by Barker (1956) but were not included in the recent list of organisms isolated from sewage sludge (Wolfe, 1971). *Methanobacterium söhngenii* is a straight or slightly curved, non-motile, Gram-negative rod that is able to ferment acetate and butyrate to methane. *Methanobacterium propionicum*, is an organism with the ability to produce methane from propionic acid. *Methanococcus mazei*, isolated by Barker (1936b), is able to ferment acetate and butyrate, and *Methanosarcina methanica* is able to ferment carbon dioxide, acetate and butyrate to methane. With the exception of *M. propionicum*, all these organisms are listed in Bergey's Manual (Breed *et al.*, 1957).

There have been numerous attempts to obtain total counts of methanogenic bacteria in digested sludge. Heukelekian and Heinemann (1939a, b) obtained counts from $600–25\,000 \times 10^3$/ml ($24–700 \times 10^6$/g solids) of acetate fermenters; $60–6000 \times 10^3$/ml ($2{\cdot}4–170 \times 10^6$/g solids) of butyrate fermenters; and $1300–250\,000 \times 10^3$/ml ($52–7000 \times 10^6$/g solids) of ethanol fermenters. In each case, the lower figure corresponds to a slowly gasifying digester, and the higher to a rapidly gasifying digester. The above workers found a close relationship between gas production and numbers of methanogenic organisms. Mylroie and Hungate (1954) found numbers between $10^5–10^8$/ml. Siebert and Hattingh (1967) also used the methods of Heukelekian and Heinemann (1939a, b), and found numbers of 500×10^3/g volatile suspended solids (VSS) formate fermenters, 50×10^3/g VSS acetate fermenters, 100×10^3/g VSS butyrate fermenters, and 300×10^3/g VSS ethanol fermenters. They showed that long periods of incubation produced higher counts for methane bacteria; their results were obtained after 83 days. Siebert *et al.* (1968) examined various media used for the enumeration of methanogenic bacteria in various habitats, including digesting sludges. They found populations ranging from $1–486 \times 10^8$/ml, which is in agreement with Smith (1966), who found that methanogenic bacteria in sludge existed in numbers greater than 10^8/ml. Hobson and Shaw (1971) examined a piggery waste sludge using the medium of Siebert and Hattingh (1967) enriched with 20% clarified digester fluid, with the addition of formate, acetate, propionate or butyrate in a carbon dioxide atmosphere. It was found that only the media containing formic and butyric acids gave rise to methane, and numbers in the order of $10^5–10^6$/ml were obtained from these digesters.

From a large number of isolates taken, all appeared to be the same and were classified as *Methanobacterium formicicum*, capable of producing methane from carbon dioxide and hydrogen, or formate. The organisms were not able to produce methane from acetate, propionate, butyrate, valerate, isovalerate, succinate, pyruvate, glucose, ethanol, propanol, butanol or isopropanol. Because the digesting sludge was not domestic, it would be difficult to conclude that these figures reinforce the findings of Mylroie and Hungate (1954). Hattingh *et al.* (1967), using the method of Siebert and Hattingh (1967), showed $84–2400 \times 10^3$/g VSS methanogenic bacteria able to ferment formic acid, while Kotze *et al.* (1968) gave counts of $18{\cdot}3 \times 10^4$/g VSS for the same organism. Thiel *et al.* (1968) examined three laboratory scale digesters fed on a synthetic substrate for a large number of parameters. The populations of methanogenic bacteria were estimated using the most probable number (m.p.n.) technique and their numbers ranged from $4–1600 \times 10^4$/ml. Attempts to correlate the numbers of methanogenic bacteria with the other parameters, in particular methane formation, were surprisingly unsuccessful. Possible correlations were indicated, however, between the numbers of methanogenic bacteria and the concentrations of the enzymes 6-phosphofructokinase and malate dehydrogenase, but no correlation was found with methane production. The authors conclude that the m.p.n. method of counting was so unreliable that it was responsible for the lack of correlation. Another factor could be that the cultural conditions only provided enumeration of some of the methanogenic bacteria, and that these comprised only an irregular portion of the total methanogenic population. It should be remembered that Smith (1966) observed that a large population of methanogenic bacteria existed which could not be isolated.

D. The Biochemistry of Methanogenesis

Despite the use of advanced biochemical methods in the examination of methanogenesis, our knowledge concerning this process is still slight and many aspects remain to be elucidated. For more detailed reviews of the biochemistry of methanogenesis, the reader is referred to Barker (1956), Stadtman (1967) and Wolfe (1971).

Söhngen showed that the methanogenic bacteria could convert mixtures of carbon dioxide and hydrogen to methane as in Eqn (1).

$$CO_2 + 4H_2 \rightarrow CH_4 + 2H_2O \qquad (1)$$

Van Neil later postulated a more general equation, inferring the reduction of carbon dioxide as being the major action. This is generalized in

Eqn (2). Compound H_2A is any substance able to be oxidized.

$$4H_2A + CO_2 \rightarrow 4A + CH_4 + 2H_2O \quad (2)$$

This hypothesis was supported by Barker (1936a) in his studies on the methane bacteria, with the observation that the amount of absorbed carbon dioxide was equivalent to the amount of methane produced. Stadman and Barker (1949) later confirmed this finding using C^{14} labelled carbon dioxide, and showed that the methane produced in this fermentation was formed entirely by reduction of carbon dioxide. Stadtman and Barker (1951a) showed that *Methanobacterium suboxydans* formed methane from carbon dioxide in the fermentations of butyrate, valerate and caproate, and *Methanobacterium propionicum* formed methane from carbon dioxide in a fermentation of propionate. *Methanobacillus omelianskii*, in the absence of carbon dioxide, produced acetic acid and hydrogen from ethanol, or acetaldehyde. The increase of hydrogen in the atmosphere inhibited the oxidation of ethanol, which was later explained by the work of Bryant *et al.* (1967) mentioned previously. Stadtman and Barker (1951b) showed that *Methanosarcina* formed methane from the methyl group of acetate and methanol rather than from carbon dioxide in the fermentations. (Eqns (3), (4), (5).)

$$C^*H_3COOH \rightarrow C^*H_4 + CO_2 \quad (3)$$

$$CH_3OH + H_2O \rightarrow CO_2 + 6H \quad (4)$$

$$3CH_3OH + 6H \rightarrow 3CH_4 + 3H_2O \quad (5)$$

$$4CH_3OH \rightarrow 3CH_4 + CO_2 + 2H_2O \quad (4)+(5)$$

These two results are inconsistent with the hypothesis of Van Neil since the methyl group is the precursor of methane and not carbon dioxide. Pine and Barker (1956) used deuterium labelled acetate and deuterium oxide to establish that the methyl group of the acetate remained intact, as illustrated in Eqns (6) and (7).

$$CD_3COOH \xrightarrow{H_2O} CD_3H + CO_2 \quad (6)$$

$$CH_3COOH \xrightarrow{D_2O} CH_3D + CO_2 \quad (7)$$

Kluyver and Schnellen (1947) showed that *Methanosarcina barkerii* could convert carbon monoxide to methane as shown in Eqn (8).

$$CO + 3H_2 \rightarrow CH_4 + H_2O \quad (8)$$

It is thought that this proceeds in two stages which are represented by

Eqns (9) and (10).

$$CO + H_2O \rightarrow CO_2 + H_2 \quad (9)$$

$$CO_2 + 4H_2 \rightarrow CH_4 + 2H_2O \quad (10)$$

Hydrogen is not necessary for the reaction and *M. barkerii* can bring the conversion about in atmospheres containing 100% carbon monoxide. In the methane fermentation of formate, Stadtman and Barker (1951c) showed that *Methanococcus vanniellii*, isolated from mud, converted formate into carbon dioxide, hydrogen and methane. However, evidence as to carbon dioxide reduction occurring in this mechanism is not conclusive. Fina *et al.* (1960) presented evidence to suggest that direct reduction of formic acid might take place, although Barker (1956) stated that formate and carbon monoxide are first oxidized to carbon dioxide which is then reduced.

Barker (1956) outlines a schematic representation of the pathways possible in the methane fermentation which is shown in Fig. 3, where it can be seen that both carbon dioxide reduction and the transmethylation of acetate and methanol are possible by this mechanism. Pine and Vishniac (1957) proposed an alternative scheme which is illustrated in Fig. 4, although this was intended for the acetate and methanol fermentations.

The inability to find intermediates such as formaldehyde or methanol in the reduction of carbon dioxide to methane, suggests that the carbon dioxide might be bound to some carrier. This was proposed by both

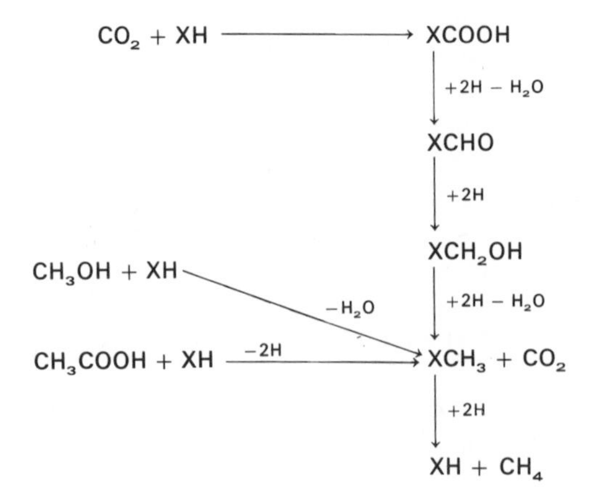

FIG. 3. Possible metabolic pathways in methane fermentation (from Barker, 1956).

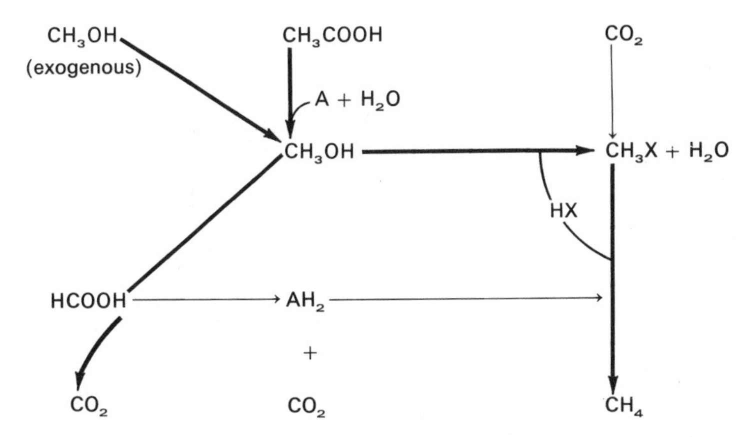

FIG. 4. Metabolic pathway in the production of methane by acetate and methanol fermentation (from Pine and Vishniac, 1957).

Barker (1956) and Pine and Vishniac (1957) in their respective schemes. The nature of this unknown compound termed "X" remains to be discovered, and also whether the compound is the same for each substrate.

Advances have been made by the use of cell-free extracts of methanogenic bacteria using known compounds as possible substrates for the production of methane from these preparations. Table IV shows the known substrates from which cell-free extracts are able to produce methane.

The first of these substrates shown to be converted to methane by cell-free extracts was pyruvate (E. A. Wolin *et al.*, 1963). These workers

TABLE IV. Compounds from which cell-free extracts of methanogenic bacteria are able to produce methane (taken from Wolfe, 1971)

Compound
C^*O_2
CH_3COC^*OOH
$C^*H_2OHCHNH_2COOH$
5, C^*H_3—H_4—folate
5, 10, C^*H_2—H_4—folate
C^*H_3—B_{12}
HC^*OOH
C^*H_3OH
HC^*HO

* Represents the C atom converted to CH_4

found that cell-free extracts of *Methanobacillus omelianskii* were able to produce methane from hydrogen, carbon dioxide, coenzyme A, and adenosine-5′-triphosphate (ATP), and furthermore that pyruvate, serine, or *o*-phosphoserine substituted for ATP and carbon dioxide. Ethanol, which was normally considered to be the substrate for *M. omelianskii*, was found to be ineffective in methane production, whereas hydrogen was shown to be essential for methane production. The compound ferrodoxin was also shown to stimulate methane formation from pyruvate, although above a certain concentration ferrodoxin became inhibitory. Wood *et al.* (1965) showed that carbon 3 of L-serine became the carbon of the methane produced in the fermentation. Blaylock and Stadtman (1963, 1964) showed that cell-free extracts of *Methanosarcina barkerii* were able to convert methylcobalamin (CH_3—B_{12}) to methane which was formed from the methyl group. In addition, the reaction was shown to depend on the amount of pyruvate added, which was used as an electron donor. M. J. Wolin *et al.* (1963) showed that the formation of methane from methylcobalamin was dependent on ATP. Pyruvate was shown to be unnecessary for the formation of methane from methylcobalamin by this organism. From the evidence of these workers it seemed that ATP was necessary for the terminal reduction of the methyl group. Adenosine-5′-diphosphate (ADP) could substitute for ATP, but adenosine-5′-monophosphate (AMP) was unable to provide energy for methanogenesis. Wolin *et al.* (1964) showed that the cobalamin (B_{12}) appeared to be spectrographically the same as vitamin B_{12}. They showed that an electron source causes the cleavage of the CH_3—B_{12} complex to form CH_4 and B_{12}, or its thiol complex, but the nature of the electron donor was not identified.

Another compound very similar to methylcobalamin is methyl-factor III isolated from *Methanobacillus omelianskii* (Lezius and Barker, 1965). Hydrolysis of this cobamide produces methyl-factor B. Both of these compounds were shown by Wood *et al.* (1965) to act as substrates for methanogenesis, but the extent to which cobamide compounds are involved in methanogenesis still remains to be discovered. If they are involved, the process for the production of methylcobamides is also by an unknown mechanism. Several other factors, some uncharacterized as yet, are involved in the complex methanogenesis process. These include a methyl corrinoid derivative acid-stable co-factor and ATP (Blaylock and Stadtman, 1966).

N-5, N-10-methylene tetrahydrofolate and N-5-methyl tetrahydrofolate have also been shown to act as substrates for methanogenesis (Wood and Wolfe, 1965) and as with many other substrates, there was a requirement for ATP. It is possible that tetrahydrofolate derivatives may play a role in methanogenesis, either as an alternative to the methyl-cobalamin derivatives, or in association with them. The requirement for ATP has so far been found necessary in all systems examined (Wolfe, 1971), but the exact role played by ATP and the mechanism by which it is produced by methanogenic bacteria remain unknown.

One other aspect in the physiology of these important anaerobic bacteria, pointed out by Toerien *et al.* (1970), is that if polysaccharide cellular material can be produced from such simple substrates as carbon dioxide and hydrogen, which are the sole requirements of methanogenic bacteria, then why should not the reverse process be possible, of carbon dioxide and hydrogen production from polysaccharide material outside of the cell with the consequent conversion to methane. From the literature, there appears to be only one indication of the existence of this metabolic route, namely that of Pretorius (1968) from Toerien *et al.* (1970). It is clear, therefore, that many details of the most interesting process of methanogenesis remain still to be elucidated.

References

Barker, H. A. (1936a). *Arch. Mikrobiol.* **7,** 404–419.
Barker, H. A. (1936b). *Arch. Mikrobiol.* **7,** 420–438.
Barker, H. A. (1940). *Antonie van Leeuwenhoek* **6,** 201–220.
Barker, H. A. (1956). "Bacterial Fermentations." John Wiley, New York.
Blaylock, B. A. and Stadtman, T. C. (1963). *Biochem. biophys. Res. Commun.* **11,** 34–38.
Blaylock, B. A. and Stadtman, T. C. (1964). *Ann. N.Y. Acad. Sci.* **112,** 799–803.
Blaylock, B. A. and Stadtman, T. C. (1966). *Archs Biochem. Biophys.* **116,** 138–158.
Breed, R. S., Murray, E. E. D. and Smith N. R. (1957). "Bergey's Manual of Determinative Bacteriology" (7th Edn). Williams and Wilkins, Baltimore, Maryland, U.S.A.
Bryant, M. P., Wolin, E. A., Wolin, M. J. and Wolfe, R. S. (1967). *Arch. Mikrobiol.* **59,** 20–31.
Fina, L. R., Sincher, H. J. and deCou, D. F. (1960). *Archs Biochem. Biophys.* **91,** 159–162.
Gaub, W. H. (1924). *New Jersey agric. Exp. Stn Bull.* **394,** 1–24.
Harkness, N. (1966). *J. Proc. Inst. Sew. Purif.* 542–554.
Hattingh, W. H. J., Kotze, J. P., Thiel, P. G., Toerien, D. F. and Siebert, M. L. (1967). *Wat. Res.* **1,** 255–277.
Heukelekian, H. (1948). *Sewage Wks J.* **20,** 490–498.

HEUKELEKIAN, H. (1957). *In* "Biological Treatment of Sewage and Industrial Wastes" (Eds Brother J. McCabe and W. W. Eckenfelder, Jr.), Vol. 2, pp. 25–43. Reinhold, New York.
HEUKELEKIAN, H. and HEINEMAN, B. (1939a). *Sewage Wks J.* **11,** 426–435.
HEUKELEKIAN, H. and HEINEMAN, B. (1939b). *Sewage Wks J.* **11,** 436–444.
HOBSON, P. N. and SHAW, B. G. (1971). *In* "Microbial Aspects of Pollution". Soc. Appl. Bact. Symp. Ser. No. 1 (Eds G. Sykes and F. A. Skinner), pp. 103–121. Academic Press, London and New York.
HOTCHKISS, M. (1924). *J. Bact.* **9,** 437–454.
VAN HOUTE, J. and GIBBONS, R. J. (1966). *Antonie van Leeuwenhoek* **32,** 212–222.
HUNGATE, R. E. (1950). *Bact. Rev.* **14,** 1–49.
KIFF, R. J. (1972). *Wat. Pollut. Control* **71,** 475–484.
KIRSCH, E. J. (1969). *Devs ind. Microbiol.* **10,** 170–176.
KIRSCH, E. J. and SYKES, R. M. (1971). *Progr. ind. Microbiol.* **9,** 155–237.
KLUYVER, A. J. and SCHNELLEN, C. G. T. P. (1947). *Archs Biochem.* **14,** 57–70.
KOTZE, J. P., THIEL, P. G., TOERIEN, D. F., SEIBERT, M. L. and HATTINGH, W. H. J. (1968). *Wat. Res.* **2,** 195–213.
LEZIUS, A. G. and BARKER, H. A. (1965). *Biochemistry* **4,** 510–518.
LIGHTHART, B. and OGLESBY, R. T. (1969). *J. Wat. Pollut. Control Fed.* **41,** R267–281.
MAH, R. A. and SUSSMANN, C. (1968). *Appl. Microbiol.* **16,** 358–361.
MAKI, L. R. (1954). *Antonie van Leeuwenhoek* **20,** 185–200.
MARA, D. D. and WILLIAMS, D. J. A. (1970). *J. appl. Bact.* **33,** 543–552.
MCCARTY, P. L. (1964). *Publ. Wks* **95,** 107–112.
MYLROIE, R. L. and HUNGATE, R. E. (1954). *Can. J. Microbiol.* **1,** 55–64.
O'SHAUGHNESSY, F. R. (1914). *J. Chem. Soc. Ind.* **33,** 3–23.
PAINTER, H. A. (1970). *Wat. Res.* **4,** 393–450.
PINE, M. J. and BARKER, H. A. (1956). *J. Bact.* **71,** 644–648.
PINE, M. J. and VISHNIAC, W. (1957). *J. Bact.* **73,** 736–742.
PIPES, W. O. (1967). *Adv. Appl. Microbiol.* **9,** 185–234.
POST, F. J., ALLEN, A. D. and REID, T. C. (1967). *Appl. Microbiol.* **15,** 213–218.
POSTGATE, J. (1959). *A. Rev. Microbiol.* **13,** 505–520.
PREVOT, A. and FREDETTE, V. (1966). "Manual for the Classification and Determination of the Anaerobic Bacteria." Lea and Fabiger, Philadelphia, U.S.A.
RUCHHOFT, C. C., KELLER, J. G. and EDWARDS, G. P. (1930). *J. Bact.* **19,** 269–294.
SALLE, A. J. (1954). "Fundamental Principles of Bacteriology" (4th Edn). McGraw-Hill, London.
SIEBERT, M. L. and HATTINGH, W. H. J. (1967). *Wat. Res.* **1,** 13–19.

SIEBERT, M. L. and TOERIEN, D. F. (1969). *Wat. Res.* **3,** 241–250.
SIEBERT, M. L., TOERIEN, D. F. and HATTINGH, W. H. J. (1968). *Wat. Res.* **2,** 545–554.
SKERMAN, V. B. D. (1967). "A Guide to the Identification of the Genera of Bacteria" (2nd Edn). Williams and Wilkins, Baltimore, Maryland, U.S.A.
SMITH, P. H. (1961). *Bact. Proc.* **61,** 60.
SMITH, P. H. (1965). *Proc. 20th Ind. Waste Conf., Purdue* 583–588.
SMITH, P. H. (1966). *Devs Ind. Microbiol.* **7,** 156–161.
SMITH, P. H. and HUNGATE, R. E. (1958). *J. Bact.* **75,** 713–715.
SOKATCH, J. R. (1969). "Bacterial Physiology and Metabolism." Academic Press, New York and London.
STADTMAN, T. C. (1967). *A. Rev. Microbiol.* **21,** 121–142.
STADTMAN, T. C. and BARKER, H. A. (1949). *Archs Biochem.* **21,** 256–264.
STADTMAN, T. C. and BARKER, H. A. (1951a). *J. Bact.* **61,** 67–80.
STADTMAN, T. C. and BARKER, H. A. (1951b). *J. Bact.* **61,** 81–86.
STADTMAN, T. C. and BARKER, H. A. (1951c). *J. Bact.* **62,** 269–280.
THIEL, P. G. (1969). *Wat. Res.* **3,** 215–223.
THIEL, P. G., TOERIEN, D. F., HATTINGH, W. H. J., KOTZE, J. P. and SIEBERT, M. L. (1968). *Wat. Res.* **2,** 393–408.
TOERIEN, D. F. (1967a). *Wat. Res.* **1,** 55–59.
TOERIEN, D. F. (1967b). *Wat. Res.* **1,** 147–155.
TOERIEN, D. F. (1970a). *Wat. Res.* **4,** 129–148.
TOERIEN, D. F. (1970b). *Wat. Res.* **4,** 285–303.
TOERIEN, D. F. (1970c). *Wat. Res.* **4,** 305–314.
TOERIEN, D. F. and HATTINGH, W. H. J. (1969). *Wat. Res.* **3,** 385–416.
TOERIEN, D. F. and KOTZE, J. P. (1970). *Wat. Res.* **4,** 315–326.
TOERIEN, D. F., SIEBERT, M. L. and HATTINGH, W. H. J. (1967). *Wat. Res.* **1,** 497–507.
TOERIEN, D. F., THIEL, P. G. and HATTINGH, M. M. (1968). *Wat. Res.* **2,** 505–513.
TOERIEN, D. F., THIEL, P. G. and PRETORIUS, W. A. (1970). Paper presented to *5th int. Conf. Wat. Pollut. Res.* II–29/1–7.
WOLFE, R. S. (1971). *Adv. Microb. Physiology* **6,** 107–146.
WOLIN, E. A., WOLIN, M. J. and WOLFE, R. S. (1963). *J. biol. Chem.* **238,** 2882–2886.
WOLIN, M. J., WOLIN, E. A. and WOLFE, R. S. (1963). *Biochem. biophys. Res. Commun.* **12,** 464–468.
WOLIN, M. J., WOLIN, E. A. and WOLFE, R. S. (1964). *Biochem. biophys. Res. Commun.* **15,** 420–423.
WOOD, J. M. and WOLFE, R. S. (1965). *Biochem. biophys. Res. Commun.* **19,** 306–311.
WOOD, J. M., ALLAM, A. M., BRILL, W. J. and WOLFE, R. S. (1965). *J. biol. Chem.* **240,** 4564–4569.

3

Fungi

T. G. Tomlinson

2 Deard's End Lane
Knebworth
Herts
England

I. L. Williams

Applied Hydrobiology Section
University of Aston in Birmingham
Birmingham B4 7ET
England

I. Introduction

The majority of records of fungi flourishing in waste-treatment plants relate to their occurrence in the film of percolating filters (bacteria beds, or trickling filters). They are generally considered undesirable as dominant members of the film community; an accumulation of a predominantly fungal film quickly causes blockages of the interstices between the stones, with consequent impedance of aeration and drainage. Surface pools of sewage may collect, a condition known as ponding, and within the filter anaerobic regions may develop, with a resultant deterioration in the efficiency of purification since this is dependent on the metabolic activity of aerobic micro-organisms. Further, heavy fungal film growths may support a very large population of fly larvae and result in a fly nuisance. Fungi have been recorded also on occasion as major constituents of the floc in activated-sludge plants, causing bulking of the sludge with a consequent fall in effluent quality. Saprophytic fungi can be as efficient as saprophytic bacteria in the removal of organic materials from sewage, but there is evidence that they produce greater biomass for weight of nutrient utilized than certain of the bacteria important in purification (Water Pollution Research, 1955). This could cause more rapid film accumulation under favourable conditions and eventually more sludge production. Further, the mycelial habit renders fungal film tougher and less readily sloughed than thick growths of predominantly bacterial film in percolating filters, and the interwoven masses of hyphae are responsible for the lowering of density of the activated sludge floc which results in bulking. Thus fungi, however efficient in purification, are a nuisance when they become dominant in the film of percolating filters, and their presence in quantity in activated-sludge plants is often disastrous. Effective control is essential to maintain effluent standards, and this depends ultimately on a knowledge of the factors favouring their development in biological oxidative treatment processes, although much empirical work has been done towards more efficient works operation and this has sometimes resulted in a notable reduction in the incidence of troublesome fungi. (Section II A3.)

Fungi may also occur in channels and pipes on treatment plants, or as growths anchored to weirs or machinery, sometimes causing blockages. The anaerobic sludge digestion tank might be thought an unfavourable environment for a group of organisms classically regarded as strictly aerobic, but Cooke (1965) has found that a number of fungi can make limited growth under these conditions. However, by contrast with their role in aerobic treatment processes, they appear to contribute little towards the process of anaerobic sludge digestion. (Section IIC.)

During the past twenty years much ecological survey work has been carried out in the United States on the occurrence of fungi in polluted waters; Cooke (1954a), in a literature review covering records from all types of polluted aquatic habitats in Europe and N. America, found only ten species named as isolates from sewage-treatment plants by earlier workers, yet his own preliminary surveys in this field had yielded over fifty species. This difference can be accounted for by a difference of approach; previous workers had isolated dominant organisms for masses of fungal film or slime causing problems in works operation, while Cooke sought to recover all fungal species present in the source material, considering all to be of potential importance in the ecosystem. Apart from fungi flourishing at the expense of dissolved organic substances in the sewage, some fungi in treatment plants live saprophytically on dead insects, others as predators of protozoa, nematodes or rotifers; their role in the ecosystem is probably a minor one as a rule. Others enter with the sewage as disseminules and may be trapped temporarily in the treatment plant, but die or leave in the effluent, never having flourished or even germinated in some cases, and thus have no significance in the ecosystem. Becker and Shaw (1955) made a quantitative study of fungi present at various stages in the treatment of domestic sewage, and concluded that the majority of entering fungi fail to multiply in sewage systems, whilst those that do flourish may make extensive vegetative growth but sporulate sparsely; thus plate counts of fungal disseminules in the effluent give no guide to the extent of the occurrence of any species in the percolating filters.

A check list of fungi isolated from sewage and polluted waters has been prepared (Cooke, 1957a) and over a hundred of these illustrated in a laboratory guide to fungi in polluted waters (Cooke, 1963). A scheme for categorizing fungi from such habitats was published in the former paper; using the prefix "lyma" (from a Greek word meaning filth) to denote organic pollution of domestic or industrial origin, four groupings were made:

lymabiont species—found growing only in the presence of such pollutants
lymaphilic species—found growing commonly in the presence of such pollutants but also found growing elsewhere
lymaxene species—growing occasionally in such habitats but more commonly elsewhere
lymaphobe species—not found growing in the organically polluted habitat

This system was based on nutrient requirements and tolerances, and the distinction was made on the percentage incidence in samples from the habitat. There was an attempt to discover the relative importance of the various species obtained from bacteria beds by a system of occurrence scoring; in 112 species isolated in a study of the Dayton (Ohio) beds by Cooke (1958a), 29 were present often enough and in sufficient numbers to be considered permanent members of the population, forming an appreciable component of filter slime, and occurrence values for each species were obtained and the fungi arranged in order of the size of these values. The occurrence value for a given species was obtained by calculating the number of colonies of that species detected by plating techniques as a percentage of the total number of colonies counted for each station, the percentage occurrences then being added together to give an occurrence value. The majority of the species were lymaxene, and it is implied that the 29 regularly occurring species were regarded by Cooke as lymaphilic, while a number of species were of doubtful position in respect of these two categories. Only two species of the 112 were identified as lymabionts: *Ascodesmis microscopica*, found only once (and with only two previous records, both from mammalian dung), and *Subbaromyces splendens*, known only from percolating filters, from which it was first isolated and named by Hesseltine (1953). Cooke himself pointed out elsewhere (Cooke, 1959) that in view of the very brief history of the percolating filter as a habitat it is highly unlikely that any species have become adapted to life in it; however, the natural habitat of *Subbaromyces splendens* remains obscure.

Although extensive survey work on the fungal populations of sewage treatment plants has been carried out in the United States, records of nuisance from excessive fungal growths are few, and ponding on percolating filters seems to be uncommon. By contrast, in the U.K. much ecological work has been carried out on fungus-dominated film in percolating filters towards improvement of operating efficiency by the control of film accumulation. This work is discussed in detail later in Section II A3. Studies on the physiology of these fungi are few, and the inclusion of some hitherto unpublished work in Section III B is an attempt to redress the balance. The notes on species in Section III C are intended to help in identification and to draw attention to literature helpful in taxonomic aspects.

II. Occurrence of Fungi in Treatment Plants

A. Percolating Filters

1. *General Considerations*

Early work in this field was largely confined to description and attempted identification of fungi dominant in the film on or near the bed surface (Rettger, 1906; Cox, 1921). Attention later turned also towards factors affecting the occurrence, distribution and accumulation of predominantly fungal film; the effect of seasonal variations was studied in some detail by Haenseler *et al.* (1923), who showed that five species of fungi regularly occurred in the percolating filters at that works in notable quantities, and all showed a repeating seasonal pattern of growth, climbing steadily during autumn to a peak in either winter or spring and disappearing or becoming scarce in mid-summer. Holtje (1943) noted a similar seasonal growth pattern. Bell (1926) showed that the grazing activity of the collembolan *Hypogastrura viatica* (*Achorutes subviaticus*) was an important factor in the control of film levels, and demonstrated by experiments that this insect would feed on fungi isolated from the film.

Most of the later work, apart from the fungal population survey work referred to in Section I, has been concerned with a study of factors affecting film accumulation at works suffering from fungal film growths reaching nuisance levels. The complexity of this problem was summed up by Hawkes (1961), who pointed out that film accumulation was a balance between rate of growth of film and rate of removal of film by autolysis, bacterial attack, and grazing fauna. Many factors have been shown to affect growth rates of film micro-organisms; physical and chemical factors such as temperature, pH, oxygen concentration, rate, method and periodicity of dosing, and the chemical nature and strength of the sewage applied. In addition, growth rates of film micro-organisms would be affected by interspecific competition, but this would be controlled by the factors already mentioned.

Some of the best documented experiments on the treatment of sewage in percolating filters were carried out between 1938 and 1953 by the Water Pollution Research Laboratory, mainly at the Minworth works of the Upper Tame Main Drainage Authority, Birmingham, England. Biological observations were made of filters operated in a variety of ways; principal among these, and the main object of the whole project, was the method of alternating double filtration (ADF) in which two filters are operated in series with periodical reversal of the order of the filters. The other methods investigated included single low rate filtration (used as a control filter), filtration with recirculation of effluent, double filtration without alternation, and filtration with controlled periodicity of dosing (Wishart and Wilkinson, 1941; Mills, 1945a, b; Tomlinson, 1941, 1946; Tomlinson and Hall, 1955).

The object of these modifications in operation and design of percolating filters was to increase the volume of sewage treated to a given standard per unit volume of filter. In most cases the accumulation of film in the interstices of the filter, particularly during the colder months of the year, limited the rate at which liquid could be applied to the filter. Investigation of the processes by which film accumulated and was disintegrated was therefore of importance.

In these particular experiments, as in many percolating filter installations in Great Britain, fungi constituted the dominant structural element in the composition of the film. Relatively few species accounted for the bulk of the fungal mycelium encountered (Section II A2). A study of the autecology of these, together with observations in the field over a number of years in relation to the operational factors mentioned above, provides a working theory of film accumulation in percolating filters.

Nutrients supporting the growth of fungi are present in domestic sewage; other materials, some organic and some inorganic, are added to the sewage in the form of used waters from industry. Painter and Viney (1959) give the composition of domestic sewage. Approximately 60% of the organic carbon is derived from faeces and urine and the remainder from food preparation, personal washing and the washing of clothes and dishes. 70% of the organic carbon is in the solid matter, half of which settles after one hour. Hence in settled sewage about half the organic matter is in solution and the rest in particulate matter. About 80% of the organic matter in solution can be assigned to known groups of compounds: 30% consists of carbohydrates, predominantly glucose, with smaller amounts of lactose and sucrose; 11% is accounted for by lower fatty acids, mainly acetic acid, 15% non-volatile acids (citric and lactic acids being identified); 11% anionic surface active compounds; and 11% free and bound amino acids. The most important constituents of the suspended solids are higher fatty acids, principally stearic, carbohydrate and protein. The reader is referred to Ch. 7 of Volume 2 for further details on the composition of sewage.

Becker and Shaw (1955) made a special survey of fungi in two (primarily domestic) sewage-treatment plants. They recovered disseminules from raw sewage and effluents at various stages of treatment by a variety of plating techniques and occasionally by baiting procedures (Section III A). They noted that the only species producing a visible fungal film was *Fusarium aquaeductuum* (at one works, during March only); however, there was no observed increase in the number of disseminules of this fungus recovered in the effluent during the limited period of visible growth. Further, the disseminule density of all species was almost always greater in the sewage than at another stage, suggesting that the initial source of the fungi was, in the main, the sewage, and that multiplication did not occur in the filters apart from vegetative growth of mycelium. Experiments with *Aspergillus niger*, used as a tracer species (as it was easily recognized and had never been isolated from that works), indicated that half the spores introduced passed through the system within two weeks, after which time no more were recovered from the effluent; a proportion were doubtless destroyed in the filters or settled out in the sludge, and no evidence was obtained that they multiplied in the filters. Becker and Shaw found no fungal mycelium in filters apart from that of *F. aquaeductuum*, and concluded that growth of fungi in sewage systems recorded in the literature could be explained by the influx of various organic industrial wastes in the sewage.

Cooke (1958a) noted that most of the fungi flourishing in percolating filters showed largely or entirely vegetative growth, spores being rare and sometimes atypical in form, which agrees with the findings of Becker and Shaw and their conclusion that plating of liquid effluent samples may not give an accurate indication of the occurrence and relative abundance of fungal species in the filters.

2. Important Fungi in the Film

In the Minworth experiments referred to in Section II A1 six fungi accounted for most of the mycelium content of the film. These were *Sepedonium* sp., *Subbaromyces splendens*, *Ascoidea rubescens*, *Fusarium aquaeductuum*, *Geotrichum candidum* and *Trichosporon cutaneum*. Their greatest abundance was on the surface and in the top 15 cm of filter medium. The impact of sewage striking the surface of the filter prevented all except *F. aquaeductuum*, and to a smaller extent *G. candidum*, growing on the surface. Under these conditions *Fusarium* is dimorphic, having an adhering system of prostrate isodiametric-celled hyphae from which the typical free hyphae grow. In the alternating double filtration (ADF) system the free hyphae disappeared from the secondary-stage filter receiving effluent treated in the primary stage, leaving behind the prostrate hyphae which then became overgrown by algae. When the filter received settled sewage in the primary stage, free hyphae grew from the cells of the prostrate mycelium. Species of *Phoma* and *G. candidum* were usually associated with *Fusarium*.

Sepedonium sp. first became established in the sub-surface trickling zone, but in heavily loaded filters, and in particular conventional single filters, the mass of mycelium firmly anchored between the surface layers of medium was able to grow over the surface and could, during the winter, attain a thickness on the surface of 2 cm or more.

Subbaromyces and *Ascoidea* occupied the same niche as *Sepedonium*, but being less vigorous were usually subordinate to it. *Ascoidea* often occurred as loose floccose masses below the *Sepedonium* zone, i.e. at depths of 15–30 cm. *Trichosporon cutaneum*, though frequently isolated, was not readily identified in samples of film, but it seemed to be associated with *Geotrichum candidum*.

Fungi which were not identified as constantly as the above six species and which occurred in much smaller amounts were as follows. *Phoma* has already been mentioned as one of the first colonists with *Fusarium* in the surface splash zone. The phycomycetes *Leptomitus lacteus* and *Pythium gracile* were only rarely seen. Many different fungi were obtained on nutrient agar plates including various hyphomycetes, species of *Mucor* and yeasts including *Rhodotorula* species, but it is unlikely that their contribution to the bulk of film present was great.

Hesseltine (1953) records notable amounts of fungal film in filters treating the waste from a pharmaceutical factory at Pearl River (N.Y.) producing antibiotics. Studies were made of the growths on stones from these filters both from the surface and at depth, over a period of two years. Only three species were regularly present, *Fusarium episphaeria* (≡ *Fusarium aquaeductuum*) which was most abundant from spring through summer to autumn, *Oospora lactis* (≡ *Geotrichum candidum*), commoner in the colder months, and *Subbaromyces splendens.* This last-named fungus, unlike the others, was regularly found deep in the filter where conditions often appeared anaerobic, and, by trapping sewage solids, caused ponding.

Very few other species of fungi are on record as causing growths of nuisance proportions in film in percolating filters. A species of *Saprolegnia* became the most abundant organism in the top layer of the primary filter of an experimental ADF installation treating wastes from a manufacturing dairy. It formed a tough persistent film at the higher rates of treatment, particularly when whey washings were present (Section II A3(b)), and unlike the other fungi present was not readily removed by a change in order of the filters (Water Pollution Research, 1941).

3. *Factors Affecting Fungal Predominance*

(*a*) *Temperature.* The effect of temperature on film accumulation is complex and is both direct, acting on the growth rate of both bacteria and fungi, and indirect, affecting the organisms responsible for grazing or lysing the film. Within the temperature range recorded in percolating filters the metabolic rate of most micro-organisms active therein will increase with increasing temperature, and thus the higher the temperature the smaller the standing crop supported when nutrient levels are limiting, as is considered to be the case with most sewages (Hawkes, 1961). Some fungi growing in filters, e.g. *Sepedonium* sp., have lower temperature optima than most other micro-organisms active in the film, and flourish at temperatures below those needed for the activity of grazing macrofauna and competing bacteria. Further, the rate of lysis of fungal film also increases with temperature (Tomlinson, 1942) and closely follows the rate of biomass increase for *Sepedonium* sp. (Hawkes, 1965a). Thus the overall effect of temperature is a tendency to restrict the accumulation of fungal growth in the film, and this may be a factor in the increased amounts of film commonly noted in percolating filters in winter (Tomlinson, 1941, 1946; Hawkes, 1957), although sewage strength is also important, as discussed later (Section II A3(d)).

Not all fungi flourishing in percolating filters have the same temperature range. In contrast to *Sepedonium* sp., *Subbaromyces splendens*, which grows in the Langley (Warks) filters during the summer and early autumn months, has been shown to have a temperature optimum between 25–27°C. Conidia failed to germinate at 5°C, and their germination rate was very poor at 10°C and still severely affected at 15°C (Williams, 1971). Thus it seems unlikely that this fungus could establish growths in the winter months, and this conclusion is supported by examination of film samples.

The limitations of the results of laboratory batch culture experiments in relation to percolating filter conditions should be appreciated (Hawkes, 1965a). In batch culture the organism is grown under non-nutrient-limited conditions but is subject to its own accumulating metabolic products, whereas in the filter, nutrient is always to some extent limiting but metabolites are continuously removed. Williams (1971) designed a continuous flow apparatus for growing percolating filter fungi under conditions more closely simulating those of the filter and permitting regular measurement of wet weight of film. Results for *Sepedonium* sp. suggested that the greater biomass increase obtained in the short term (2–3 weeks) at high temperatures was due to the shorter lag period needed for the growth to establish. The growth rate at 5°C in the linear phase was the same as that at 20°C, although the biomass accumulated in three weeks was only 60% of that at 20°C.

(*b*) pH. The effect of pH on micro-organisms seen in culture is recognized to be complex in that its effect is exerted through a number of other factors such as enzyme activity, metal solubilities, and entry of substances into the cell; further, these factors may be effective in one part of the hydrogen ion concentration range only, causing complex pH/growth rate relationships (Cochrane, 1958). In practice it is not an important factor, as the pH of domestic sewage is usually near to neutral point due to the presence of buffering substances, and any industrial wastes likely to move the pH appreciably from this point would be pre-treated to prevent possible damage to plant and equipment as well as to avoid upset of the ecological balance of biological treatment processes.

However, pH control may fail occasionally, and periodic flushes of acid sewage could encourage certain fungi. Reynoldson (1942) suggested that the prevalence of *Geotrichum candidum* at Huddersfield might be associated with the occasional fall of the pH of the settled sewage below 5·0. Amounts of *Saprolegnia* sp. in filters treating milk processing wastes by ADF increased when whey washings were included and the pH fell to 4·8 (Water Pollution Research, 1941).

(*c*) *Nature of sewage.* Sewage containing a high proportion of industrial effluent is frequently associated with fungal growths in percolating filters. Hawkes (1965b) suggests that the carbon to nitrogen ratio may be among the factors affecting competition between bacteria and fungi in film, noting the relative rarity of fungal growths in film on filters treating purely domestic sewages. A specific instance is reported earlier (Hawkes, 1957) from studies on three works in the Birmingham area, two treating domestic sewage showing predominantly bacterial film whilst at the third works (Minworth) the sewage was about 50% industrial effluent and the film fungal in character. Many industrial waste waters are organic in nature, e.g. wastes from dairies, canneries, distilleries, pharmaceutical and chemical manufacturers, and carbonization processes; most of them show a high carbon to nitrogen ratio. This ratio for domestic sewage is relatively low because of the high proportion of nitrogenous compounds in human faeces and urine.

Hesseltine (1953) records ponding as a result of heavy fungal growths in filters treating wastes from antibiotic manufacture, and the presence of phenolic substances and tar acids was shown to be the cause of excessive fungal film and consequent ponding at Keighley, Yorks (Watson *et al.*, 1955). Reynoldson (1942) records a film composed largely of *Geotrichum candidum* (reported as *Oospora lactis*) in filters treating a sewage containing a very high proportion of textile and dye-stuff wastes, although he considered that periodic drop in pH of the sewage was an important factor here, as reported in the previous section. Several other workers have reported an association between industrial waste content of sewage and fungal growth (Sladka and Ottova, 1968; Feldman, 1955, 1957), but in most cases where ponding of percolating filters has been attributed to a high concentration of industrial wastes in the sewage, the offending film has apparently not been examined.

(*d*) *Sewage strength.* The fact that strong sewages favour fungal growths in film has been noted several times. Hawkes (1965a) carried out experiments in which inocula of *Sepedonium* sp. were grown on gauzes drip-fed with Minworth sewage for three weeks at 20°–22°C, the experiment being repeated each month for two years. It was found that there was a positive correlation between sewage strength (as measured by PV and ammoniacal N. values) and the amount of growth obtained. Precautions were taken to exclude grazing macrofauna. Tomlinson (1946) measured weights of film accumulating on experimental vessels set in the surface layers of two filters, one treating undiluted Minworth sewage and the other a mixture of sewage and effluent with a BOD slightly over half that of the sewage. It was arranged for each set of vessels to receive the feed at the same rate of flow. The rate of growth of the predominantly fungal film was five times greater on the filter receiving undiluted sewage.

(*e*) *Methods of filter operation.* Certain modifications of the method of dosing percolating filters to improve efficiency appear to have worked largely by controlling film accumulation, particularly where fungi dominated the film. The three most successful methods have been recirculation, alternating double filtration (ADF) and low frequency dosing.

Recirculation of a proportion of effluent gives a reduction in feed strength combined with an increase in hydraulic load. The former limits fungal growth at the surface but the latter increases the percolation rate; the rate of BOD reduction with depth is less, and film will accumulate to a greater depth, than in a single filter operated conventionally. Thus, although the total amount of film may be the same, the more even distribution will reduce the risk of ponding. The importance of the correct recirculation ratio in relation to sewage strength has been emphasized by Lumb and Eastwood (1958), and the failure sometimes attending the common practice of varying the amount of recirculated effluent inversely with the sewage flow to the works has been explained by Hawkes (1961) by reference to microbial growth curves. He showed that the degree of dilution with this regime could be insufficient to control surface growths.

Alternating double filtration leads to an actual reduction in the amount of film in the filters, as much of the film built up during the primary period breaks down during the secondary period provided that the concentration of organic matter (carbonaceous BOD) in the primary effluent has been reduced below the level that will support film growth; if these conditions are obtained, endogenous respiration will occur during the secondary period and in the long term the amount of film growth will be smaller (Hawkes, 1961).

In the early experiments at Minworth the effect of alternation of the filters on the film composition could be seen quite clearly at the filter surface in spring, summer and autumn. The green colour prevailing on the surface of the filter at the end of the secondary period, due to algal growths, gradually gave place to a grey colour as bacterial and fungal growth covered the algae during the period of operation as a primary filter, until just before the change the surface was orange-pink with *Fusarium aquaeductuum*. As the secondary period proceeded this slowly disappeared, the colour change being reversed until the algae were again uncovered (Tomlinson, 1941). The successful re-establishment of *F. aquaeductuum* in each primary period was related to the nature of the holdfast system established on the stones (see Section II A2). In the winter period the orange-pink film of *F. aquaeductuum* persisted all the time, masking the algal growth still living beneath. There was some sub-surface film consisting mainly of *Sepedonium* sp. with some *Geotrichum candidum*, but only occasionally did this grow up to the surface, and there was never serious ponding. By contrast, on the control filter, operating on single filtration at conventional rates, fungal film persisted for most of the year, *Sepedonium* sp. growing up from the sub-surface film and covering the *Fusarium* film during the winter. Ponding was extensive, with anaerobic conditions prevailing beneath the ponded areas. Studies made of fungal growth on glass slides placed on the filter surface (Tomlinson, 1942) showed that *G. candidum* as well as *F. aquaeductuum* could establish on a substratum directly subject to sewage impingement at the high rates of dosing used in ADF, but that *Sepedonium* sp. was unable to do so. The work also revealed the effect of nutrient starvation on the fungi during secondary periods of operation; growth became concentrated in certain portions of the hyphae at the expense of the rest of the mycelium, leaving resting cells separated by dead regions. A further experiment in the laboratory, feeding both sterilized and unsterilized filter effluent to pure cultures of *Sepedonium* sp. and *F. aquaeductuum*, showed that disintegration of starving mycelium in sterilized effluent did not occur in 20 days, but was rapid, occurring within a few days, in the presence of bacteria. Only dead portions of the hyphae were affected; resting cells and chlamydospores were not attacked and retained their viability.

The efficiency of ADF is limited by the quality of the primary effluent produced, for if this allows a positive growth rate in the film of the secondary filter, accumulation of film will occur over a period. Low frequency dosing, in which a heavy dose of sewage is fed intermittently to the filters, gives rather a different nutritional regime. Most of the sewage applied will be held in the interstices of the film until flushed out by the next dose some minutes later. During this time it will be reduced in strength by microbial oxidation until the concentration falls below that capable of supporting growth of the fungi. An endogenous period will follow, lasting until the next dosing, and in this period some regression of the film will occur. Thus film accumulation is limited within each dosing cycle and large amounts of film do not develop. The theory of the process has been discussed at length by Hawkes (1961).

Lumb and Barnes (1948) first drew attention to the value of the process when describing a series of experiments they carried out at Halifax (Yorks). Tomlinson and Hall (1955) reported the results of operating the Minworth experimental plant as two pairs of filters on ADF, one pair with normal continuous dosing and the other geared for low frequency dosing and operated on a series of dosing cycles from 2–14 min, each cycle being tried out for several months. Ponding was not seen on the filters operating on low frequency dosing and they accumulated much less film than the other pair, which regularly ponded. Hawkes (1955) followed film accumulation closely in the later stages of this experiment, and showed that film amounts in the upper 0·75 m of the intermittently dosed filters were consistently small in amount, whereas the continuously dosed filters showed heavy film accumulation in winter and spring, with septic conditions in the film accompanying ponding. Performance was affected to some extent, particularly in respect of nitrification, which was notably better in the filters on low frequency dosing. Maximum efficiency was considered to have been achieved at Minworth with a dosing cycle between 4 and 8 min, but small scale filters at Coventry gave results suggesting that a longer dosing cycle (12 min) would be better; these filters were operating at higher hydraulic loadings and treating a different sewage.

In a comparison of percolating filters operating on ADF and double filtration (DF) without alternation, Hawkes and Jenkins (1964) (working at Minworth) found that controlled frequency of dosing enabled the primary filter in DF to be maintained in a satisfactory condition for the effective removal of organic load in the primary stage, permitting suitable conditions for nitrification to prevail in the secondary filter. The primary filter did accumulate appreciable quantities of film but heavy fungal growths did not occur in the upper layers, and the more even distribution of film prevented clogging. With low frequency dosing, retention time of liquid in the filter might be expected to fall with increasing length of dosing interval because of the high instantaneous rate of application, and therefore purification efficiency might be reduced. These workers, using lithium salts as tracers, showed that the overall retention time in the filter, at the highest rate of dosage used in DF, did not decrease within the range of 3 to 20 min per revolution of the 4-arm rotating distributors.

It cannot be assumed that the optimal dosing cycle is the same for all sewages, as sewage nature determines the dominant fungus of the film and fungi differ in their responses to dosing periodicity. Experimental work using the continuous flow apparatus previously mentioned (Williams, 1971) showed that dosing once every ten minutes severely affected the growth of *Sepedonium* sp. but had little effect on *Ascoidea rubescens* (Fig. 1).

(*f*) *Other factors.* Fungi are essentially aerobic and the supply of oxygen to the film could be an important factor in the accumulation of fungal film, although Cochrane (1958) points out that many fungi can grow in

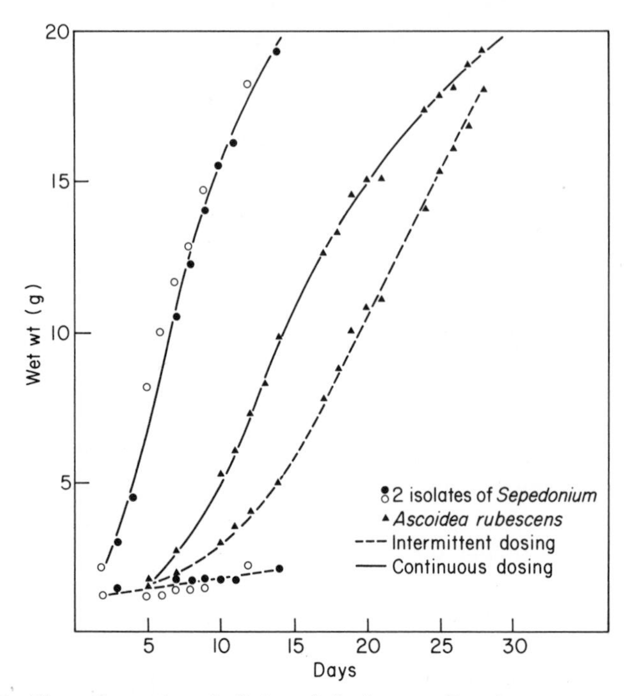

FIG. 1. The effect of periodicity of dosing on *Sepedonium* sp. and *Ascoidea rubescens* in culture.

quite low oxygen concentrations. Thick growths of fungal film may become anaerobic beneath, when they block the filter and cause ponding. This will lead ultimately to sloughing of the film (Hawkes, 1957). But quite thick layers of fungal film on percolating filters may remain aerobic; sometimes this is due to the open texture of the film which could allow air penetration, but film is best considered as a layer completely permeated by liquid through which dissolved oxygen must penetrate. Tomlinson and Snadden (1966) studied the oxygen supply to films of micro-organisms in a Gloyna pattern rotating tube, and calculated the diffusion coefficient for oxygen penetrating a bacterial film to be about two-thirds of the value of oxygen diffusion through a water film, and the depth of film to which oxygen could be expected to penetrate to be about 0·2 mm. The value obtained for fungal film was, however, about ten times the value for the passive diffusion of oxygen through water, indicating that diffusion must be assisted by some other process. The above workers suggested that protoplasmic streaming was probably responsible for efficient oxygen transport through fungal films of considerable thickness. Such streaming can be readily observed in *Sepedonium* sp.

The importance of the effect of mechanical scour due to the force of impingement of sewage on the film has been variously assessed. Some consider it an important factor in the control of film accumulation, while others consider it unimportant compared with biotic factors such as grazing macrofauna or lytic bacteria (Hawkes, 1965a). Much depends on the nature and condition of the film; a bacterial film is more easily flushed off than a fungal film, and starving fungal film is subject to bacterial lysis and would then readily disintegrate (Tomlinson, 1942; Hawkes, 1961, 1965a). Jet scour was studied by Hawkes (1959) in the evaluation of the role of various types of jet in film accumulation. Under conditions promoting heavy fungal film growths (high organic loading in cold weather), surface growth was limited under the more powerfully impinging types of jet, but a very tough, leathery sub-surface film formed, leading to ponding and anaerobic conditions. Percolation of sewage was hindered and lateral spread encouraged upward growths of film which eventually covered the filter surface. The dominant fungus was *Sepedonium* sp. Under these jets the macrofauna was adversely affected, probably by the ponded conditions, thus encouraging film accumulation. In contrast, jets providing a more even distribution of the sewage produced a more open-textured film supporting a large and varied macrofauna. Tomlinson (1942) showed, with glass slides placed in the jet-track on ADF filters with high hydraulic loading, that certain fungi could establish growth (see previous section). Thus it seems that scour is unimportant in the control of fungal film in percolating filters.

There are few records of the effect of light on percolating filter fungi. Long-term experiments involving drip-feeding *Sepedonium* sp. with sewage under controlled conditions in both dark and light revealed no difference in the amount of growth produced (Hawkes, 1965a). Tomlinson (1942) found *Fusarium aquaeductuum* dominant on glass slides placed on the filter surface exposed to light, while *Geotrichum candidum* was dominant on those slides shielded from the light. *Fusarium* was able to compete successfully with algae such as *Stigeoclonium* for a place of attachment on the stones in the light, but *Geotrichum* was unable to do so, though it became dominant over *Fusarium* in the absence of algae.

The nature of the medium exerts its effect both through its physical characteristics (void space, surface per unit volume) and indirectly by influences on the grazing population. From two full-scale trials, Hawkes and Jenkins (1955, 1958) concluded that no one type of medium was superior under all operating conditions, but that where film accumulation was a problem, as with sewages that promote fungal growths, larger media gave a better overall performance. The larger voids accommodated better the film growths liable to build up, and reduced the risk of ponding.

Recently there has been a considerable development in the use of plastic modules, in various conformations such as sheets and tubes, as a substitute for conventional random mineral and coke media (Bruce *et al.*, 1970). The spacing of the surfaces in these filters is such that fungal growth is unlikely to block the filter. As far as is known the effect on the flora is quantitative rather than species-selective. Small plastic units designed for random packing have also been marketed in recent years. These media have similar properties to the sheet and tube modules but are more easily packed into the filter shell. In a sample of fungal film taken from a filter using random pack plastic medium and treating an industrial sewage, *Fusarium aquaeductuum* and *Geotrichum candidum* were found to be the predominant fungi; *Phoma* spp. were also common. Although the film was considerable in amount, no problems with blockage of the filters had occurred.

The outcome of competition between living organisms in the film is largely a result of the differential effect on them of the various factors already discussed, and it may in certain cases favour fungal growths. Food preferences of the grazing macrofauna of filters have been studied little, but most appear to graze fungal growths. Bell (1926) described in detail the grazing activity of the apterygote insect *Achorutes viaticus* on filter film. Without apparently understanding fully the role of film, he nevertheless appreciated that these insects could control excessive growths given the right conditions; they were unable to tolerate immersion or shortage of air. They were thus favoured by distribution jets that produced discontinuous tracks of growth, and discouraged by a continuous even spray. An earlier excursion into the field of biological control by distributing these insects to sewage works in many parts of the world is recounted in the above paper. The influence of the style of dosing jet on the composition of the grazing fauna is dealt with fully by Hawkes (1959).

Reynoldson (1942) showed that the strength, nature and pH of the sewage, together with seasonal temperature fluctuations, all played a part in the complex interactions between the growth of fungal film and the grazing fauna in the double filtration plant at Huddersfield. The toxic nature of the sewage virtually restricted the grazers to the larvae of the fly *Psychoda alternata* and the film was dominated by *Oospora* sp. (probably *Geotrichum candidum*), and this restricted biota showed clear-cut interactions. A strong sewage and a fungus resistant to toxic compounds ensured that food was never limiting for either, but toxic compounds affected the *P. alternata* populations at times and allowed the fungus to develop excessively, with attendant ponding and septicity, which further adversely affected the fly larvae. Although both were inhibited to a large extent at low winter temperatures, the fungus recovered more quickly so that spring peaks of film accumulation were experienced.

Hawkes (1957, 1961, 1965a) has shown by field observations and experiments that although the grazing organisms are important in controlling amounts of film in filters, they are not the factor responsible for seasonal fluctuations in the standing crop of fungus; this is accounted for by seasonal changes in temperature and strength of sewage. The amount of film has more effect on the grazing population than the latter has on the amount of film. Spring unloading of film from filters will occur in the absence of grazers, as a result of starvation following increase of temperature coupled with a reduction in sewage strength (as with spring rains). The growth rate of the fungi increases but not enough food is available to support it, and autolysis and bacterial attack on the starving mycelium cause disintegration as described earlier. Several instances of bacterial/fungal and interfungal competition have previously been quoted while discussing physical and chemical factors.

B. Activated Sludge

Fungi are not normally found as dominant organisms in the activated-sludge process, except for the occasional occurrence of heavy growths of predacious fungi (Section II E). Farquhar and Boyle (1971) found fungus dominant only once in a total of 16 samplings from 10 plants. Cooke (1970), reporting results of colony counts of fungi and bacteria from all stages of treatment by the activated-sludge process at Lebanon, Ohio, showed that while there was a notable increase in numbers of fungi in mixed liquor in the aeration tanks compared with the numbers in the sewage, the limitation of the enumeration methods used made assessment of the true extent of this increase very difficult. Fungal filaments were detected in much of the floc material by microscopical examination.

When growth of bacteria is inhibited, fungi may grow profusely; at one works (Yardley, Birmingham) sudden discharges of acid plating wastes lowered the pH below 3·0 and subsequently *Geotrichum candidum* grew abundantly (Hawkes, 1963). In general the appearance of fungi in activated sludge is associated with industrial effluent present in the sewage, and when they flourish fungi may adversely affect the settling qualities of the sludge, causing bulking. Such sludges may be quite efficient in purification, as Hawkes (1963) has pointed out, but generally do not settle well and consequently impair the effluent quality. Extensive fungal growths do not, however, necessarily cause bulking. During the operation of a large-scale experimental plant investigating the oxidation of lactose, after a few weeks the sludge, previously dominated by bacteria, became very dense and dark in colour. The well-defined, thick flocs were dominated by *Pullularia pullulans* (≡ *Aureobasidium pullulans*). This condition persisted but bulking did not occur (Jenkins and Wilkinson, 1940).

In laboratory trials studying the feasibility of treatment of effluent from synthetic resin manufacture by the activated-sludge process, Harkness and Jenkins (1964) found the flocs became dominated by filamentous fungi when a toxic and highly acidic discharge was received. On return to normal conditions the fungus disappeared. In conditions of low nitrogen supply fungal filaments became very extensive and dominated the floc. This (unidentified) fungus grew in the presence of 400 mg/l of phenol, which it tolerated but did not use, but 200 mg/l formaldehyde proved toxic. When the full-scale plant was installed at the factory, difficulties were experienced with the settling characteristics of the sludge, and Holmes (1970) found a fungus present which he identified as the yeast *Candida tenuis*. It was the dominant organism and produced filamentous growths in the floc intermittently, but its large single cells could always be seen. A count of yeasts was made frequently during the six weeks of the project and gave results consistently in the range 15 000 to 23 000 colonies/ml.

The occurrence of fungi in activated-sludge plants was specifically studied by Pipes and Cooke (1969). They examined liquors at all stages of treatment (including raw sewage) in 18 installations, only four of which received more than 10% of industrial effluent. The colony count method was used to estimate numbers of fungal disseminules present, and over 90% of fungi recovered belonged to four common genera: *Penicillium* spp., *Cephalosporium* spp., *Cladosporium cladosporioides* and *Alternaria tenuis*. But the most consistently occurring species, present in almost all the plants, were *Geotrichum candidum* and species of *Trichosporon*. These were present all the year where the others tended to be scarce or absent in winter. There was no indication of problems with bulking in any of these works or in that studied by Cooke (1970).

Few studies have been made either of the importance of fungi in activated sludge or of the identity of filamentous organisms reported to be associated with bulking sludges. The latter are commonly assumed to be filamentous bacteria (such as *Sphaerotilus natans*) rather than fungi, but in at least one case this assumption proved ill-founded (Pipes and Jones, 1963). Jones (1964, 1965) subsequently studied the conditions affecting the growth of *Geotrichum candidum* in activated sludge, and concluded that it held a competitive advantage over other micro-organisms in N and P deficient wastes. The contribution of filamentous micro-organisms to sludge bulking has long been a matter of controversy, and Finstein and

Heukelekian (1967) attempted to quantify this relationship. They found that to a large extent filamentous material governed the sludge volume index (SVI), notably in plants where flocs were consistently filamentous, but the methods used did not allow distinction between fungal and bacterial filaments.

The origins of sewage will ensure that disseminules from a wide range of fungal species will be present in it, and in view of the aerobic nature of activated sludge and the organic nutrients present it would be surprising if some of them did not grow in it to some extent. The fact that they are not normally dominant in the flocs, at least where a basically domestic sewage is being treated, suggests that in this ecosystem they cannot usually compete successfully with bacteria, and this may be related to the relatively high saturation constants of fungi as compared with bacteria (Section III B). From work discussed above it seems that C/N ratio, pH, and the presence of toxic substances are also important. But clearly there are yet other factors. Although in many ways the activated-sludge process and the percolating filter are basically similar habitats, some organisms present in the latter find the former inhospitable: macrofauna such as oligochaete worms and insect larvae, for example, are usually absent. This could also be true for some fungi, as at least four of the troublesome species in filter film are not reported from activated sludge. A fungus like *Fusarium aquaeductuum*, which grows firmly attached to the medium in the filter and repeatedly recolonizes from a prostrate system of hyphae, could find the small, loose, moving flocs unsuitable, although it might be found on walls and pipework not normally sampled.

C. Sludge Treatment

In a survey of all parts of the Dayton (Ohio) sewage treatment plant, the range of fungal species occurring frequently in the raw and digesting sludges was found to be similar to that in filter film (Cooke, 1958a). Viable fungi were even recovered from vacuum filter cake after conditioning of the sludge with $FeCl_2$ and lime. Apart from yeasts (which were the most abundant forms but were not separately identified), *Geotrichum candidum*, *Fusarium aquaeductuum*, *Margarinomyces heteromorphum*, *Penicillium ochrochloron* and *P. lilacinum* were the commonly occurring species in most of the sludges sampled, with *Pullularia pullulans* prevalent in raw sludges. *Trichosporon cutaneum* was very abundant in sludge from drying beds, but was hardly recorded elsewhere. Later Cooke (1965) recovered quite large populations of fungi (by various plating and other techniques) from digesting sludge in an experimental plant. To ascertain whether they were growing there or merely surviving, actively digesting sludge in one digester was subsequently fed only with sterilized fish meal, and in the other only with unsterilized fish meal. Fungal populations were assessed regularly through seven digestion cycles, and the numbers recovered by plating showed considerable fluctuations in population size in both digesters, indicating that some growth was taking place. During this experiment 45 species of fungi were isolated, approximately 75% of which were filamentous forms.

It is generally considered that fungi only grow successfully under aerobic conditions, and Cochrane (1958) points out that while many fungi can grow in quite low oxygen concentrations, none are known to grow better than at normal atmospheric levels. Tabak and Cooke (1968) tested 13 species of fungi isolated from sewage sludge and other habitats; although growth under anaerobic conditions was considerably poorer than under aerobic, all showed some growth. *Geotrichum candidum*, *Fusarium oxysporum*, *F. solani* and *Mucor hiemalis* were particularly successful.

Although many fungi can survive and make some growth under conditions prevailing in sludge digesters, there seems no evidence that they play any appreciable role in the process. However, Becker and Shaw (1955) noted that *Geotrichum candidum* was so plentiful in digesting sludge that its possible significance in the process should be further investigated.

D. Fungi in Sewers and Channels

Provided aerobic conditions prevail, fungi will grow in sewers, and their presence is brought to the notice of the biologist when growth is so profuse that the sewer becomes blocked or pieces of fungal mycelium cause trouble in mechanical equipment at the sewage works, as for example by blockage of screens or distributor jets. In one particular case, blockage of small bore sewers carrying domestic sewage and industrial effluent in a factory resulted from such growths of fungus; the relatively large volume of effluent kept the sewage sufficiently aerobic to encourage the growths.

Fungi in these particular habitats have been little studied, but seem to belong to the same range of species as that occurring in percolating filters. *Leptomitus lacteus* was found in profuse quantities in the feed channels at one works, and detached growths blocked distributor arms, but the fungus did not establish itself in the filters (Hawkes, 1963). *Sepedonium* belonging to a strain similar to that occurring profusely in percolating filters on the Minworth works was also found growing in the vegetation at the edge of the effluent channel and in the receiving stream (R. Tame).

E. Occurrence of Predacious Fungi

Fungi which capture small animals, such as nematodes, rotifers and protozoa, for food are well known from soil, plant remains and manure; representatives occur in several families of fungi and have been extensively studied (e.g. Dreschler, 1935, 1959; Duddington, 1956, 1957). Since prey is abundant in activated sludge and in film of percolating filters it might be expected that such fungi would be common in these habitats also. Although there are few records of their presence in abundance in sewage treatment plants, many workers have reported encountering them. The mycelium of nematode-trapping fungi is frequently seen in activated sludge rich in nematodes, the fungus *Dactylella bembicoides* Dreohski being one species positively identified from this habitat. Tomlinson and Snaddon (1966), investigating the oxidation of sewage in a rotating tube apparatus, found that *Arthrobotrys oligospora* and species of *Trichothecium* captured large numbers of nematodes. Another fungus parasitic on nematodes which was occasionally found was *Harposporium anguillulae*.

Cooke and Ludzack (1958) described a heavy growth of a fungus identified as *Zoophagus insidians* in a laboratory activated-sludge tank. The fungus was feeding on the loricate rotifer *Monostyle* (≡ *Lecane*) (which in its turn was the principal predator of the bacteria) and caused bulking of the sludge. Control of the fungus was achieved by the use of chlorine but this upset the ecological balance; in particular, nitrification almost ceased. Pipes (1965) records the same fungus in an activated-sludge pilot plant treating domestic sewage, and suggested that flotation of the fungus was largely a result of gas produced by bacterial action in the impaled empty loricas of the dead rotifera, since small masses of mycelium seen to settle either carried few trapped rotifers or had rotifers with body contents invaded by haustoria still visible. Pipes pointed out that this fungus is an obligate parasite, but many predacious fungi found in treatment plants are capable of saprophytic nutrition and thus could flourish in a wider range of conditions. Several species of *Arthrobotrys* with different types of trapping nets were isolated in this work, and these were used in an experiment on sludge settlement. When the fungus was grown in aerated sewage with bacteria and protozoa only, it formed small, compact masses each of which was the basis of a floc, and the sludge settled well. However, when nematodes were added the fungus developed much more extensive growths and there was bulking, the SVI rising by 15–50%. It was suggested that this work illustrated the doubtful applicability of much of the theoretical work on activated sludge, especially while so little is known of ecological factors which can drastically alter the composition, and thus the behaviour, of the living population of the sludge.

III. Studies on Isolates of Predominant Fungi

A. Isolation and Cultural Methods

A paper on isolation techniques was published by Cooke (1954b) in which a number of media, mostly modifications of Martin's rose bengal–streptomycin agar (Martin, 1950) employing various concentrations and combinations of nutrients, colony growth inhibitors and anti-bacterial antibiotics, were compared to determine the most suitable medium for general use in the isolation and enumeration of fungi from polluted habitats. A dextrose/soy-hydrolysate/mineral-salts/rose bengal/aureomycin agar was found most satisfactory and full details for its use, together with other techniques for the recovery of aquatic phycomycetes and yeasts, are given in the laboratory guide referred to earlier (Cooke, 1963). For continuous sampling of bacteria bed populations Cooke (1958b) used glass slides placed on the surface of the beds, a method used previously by Tomlinson (1941). These were removed successively at intervals from one day to one month for weighing of air-dried growth and for isolation and enumeration of fungi by plating this material on the rose bengal–aureomycin agar developed by him. Earlier, Harvey (1952) had described the use of hemp-seed baiting techniques in a search for fungi in polluted streams, and concluded from his work that few fungi were regularly found growing in association with polluted conditions. Although standard dilution and plating techniques were tried in addition, using corn-meal agar and potato-dextrose agar sometimes with antibiotics (streptomycin, penicillin, or actidione) to inhibit other types of micro-organisms, only three genera of deuteromycetes were recorded: *Cephalosporium*, *Trichoderma* and *Geotrichum*, the majority of fungi found being aquatic phycomycetes. Cooke (1963) in his laboratory guide recommends the baiting techniques as the best method for the recovery of aquatic phycomycetes, but his development of plating techniques has shown that a large number of deuteromycete species and some ascomycete species occur regularly in polluted aquatic habitats.

The use of the ring-plate method in our experience has proved very useful for isolating fungi from slimes, film or floc collected from various parts of the treatment plants. Small lumps of inoculum may be washed in sterile water or saline and examined microscopically to allow the removal of small invertebrate animals which might cause contamination of the isolation plates. Suitable rings may be made from 3 mm lengths of thin-walled glass tubing of about 20 mm diameter with three blobs of glass fused on to one edge to serve as feet. Plates should be poured with the sterile rings standing in each, sufficient agar being used to cover the feet and lower half of the ring only. When set the prepared inoculum is placed on the agar surface in the centre of the ring; bacterial growths developing are usually confined by the ring while the fungal hyphae grow through the medium and out between the feet to invade the peripheral region. This method is particularly valuable for fungi such as *Subbaromyces splendens* and species of *Sepedonium* which may be totally inhibited by the concentrations of rose-bengal and antibiotics in general use.

Carefully chosen media may assist the isolation procedure; for example, the use of modified Czapek-Dox agar (Oxoid C.M. 97; Oxoid Manual, 1969) is successful for the isolation of *Fusarium aquaeductuum* in ring-plates as it both restricts bacterial growth and allows recovery of this fungus in the presence of *Geotrichum candidum* for which fungus its nutrient status is unsuitable. In some cases rings are unnecessary, as with the isolation of *Sepedonium* sp. which produces rapidly extending hyphae in agar without added nutrients, on which medium bacterial growth is negligible. Portions of the agar containing the tips of these hyphae may be cut out aseptically and removed to a richer medium.

The temporary anaerobic conditions liable to occur if samples are stored overnight in a closed jar at room temperature are unlikely to harm the fungi required but will kill many of the animals causing problems in isolation procedures. There is, however, an element of risk in that certain components of the film may be grazed to extinction if the animals do not become inactivated.

Problems of identification are complicated by the tendency of fungi flourishing in habitats associated with sewage treatment to occur in the vegetative condition, and some species are unreliable sporulators in culture also. Although colony form on agar media is not generally acceptable as a character for identification purposes because of its variability under different environmental conditions and in different strains, as in *Geotrichum candidum* and species of *Fusarium*, it is a useful aid to the separation of superficially similar species where it is consistent, as in the separation of *Subbaromyces splendens* and *Ascoidea rubescens*. Section III C of this chapter gives some guidance on the identification of fungi causing problems in percolating filters and gives sources of reference.

Methods of enumeration of fungi in water, mud, slimes and similar materials in general, make use of a medium similar to that of Cooke (1954b) described above, with rose-bengal to restrict the rate of growth of the faster growing fungi and some antibiotic chosen to inhibit bacteria. Burman *et al.* (1969) recommend Kanamycin as antibiotic, and describe the use of a membrane filtration modification of the above method suitable for water in which the number of fungi is relatively low.

The degree of accuracy of fungal counts as an estimate of populations in bacteria beds was discussed by Cooke and Hirsch (1958). They pointed out that if the film was broken up by a homogenizing procedure in an ideal way, every other cell of a mycelial hypha could give rise to a colony (theoretically), the interposed cells being broken up, while every yeast cell could produce a colony. Whilst it is unlikely that any blending procedure would be so efficient, it is probable that figures produced by such counts are sufficiently accurate to correlate reasonably with the degree of incidence of any given fungus, presuming of course that the medium will permit the recovery of that fungus.

B. Physiological Studies

A study was made of the comparative physiology of the six dominant species with a view to elucidating the fungal ecology of the percolating sewage filters at Minworth and Langley, Birmingham. In the following sections the carbon and nitrogen nutrition is described first, followed by a section on growth rates in liquid media including sewage and the effect of nutrient concentration as predicted by the Michaelis–Menten theory. The lowest concentration of sewage necessary to provide energy of maintenance is discussed. The relative rates of growth on solid media provide data for a comparison of the lengths of hypha contributing to apical growth, and the rapid falling off with depth in the filter of the weight of fungus is discussed in the light of the data obtained. Other aspects dealt with are the effects of pH value, ammonia, anionic detergent, substrate concentration and interspecific inhibition.

1. *Nutrition*

(*a*) *Nitrogen requirements.* Painter (1954) found that *Sepedonium* sp. grew best on complex organic sources of nitrogen. It grew poorly on nitrate, ammonia and glycine. *Geotrichum candidum* and *Trichosporon cutaneum* grew on ammonia as well as organic nitrogen and *Fusarium aquaeductuum* was able to grow on nitrate as well as ammonia and organic nitrogen. Table I shows that *Subbaromyces spendens* and *Ascoidea rubescens* when grown on Painter's base medium to which different sources of nitrogen were added in concentrations equivalent to 0·035% N required organic sources of nitrogen for growth. *Sepedonium, Ascoidea* and *Subbaromyces* fall into Category IV of Robbins' (1937) classification

TABLE I. Growth of fungi (mg dry wt.) on various nitrogen sources

	Ascoidea rubescens	*Subbaromyces splendens*
Ammonium sulphate	nil	nil
Potassium nitrate	nil	nil
Sodium glutamate	nil	nil
Asparagin	nil	nil
Hydrolysed casein	7·82	0·73
Proteose peptone	0·50	4·5

since they require an organic source of nitrogen. On this basis *Fusarium* would come in Category II (NO_3, NH_3 and Org. N). and *Geotrichum* and *Trichosporon* in Category III (NH_3 and Org. N).

(*b*) *Growth on specific compounds.* Preference for certain types of nutrients was investigated by comparing rates of radial extension of colonies and density of growth on sewage-agar plates to which the individual compounds were added in a concentration of 0·1% w/v with growth on sewage agar without any added nutrient. In a second experiment the increased weight of mycelium produced by adding each compound in a concentration of 200 mg/l to sterile sewage was measured. (Table II.)

Glucose, acetate and butyrate were good sources of carbon for all species. In general the high molecular weight compounds, starch, casein and "Complan" promoted growth of the leading hyphae, and sugars and fatty acids promoted branching. "Complan" (Glaxo Ltd) is described as a blend of dried skimmed milk, arachis oil, calcium caseinate, maltodextrin, cane sugar, vitamins and trace metals and contains 31% protein, 16% fat and 44% carbohydrates.

Growth of *Subbaromyces* on sewage was not stimulated by the addition of glucose. *Ascoidea, Fusarium, Geotrichum* and *Trichosporon* share the capacity to utilize sucrose and arabinose (except *Geotrichum*) and glycerol as well as glucose. They may be classed as "sugar fungi". *Sepedonium* and *Subbaromyces* grew well on the complex organic compounds casein and "Complan" and were notably successful on sewage; *Sepedonium* grew well on starch, having strong amylase activity.

The effect of the strength of sewage on the growth of fungi is shown in Table III. In these experiments a strong detergent-free sewage was diluted to give BOD values ranging from 800 to 50 mg/l and ammonia ranging from 450 to 50 mg/l. The pH value was in the region of neutrality.

TABLE II. Growth of fungi on various carbon sources

	Sepedonium			*Subbaromyces*			*Ascoidea*			*Fusarium*			*Geotrichum*			*Trichosporon*		
	E	D	W	E	D	W	E	D	W	E	D	W	E	D	W	E	D	W
Starch	+		+	+						+			+					+
Sucrose									+		+	+		+		+	+	+
Lactose																	+	
Glucose		+	+				+	+	+	+	+	+		+	+	+	+	+
L-arabinose									+		+	+					+	+
Glycerol									+			+			+			+

Acetate		+	+		+	+	+	+	+		+	+			+			
Butyrate		+	+		+	+	+	+	+		+			+	+		+	+
Glycerol monostearate	+		+							+			+					
Glycine																		
Asparagin												+						
Glutamate			+															+
Soluble casein	+		+	+		+				+			+					
Hydrolysed casein		+	+												+		+	+
Bacto-peptone		+	+							+					+		+	
"Complan"	+		+	+		+				+		+	+			+	+	+

The letters E, D and W refer respectively to radial extension, growth density and weight of mycelium.

TABLE III. Growth of fungi (mg wt.) on sewage of differing strengths

Experiment 1

BOD mg/l	*Sepedonium*		*Fusarium*	*Geotrichum*		*Trichosporon*	
	Time (h)						
	76	98	51	45	69	45	69
800	3·3	4·3	14·8	1·6	1·3	19·2	6·1
400	2·2	1·5	7·1	0·4	0·6	8·8	8·1
200	0·9	1·0	1·2	1·0	nil	4·3	4·5
100	0·18	0·14	0·39	nil	nil	1·6	1·5

Experiment 2

BOD mg/l	*Sepedonium*	*Subbaromyces*	*Geotrichum*	*Trichosporon*
	Time (h)			
	92	92	65	96
460	12·2	8·3	8·4	7·2
230	5·7	5·3	3·3	3·2
103	2·2	nil	1·5	1·4
50	0·5	nil	0·45	0·4

The values in Table III are about of the same order of those expected from synthetic media of the same strength. Very little growth occurred in sewage having a BOD of about 50. It is shown elsewhere that the limiting concentration for maintenance of *Sepedonium* is about 50 mg/l. Sewage would appear to be as favourable for growth of filter fungi as synthetic media of the same strength. This is consistent with their utilization of sucrose, glucose, volatile fatty acids and complex nitrogenous compounds.

About half the organic matter in sewage treated in filters is in the form of particulate matter. The only evidence obtained so far of the utilization by the fungi of this type of material was the proliferation of the hyphae of *Sepedonium* seen around sewage particles in sewage agar plates.

(*c*) *Vitamin requirements.* Painter (1954) studied the vitamin requirements of four of these fungi, and found that *Fusarium aquaeductuum* and *Geotrichum candidum* were independent of an external supply of vitamins of the "B" group, *Trichosporon cutaneum* required thiamine for satisfactory growth, and *Sepedonium* sp. needed both thiamine and biotin from an external source. Four vitamins were used by Cooke (1957b) in a study of nutritional requirements of nine fungi from treatment plants; none of them was required by *G. candidum* but his strain of *F. aquaeductuum* proved deficient for thiamine.

2. Rate of Growth

(*a*) *Calculation of growth parameters.* Difficulty in assessing the rate of growth of *Sepedonium* sp. arises from the absence of conidia or other suitable disseminules. Small pieces of preformed mycelium used as inoculum may give rise to logarithmic growth in the early stages of growth but once a sizeable pellet has formed the diameter increases at a constant rate (Pirt, 1966). Under conditions of logarithmic growth it is found that the relation between rate of growth, μ, and substrate concentration follows the Michaelis–Menten expression as applied to bacterial growth by Monod (1942), namely $\mu = \mu_m\left(\frac{s}{K_s + s}\right)$ where μ_m is the maximum rate of growth when substrate concentration is not limiting, s is the substrate concentration and K_s is the saturation or Michaelis constant equal to the substrate concentration at which the rate of growth is half the maximum rate.

Figure 2 shows the results of an experiment in which the dispersed mycelium of *Sepedonium* sp. as inoculum was grown at 25°C on a medium consisting of 3 g casitone (Difco) peptone, 2 g yeast extract and 5 g glucose dissolved in water at pH 7·2 (CYG medium). The dry weight of the inoculum was 0·1 mg and the fungus grew logarithmically to a final yield of 30 mg in 150 ml of medium. Different concentrations of total substrate were made up by dissolving the above weights of medium

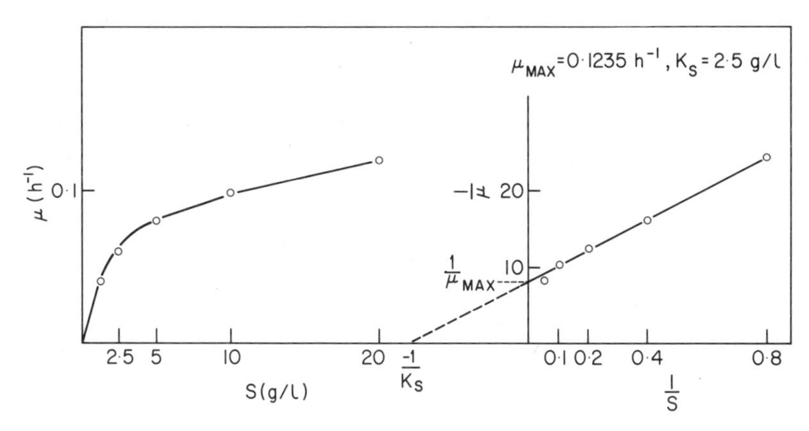

FIG. 2. Rate of growth (μ) of *Sepedonium* sp. in relation to substrate concentration (s).

TABLE IV. Maximum theoretical growth rates (μ_{max}), doubling times (t_d) saturation constants (K_s) of 6 filter fungi

Fungus	Medium	μ_{max} (h^{-1})	t_d (h)	K_s (g/l)	K_s (BOD. mg/l)
Sepedonium sp.	CYG	0·124	5·6	2·5	1850
	CYGS	0·167	4·2	1·25	900
Subbaromyces splendens	CYG	0·0628	11·0	3·84	2460
	CYG	0·077	9·0	1·54	990
Ascoidea rubescens	CYG	0·0887	7·8	1·16	1000
	CYG	0·053	13·1	1·54	1375
Fusarium aquaeductuum	CYG	0·188	3·7	0·435	278
Geotrichum candidum	CYG	0·30	2·3	0·875	560
Trichosporon cutaneum	CYG	0·22	3·2	1·0	640
	"Complan"	0·178	3·9	0·182	100

Note: All fungi grown in batch culture in liquid media at 25°C, the flasks continuously shaken.

constituents in different volumes of water. Growth rates of the mycelium were estimated in a range of substrate concentrations and the results obtained are plotted in the linear and reciprocal forms in Fig. 2. The values of the kinetic constants obtained by these methods are given in Table IV. The BODs of the medium at different concentrations were estimated and this enabled the conversion of the value of K_s into terms of BOD (Table IV). If it is assumed that one can directly relate the BOD of a laboratory growth medium with that of a settled sewage then it is possible, from the values given in Table IV, to calculate that the doubling time of *Sepedonium* sp. in a settled sewage of BOD 300 mg/l would be seven times longer than the minimum found for CYG medium. A similar experiment in which the inoculum consisted of 6 mm-diameter discs cut from a culture of *Sepedonium* sp. on agar and where the medium contained 3 g casitone peptone, 2 g yeast extract, 3·3 g glucose and 1·7 g/l soluble starch (CYGS medium), gave rather different kinetic constants and these are given in Table IV.

The above method was also used to calculate the growth constants of *Subbaromyces splendens* in two experiments on CYG medium in shake flasks and the values are given in Table IV.

Two determinations of the growth of *Ascoidea rubescens* at 25°C on CYG medium gave approximately equal values and these are given in Table IV. The fungi *Trichosporon cutaneum*, *Geotrichum candidum* and *Fusarium aquaeductuum* have similar growth characteristics that are distinct from the previous three species discussed above. The kinetic properties of all six organisms are compared in Table IV.

The absolute values obtained with these media are not so important in the context of sewage filters as the differences between the rates of growth and saturation constants of the six species. The three "colonial" fungi *Fusarium*, *Geotrichum* and *Trichosporon* have relatively high rates of growth and low saturation constants. Of the three "spreading" species, *Sepedonium* grew fastest and *Subbaromyces* and *Ascoidea* grew at approximately half the rate of *Sepedonium*. On average, the saturation constants were equivalent to a BOD of 1000 to 2000 mg/l. i.e. much higher than those of the three "colonial" species (BOD from 100 to 600 mg/l).

It is interesting to calculate growth rates expected at a BOD corresponding to the strength of sewage at Langley and Minworth (Birmingham), say 250 mg/l. At this value the doubling times for the six fungi obtained by the expression $t_d = t_{d_{min}}\left(1+\frac{K_s}{s}\right)$—where $t_{d_{min}}$ is the minimum doubling time shown in the table and s is 250 mg/l—would be *Sepedonium* 33 h; *Subbaromyces* 80 h; *Ascoidea* 58 h; *Fusarium* 8 h; *Geotrichum* 7·5 h; *Trichosporon* 7·8 h. These values refer to growth at 25°C but it is possible using the data in Fig. 3 to calculate the theoretical doubling times at a temperature of 12°C, which is an average winter temperature: *Sepedonium* 79 h; *Subbaromyces* 216 or 560 h; *Ascoidea* 135 h; *Fusarium* 13 h; *Geotrichum* 9 h; *Trichosporon* 22 h.

(*b*) *Maintenance energy*. Tomlinson (1941) showed that when sewage (BOD 210 mg/l) was allowed to drip on to pieces of filter film composed predominantly of *Sepedonium* sp. the weight doubled in about 5 days. When tap water was substituted for the sewage, there was a gradual decrease in weight. With sewage diluted by an equal volume of tap water the doubling time was over 8 days, and when diluted by four volumes of tap water the weight increased only very slowly. The concentration of sewage necessary to supply the energy of maintenance was equivalent to a BOD of about 40 mg/l.

Results obtained with *Sepedonium* and *Subbaromyces* are given in Table V. Energy of maintenance of *Sepedonium* was provided by sewage

TABLE V. Change in weight of *Sepedonium* and *Subbaromyces* after 3 days at 25°C in sewage diluted with water

BOD mg/l	*Sepedonium* Initial wt. (mg) 3·54	*Subbaromyces* Initial wt. (mg) 2·46	BOD mg/l	*Sepedonium* Initial wt. (mg) 0·62	*Subbaromyces* Initial wt. (mg) 0·67
	Wt. after 3 days			Wt. after 3 days	
200	4·40	2·46	220	2·93	2·60
100	*3·85*	2·10	110	1·71	2·46
50	3·15	*2·24*	55	*1·02*	1·35
25	3·30	1·88	27·5	0·45	*1·00*
12·5	2·60	1·80	14	0·32	0·66
			0	0·39	0·55

diluted to a BOD value 50–100 mg/l in the first and 28–55 mg/l in the second experiment. Slightly lower concentrations were sufficient to prevent loss of weight of *Subbaromyces.*

It is interesting to compare the estimated doubling times of these fungi on sewage with those obtained for zoogloeal bacteria by allowing sewage to drip on weighed microscopical cover glasses at 25–28°C. Four values in the region of 10 h were obtained, three in the region of 15 h and two about 25 h. These values are intermediate between the "spreading" and "colonial" groups of fungi. With regard to competition between fungi and bacteria the latter have much lower saturation constants. For example, Monod (1942) found values for *Escherichia coli* grown in glucose, mannitol and lactose at 37°C of 4·0, 2·0 and 20·0 mg/l. Painter *et al.* (1968) found values for organisms in activated sludge removing acetate, glucose and sucrose of 2·0 and 5–10 mg/l. Montgomery *et al.* (1971) found a value of 15 mg/l for a sewage bacterium oxidizing glucose. This means that the rate of bacterial growth would not be greatly affected until a BOD level below that necessary for the maintenance energy of the fungi had been reached.

Hawkes (1965a) correctly argued that the lower average strength of the Minworth sewage in summer and autumn would favour bacteria and limit the fungi and that the stronger sewage and lower temperatures in winter would encourage the growth of fungi. He also pointed out that sewage in the U.S.A. is generally weaker than in this country, and this probably explains why American workers rarely mention blocking of filters by fungus mycelium.

(*c*) *Radial growth rate of fungi on solid media.* In some respects the growth of a cushion of mycelium and particularly that of *Sepedonium*, which grows to a thickness of over 2 cm, is similar to that of a colony on an agar plate. Fungal hyphae increase in length only at their apices but protoplasm in a much longer portion of the hypha increases in volume in order to supply the advancing tip. Thus the rate of extension of a hypha is a function of the length of the terminal portion of the hypha contributing to apical extension. The length of the peripheral growth zone of a colony may be determined experimentally by severing hyphae at different distances from their apices and measuring the shortest hyphal length at which the rate of growth is undiminished (Trinci, 1971). It was found that the radial growth rate (K_r) of colonies was the product of the length (w) of the hyphae spanning the colony's peripheral growth zone and the specific growth rate (μ) of these hyphae, or $K_r = \mu w$. The value of μ appeared to be identical to the organism's maximum specific growth rate in submerged culture.

The determination of the length of the hyphal growth zone of the six fungi by the method of Trinci (1971) was not entirely satisfactory, but approximate values together with values of the growth rate μ_{max} from Table IV and the product $w\mu_{max}$. The rates of radial growth are given below (Table VI). There is fair agreement between K_r and the observed rates of radial growth. It may be concluded therefore that the values of w are approximately correct. The corollary to this is that the three "spreading" fungi *Sepedonium*, *Subbaromyces* and *Ascoidea* can draw on the protoplasm from 1–4 mm of hypha for apical growth; the corresponding length in the three "colonial" species is about 0·3–0·4 mm.

TABLE VI. Radial growth rates of fungi on solid media

	Length of hyphae (w) contributing to apical growth	$\mu_{max}(h^{-1})$	K_r $w\mu_{max}$ ($\mu m \cdot h^{-1}$)	Observed radial growth ($\mu m \cdot h^{-1}$)
Sepedonium	3–5 mm	0·145	580	590
Subbaromyces	1 mm	0·07	70	140
Fusarium	270 μm	0·19	51	60
Geotrichum	400 μm	0·30	120	130
Ascoidea	1 mm*	0·07	—	80
Trichosporon	375 μm*	0·20	—	75

* Calculated from observed rate of radial growth and μ_{max}
$\mu m = cm^{-4}$

TABLE VII. Colony growth characteristics of *Sepedonium* and *Subbaromyces*

	Fungus	BOD (mg/l) 0	70	140	270	540	1080
Rate of hyphal extension (mm/day)	*Sepedonium*	3·7	3·0	3·3	4·7	5·7	6·0
	Subbaromyces	1·5	1·3	1·6	1·5	1·5	1·7
Hyphal density at periphery of colony (hyphae/cm)	*Sepedonium*	3	12	62	83	110	200
	Subbaromyces	16	70	100	97	143	180

The greater "mass" growth rates of the three colonial species are due to their high rate of branching. This produces a very dense colony, of limited size. Limitation of radial growth has been shown to be due either to staling products, growth hormone or similarly inhibiting metabolite (Gottlieb, 1971).

Hawker (1957) emphasized the survival value of rapid linear growth of the mycelium under starvation conditions. This is a means of migration to a more favourable habitat. Species of *Mucor* and allied genera, species of *Sordaria*, *Chaetomium*, *Pyronemia confluens*, and facultative parasites of the *Pythium* type, possess this advantage. It is of great value where the food supply is scattered, as in soil where the distribution of dead plant and animal material is not uniform.

Both *Sepedonium* and *Subbaromyces* are able to produce "starvation growth"; indeed this characteristic may be used to isolate them from contaminating bacteria and other fungi on water agar. The following Table VII shows that the rate of hyphal extension is not greatly affected by the nutrient concentration but the density of branching is. The dominant fungi in percolating filters usually have this property of rapid hyphal extension; for example, *Saprolegnia ferax* in filters treating milk factory waste and *Mucor hiemalis* in filters treating skimmed milk waste.

3. *Effect of Temperature*

The effect of temperature on the radial growth of colonies of the 6 fungi is shown in Fig. 3. Growth increased linearly with temperature. The rates of growth at 25°C (maximum summer temperature) to that at 12°C (average winter temperature) were as follows: *Sepedonium* 2·4; *Subbaromyces* 2·7 and 7·0; *Ascoidea* 2·3; *Fusarium* 1·7; *Geotrichum* 2·7 and 3·2; *Trichosporon* 2·2 and 2·8.

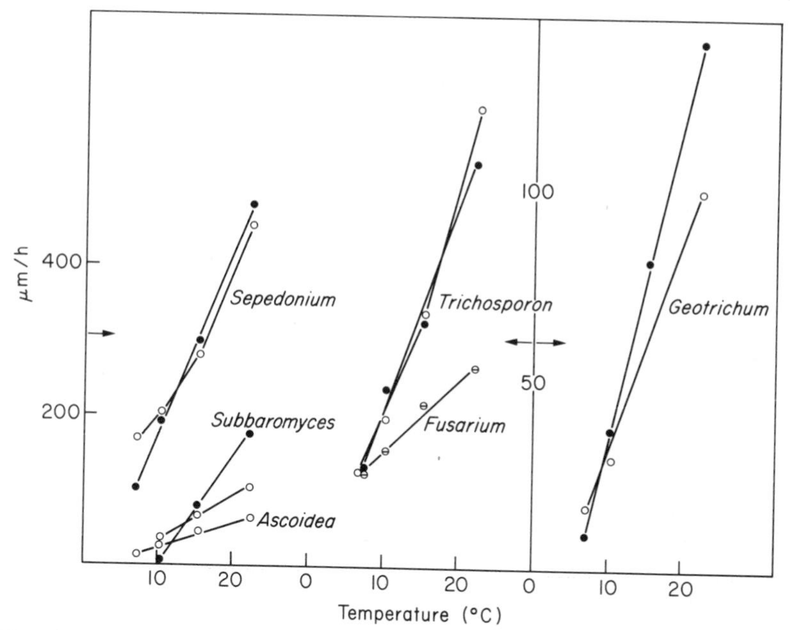

FIG. 3. Rate of hyphal extension in µm/h in relation to temperature.

4. Effect of pH

Results obtained by Painter (1954) on the effect of pH on the growth of *Sepedonium* sp., *Fusarium aquaeductuum*, *Geotrichum candidum* and *Trichosporon cutaneum* were probably influenced by the tendency of the organism to raise the pH value during the course of growth in an unbuffered medium. Painter found that the three latter fungi gave approximately the same yield over a pH range of 4·0 to 8·0 whereas *Sepedonium* sp. had a sharp peak in the region pH 6·5–8·5.

Albert (1960) cites a number of workers who have found that the neutral molecule of normal fatty acids is more inhibitory than the corresponding ion to the growth of common moulds, yeasts and some other fungi. For example, Hoffman *et al.* (1940) showed that the concentration of acetic acid (pKa = 4·8) necessary to prevent the growth of a number of common moulds at pH 6·0 was about four times the concentration necessary at pH 4·0, and above pH 6·0 acetic acid was virtually non-toxic. At pH = pKa, 50% of the molecule is ionized. A difference of 2 pH units above or below the pKa value of anionic or cationic groups increases the ionization to about 99% or the percentage of the molecule in the neutral form to 1%. However, in some cases the ionized form contributes to the inhibitory action.

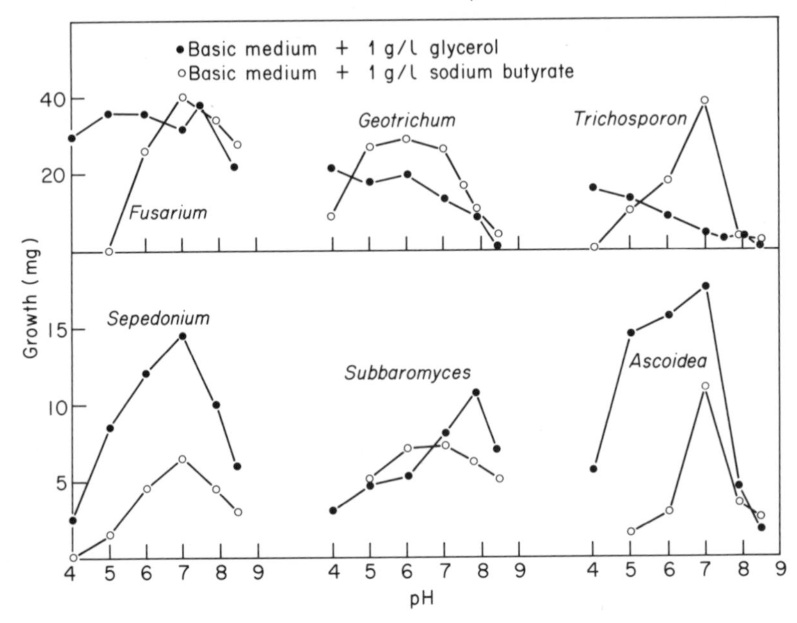

FIG. 4. Growth of fungi in relation to pH and composition of liquid medium.

Figure 4 shows that the effect of pH in the case of *Fusarium*, *Geotrichum* and *Trichosporon* depends on the composition of the medium; substituting butyrate for glycerol in a basic liquid medium containing 1 g glucose, 2 g hydrolysed casein, 0·5 g yeast extract, trace metals, and 5 g Tris buffer per litre, limited growth at lower pH values. This was probably due to the inhibitory action of undissociated butyric acid which has a pKa value at 25°C of 4·82. *Sepedonium* had a growth peak in both types of media at pH 7·0. *Ascoidea* tolerated a wider pH range in the medium containing glycerol and *Subbaromyces* showed tolerance of a wider pH range in the medium containing sodium butyrate. The six fungi can be classified into the "colonial" species which can tolerate low pH values but are susceptible to inhibition by undissociated organic acids, and the "spreading" species which have a pH optimum in the neutral range of pH.

5. *Inhibition of Growth*

(*a*) *Anionic detergent.* Anionic detergent used domestically has risen in amount from 1949 to the present day and is an important constituent of domestic sewage. The mean concentrations in the settled sewage treated at the Minworth and Langley sewage works of the Upper Tame Main Drainage Authority during 1969 were equivalent to 6·7 and

TABLE VIII. Effect of JNQ on the rate of radial growth (mm/day) of fungi

	Concentration of JNQ (mg/l Manoxol OT)						
	0	5	10	20	30	40	50
Sepedonium	12·0	10·5 (2·5)	14·0 (2·3)	2·0	1·2	0	0
Subbaromyces	2·4	2·5	2·6	2·1	1·5	0·4	0·1
Ascoidea	0·9	0·96	0·8	0·9	0·8	0·9	0·8
Fusarium	1·0	0·5	0·75	0·6	0·5	0·25	0·1
Geotrichum	2·6	2·4	2·3	1·8	1·3	0·6	0
Trichosporon	1·4	1·5	1·7	1·5	1·4	1·0	0·8

25·6 mg/l Manoxol OT respectively. During 1972 the concentration in the Stevenage sewage was in the region of 35 mg/l.

Two methods were used to measure the effect of different concentrations of anionic detergent on the growth of fungi. By the first method the rate of radial growth of a colony on agar containing CYG medium was measured, and by the second method the weight yield of three of the fungi was obtained in detergent-free sewage to which different concentrations of detergent were added. The detergent used was Dobane JNQ, a soft anionic detergent. (Table VIII.)

Ascoidea was least affected by detergent; *Trichosporon* was not inhibited by concentrations up to 30 mg/l. At this concentration hyphal growth of *Fusarium*, *Geotrichum* and *Subbaromyces* was inhibited by about 50%. The greatest effect occurred with *Sepedonium*. Even at 5 mg/l some of the mycelium became highly branched with little extension in growth (Table VIII, figures in parentheses). Other hyphae grew in length but were less branched. At 20 mg/l there was general inhibition to the extent of about 80 per cent.

In sewage *Sepedonium* had an average yield of 0·8 mg/day (dry wt. mycelium) in the control flask with no added detergent and 0·7 mg/day at a detergent concentration of 18 mg/l. At 38 mg/l of JNQ the mean yield was only 0·23 mg/day. Similarly 5 mg/l JNQ had no effect on the growth of *Subbaromyces* but 40 mg/l inhibited growth by about 90%. 40 mg/l JNQ in sewage had little effect on the growth of *Ascoidea*. The susceptibility of *Sepedonium* to inhibition by detergent in concentrations above 10 mg/l may explain its absence from the filters at Langley where *Subbaromyces* predominates. The results of these experiments are summarized in Table IX.

Manns (unpublished) in a study of the effect of the detergent Dobane JNX on*Subbaromyces splendens* in pure culture showed that colony extension rates were affected by concentrations in excess of 5 mg/l, noting a

TABLE IX. Effect of detergent concentration on mean daily yield of fungi grown in sewage

Detergent concentration mg/l	*Sepedonium* mg/day	Detergent concentration mg/l	*Subbaromyces* mg/day	*Ascoidea* mg/day
0	0·8	0	2·2	0·17
18	0·7	15	2·3	0·15
38	0·23	40	0·2	0·14

progressively increasing effect to 25 mg/l, the highest value tested (Table X). At this latter concentration conidia production was inhibited, and this may explain why conidia are rarely seen in filter film at Langley. At the higher detergent concentrations, increase in the extent of branching in the mycelium was seen, and this could be a factor contributing to the markedly slower rate of colony extension.

(*b*) *Antibiotic interaction between pairs of fungi.* Inhibition of growth of one fungus by another can be demonstrated by placing pairs of agar discs containing different species on nutrient agar in proximity and observing the presence or absence of antibiotic activity. If there is no interaction the mycelium of the two mix and there is no demarcation boundary between the two colonies. On the other hand an inhibitory species will expand and prevent the growth of the other species in the vicinity. Table XI shows the results obtained when five species of fungi were tested against each other

TABLE X. Effect of Dobane JNX on *Subbaromyces splendens* colony extension rate (Manns, unpublished data)

Detergent concentration (mg/l as Manoxol OT)	Average diameter of colony[a] (mm) after days 2	5	7	12	18	22	25
0	3·7	14·6	28·0	55·5	66·2	76·0	79·8
5	3·0	10·8	22·0	44·8	62·0	72·0	76·8
10	2·8	11·0	19·2	36·3	52·7	62·8	65·8
15	2·2	10·2	18·0	33·0	45·3	53·0	61·2
20	0	6·5	11·0	22·8	31·2	38·2	41·2
25	0	2·6	6·2	11·8	20·2	25·3	27·2

[a] Mean of two diameters for each of 3 colonies

TABLE XI. Interactions between five species of fungi on nutrient agar plates

	Se	*Su*	*G*	*F*	*T*
Sepedonium (*Se*)		=	=	>	=
Subbaromyces (*Su*)	=		=	>	=
Geotrichum (*G*)	=	=		<	<
Fusarium (*F*)	<	<	>		<
Trichosporon (*T*)	=	=	>	>	

= indicates no interaction, > indicates that organism in vertical column was inhibited by organism in horizontal column, < indicates that organism in horizontal column was inhibited by organism in vertical column.

as described above. *Geotrichum* inhibited growth of *Fusarium* and *Trichosporon* but not *Sepedonium* or *Subbaromyces*, whereas *Fusarium* inhibited the latter two species as well as *Trichosporon*.

(*c*) *Substrate inhibition.* Lamb *et al.* (1964) classified substrates into four categories: (1) not used, not toxic, (2) used, not toxic, (3) used, toxic, and (4) not used, toxic. Neutral compounds such as glucose are in Category (2) and volatile fatty acid salts come in Category (3). Growth of *Sepedonium* on different concentrations of a medium (Fig. 5) containing 2 g glucose, 2 g sodium butyrate, 6 g hydrolysed casein, 1 g yeast extract, 5 g Tris buffer per litre, pH 6·9, was inhibited at concentration of butyrate greater than 200 mg/l.

In the case of *Subbaromyces* the rate of growth increased with substrate concentration up to a value of 600 mg/l sodium butyrate and remained

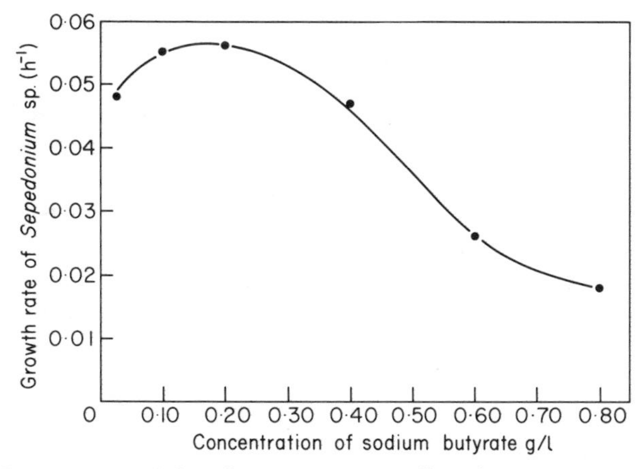

FIG. 5. Growth rate of *Sepedonium* sp. on sodium butyrate.

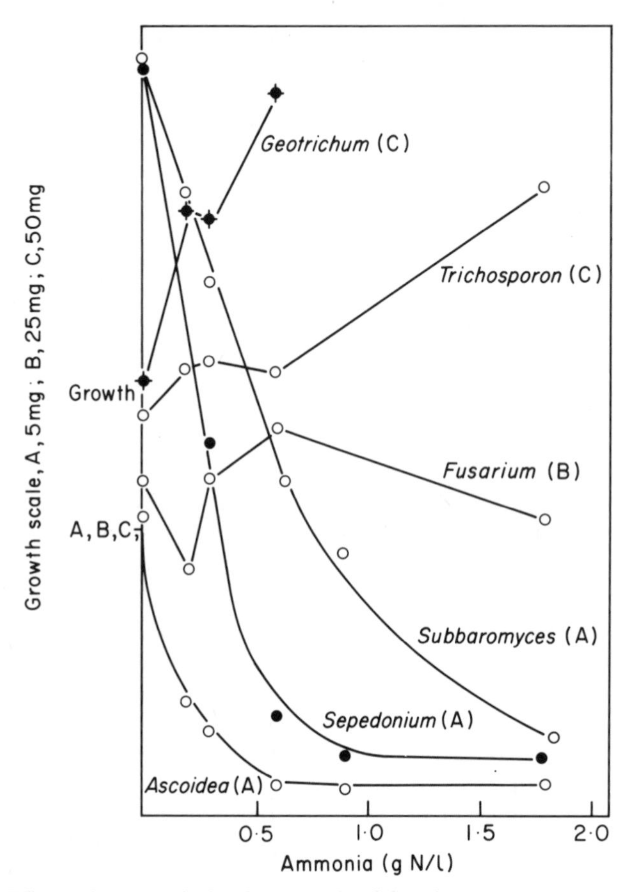

FIG. 6. Effect of ammonia in the growth of fungi.

constant over the range 600–1600 mg/l sodium butyrate. In the case of *Ascoidea* the rate of growth increased up to a substrate concentration of 600 mg/l sodium butyrate. At concentrations of 800 and 1000 mg/l the rate was much lower but at 1200 and 1600 mg/l the rate was as high as at 600 mg/l.

(*d*) *Ammonia.* Figure 6 shows the effect of the concentration of ammonium ion on the growth of the six species. The medium contained 1 g glucose, 1 g sodium butyrate, 2 g soluble casein, 0·1 g yeast extract per litre (pH 6·5). *Fusarium*, *Geotrichum* and *Trichosporon* were not inhibited by concentrations of ammonia equivalent to 2000 mg N/1 but *Sepedonium*, *Ascoidea* and *Subbaromyces* were inhibited at much lower concentrations. In this connection it may be significant that *Fusarium*,

Geotrichum and *Trichosporon* can grow on ammonia as the sole source of nitrogen whereas the other three fungi need an organic source of nitrogen.

6. Variations in Growth with Depth in Filter

Mills (1945a) produced data of the reduction in BOD of sewage with depth in percolating filters at Minworth. During the winter of 1942–43 the BOD was reduced from 132 to 50 mg/l in the top 30 cm of medium. Assuming this is a logarithmic reduction with depth it may be deduced that the BOD was halved by passage through a depth of filtering medium equal to 22 cm (0·72 ft). The rapid decrease in nutrient concentration explains the concentration of fungi in the top 30 cm of filtering medium. It also explains the efficiency of recirculating effluent in controlling growth of fungi.

C. Notes on Taxonomic, Cultural and Ecological Aspects of Named Species

Ascoida rubescens Brefold
(Ascomycotina–Hemiascomycetes–Endomycetales–Ascoidaceae)

Although there are very few records of the occurrence of this fungus in sewage treatment plants, it was abundant for a period of several years at Minworth Works (Hawkes, 1963), as foamy masses in the interstices between stones of the percolating filters but not as surface mat growths. It has not been recorded there recently, probably because of the changes that have taken place both in the nature of the sewage and in the mode of operation of the filters. This fungus has also been found at Rochdale (Lancs), where the perfect stage was seen in filter film (Fig. 7). It is not recorded from this habitat in the U.S.A. (Cooke, 1963).

Isolates from the Minworth works have produced plenty of conidia, including occasional endoconidia, but although some structures resembling asci have been seen to form, the characteristic hat-shaped ascospores were not produced. An isolate was identified by the Commonwealth Mycological Institute as *Ascoidea rubescens*, and all features observed in culture conform closely to published descriptions of this fungus (Walker, 1931, 1935). The colony form on agar plate culture is consistent on media supporting a good growth, showing a white aerial mycelium of a coarse, woolly appearance and up to 5 mm tall. Conidia do not usually form before seven days of incubation on plates; they are produced successively from the tips of hyphae or on short lateral branches, but the conidiophores are themselves unbranched. (Fig. 8). The mycelium is fairly coarse, with hyphae 15–30 μm in diameter, and shows frequent, regular

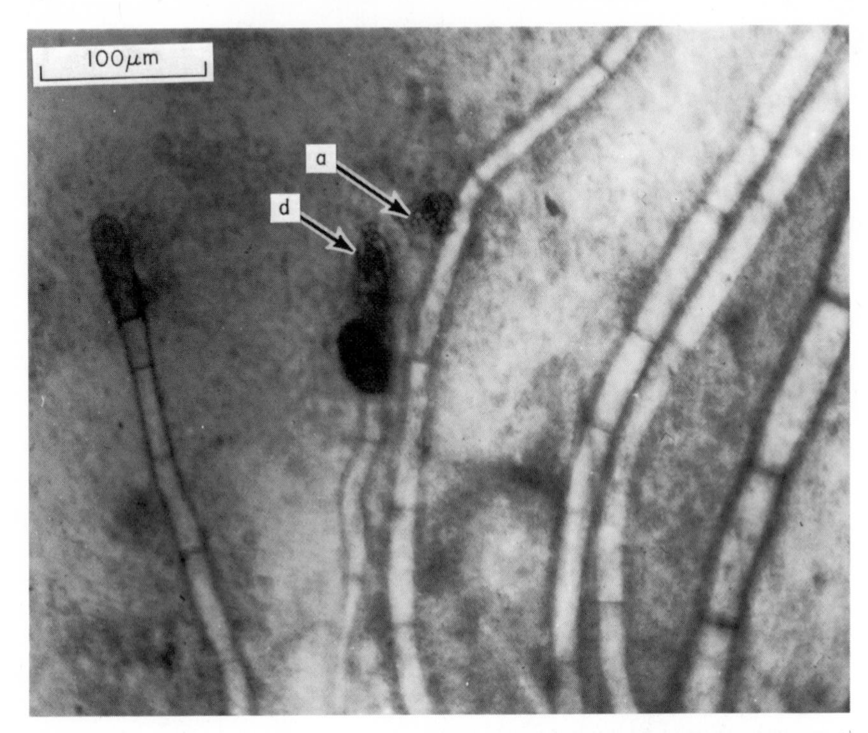

FIG. 7. Ascospore production in *Ascoidea rubescens*. (a) Mycelium with asci, in filter film. ×230. (b) Detail of ascospores. ×1000. a = ascospores, d = dehisced ascus.

branching. Luxuriant growth is only obtained on media rich in organic substances and the growth responses on various media reflect its organic nitrogen requirement. Fungi of the class Hemiascomycetes produce asci without any covering layer of sterile hyphae; those in the family Ascoidaceae are noted for the multisporous ascus and well-developed true mycelium, while *Ascoidea* species are unusual in that the asci are believed to be produced parthenogenetically (Walker, 1935).

Fusarium aquaeductuum (Radimacher and Rabenhorst) Saccardo
(= *Fusarium episphaeria* Snyder and Hansen)
(This fungus is the imperfect stage of *Nectria episphaeria*)
(Deuteromycotina–Hyphomycetes–Tuberculariaceae)

The most familiar of the percolating filter fungi, this produces orange to pink growth on the surface, closely adherent to the stones by virtue of the prostrate system of hyphae previously mentioned (Fig. 9a). The macroconidia, which are characteristic of the genus in being large, multicellular and crescentic, are in this species produced from the general surface of

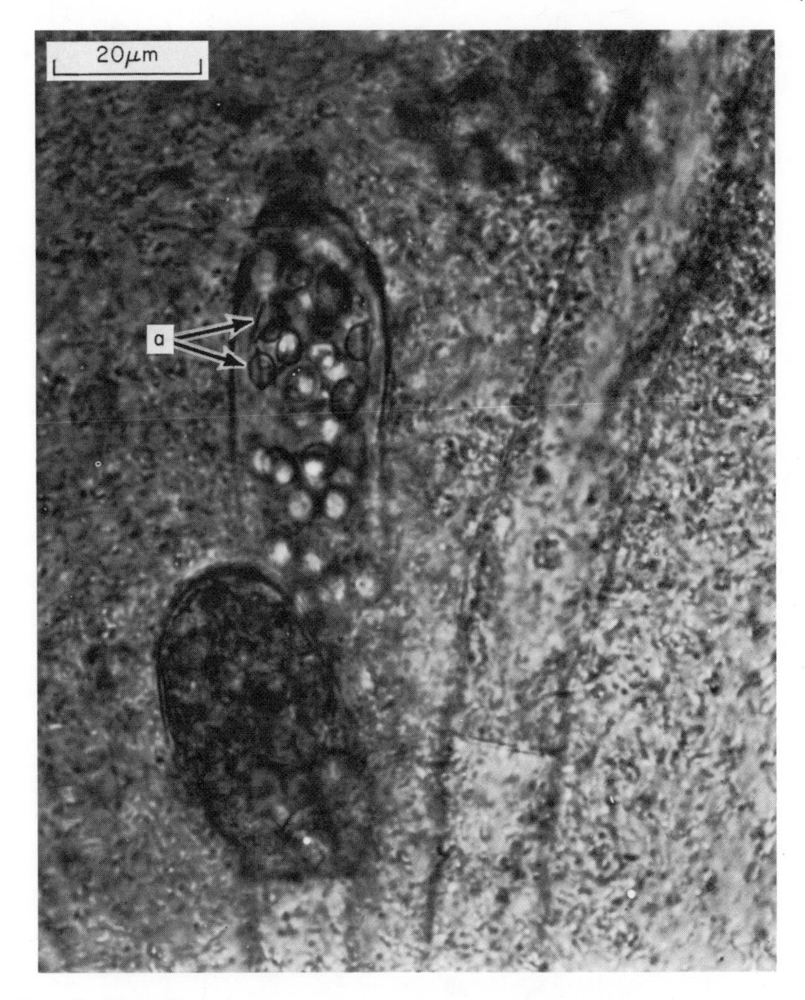

FIG. 7. (*Contd.*)

the mycelium and not in special structures (Fig. 9b). Microconidia are not formed in this species but chlamydospores are seen occasionally. Isolates from Minworth have given conidia 25–35 μm long by 4–5 μm broad, modally 3-septate, but much variation is recorded; they form on most media supporting good growth, but very few are formed in cultures grown in total darkness. Both conidia formation and pigmentation are light-dependent; the pigment is a photo-induced carotenoid (Rau and Zehender, 1959; Rau, 1967).

Colony form on agar is very variable, as is usual in this genus, but it

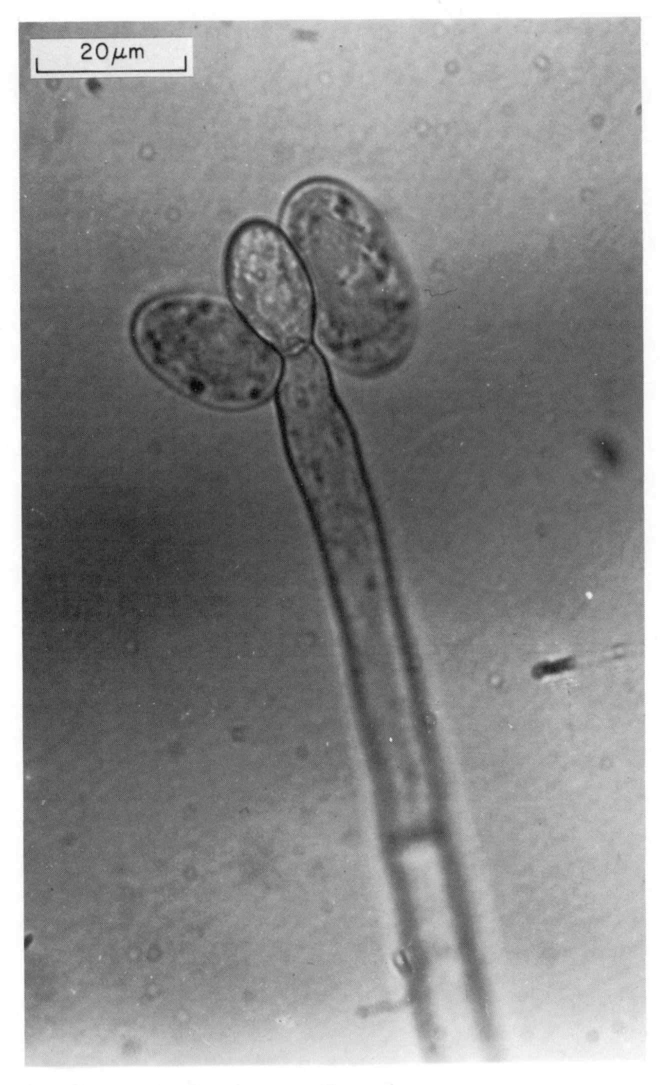

FIG. 8. Conidia production in *Ascoidea rubescens*.

frequently produces a ruckled, slimy mat which grows slowly and eventually turns pink, orange or yellow. A tuft of coremial growth characteristically appears in the colony centre. This fungus is recorded most often from works treating a sewage containing some industrial effluent, and sometimes occurs in great profusion, colouring the entire filter area pink (Winsor, 1973). Further taxonomic detail is to be found in the work of Wollenweber and Reinking (1935) and Toussoun and Nelson (1968).

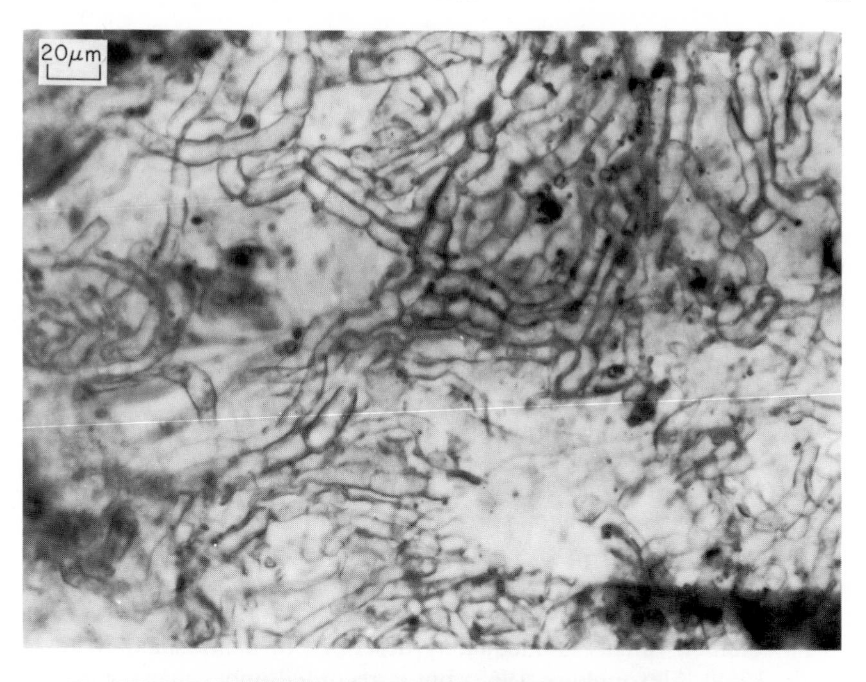

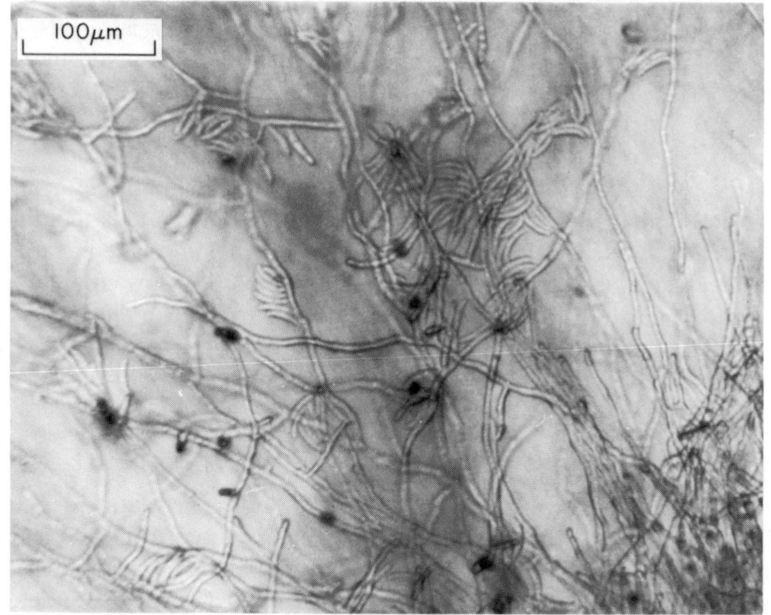

FIG. 9. *Fusarium aquaeductuum* in filter film. (a) Prostrate system of hyphae. (b) Erect system of hyphae showing conidia production.

Geotrichum candidum Link.
(Deuteromycotina–Hyphomycetes–Moniliales–Moniliaceae)

Since this fungus is world-wide in distribution, occurring as a common fungal contaminant in the dairy industry, in soil, on fruits, and with some strains pathogenic to plants or causing mycoses in man, its presence in sewage is to be expected. It has been found at most works where the fungal flora has been investigated, and heavy growths reported have usually been associated with the presence of industrial effluent. Becker and Shaw (1955) recovered it from 70% of sewage samples tested from two works, each of which was treating a primarily domestic sewage with a small proportion of food industry effluent. The incidence of this fungus remained high in samples of effluent at all stages of treatment. At Minworth it can be found in the filter film at all times of the year; on the experimental plant referred to earlier it succeeded *Fusarium aquaeductuum* in the late autumn but was itself succeeded as the main component of surface film by *Sepedonium* sp. as the temperature fell further. Hawkes (1959) has recorded it as co-dominant with *Sepedonium* sp. Film dominated by this fungus is grey to brown in colour, soft in texture, and loosely follows the contours of the stones.

This common fungus is adequately described in many standard texts (Smith, 1969; Alexopolous, 1962) but is recorded under many different names in older literature. The taxonomy, with a list of synonyms, and morphology (including colonial morphology) are fully discussed by Carmichael (1957), while Painter (1954), Brancato and Golding (1953) and Cooke (1957b) have investigated the factors affecting the growth of this organism in culture. The characteristic truncate arthrospores are freely produced in percolating filter film (Fig. 10), and the dichotomous branching of the hyphal tips (a common feature of this fungus) aids identification. Colony form on agar is very variable but colonies are invariably white or cream and aerial mycelium is very short; slimy colonies are common, particularly under extreme conditions or rich media, and the mycelium may become totally arthrosporic. A perfect stage has recently been confirmed and named (*Endomyces geotrichum*) by Butler and Petersen (1972).

Leptomitus lacteus (Roth) Aghardh

This fungus has been found in various situations in treatment plants. Cooke (1959) includes it in a long list of fungi isolated from percolating filters, and it has been found in filters treating industrial sewage (at Minworth), milk wastes, and waste waters from beet sugar manufacture. Profuse growths are not generally recorded from this habitat, but do sometimes occur in channels or streams carrying used water from certain

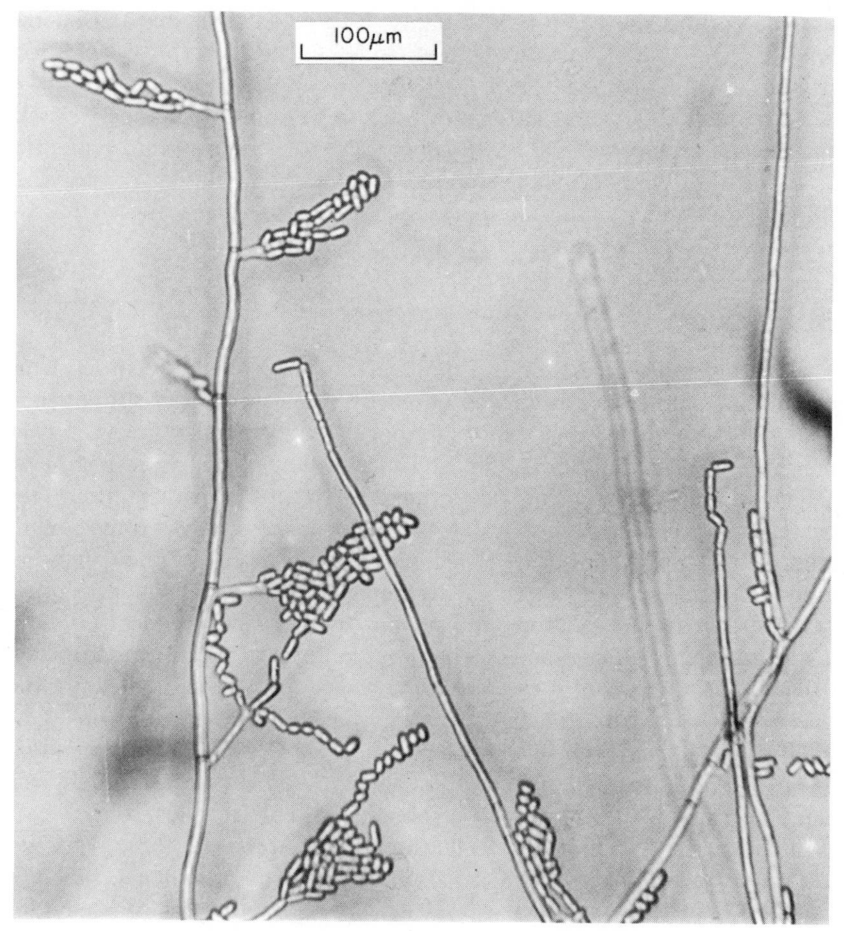

FIG. 10. *Geotrichum candidum* mycelium with arthrospores.

industries; Hawkes (1963) suggests that this may be related to the fact that it is a true aquatic fungus. Sladka and Ottova (1968) isolated it from a brook receiving dairy waste and also from an activated-sludge plant treating brewery and malthouse wastes. It has also been found in the River Dee (England) a short distance below the outfall of a chemical factory.

Cantino (1966) states that *Leptomitus* has lost the capacity to utilize both ammonia and nitrate; Zehender and Boek (1964) also report its inability to utilize inorganic nitrogen sources, finding high molecular weight nitrogen compounds essential for it. These workers also record optimum growth at pH 6·0 and 8°C, and found organic acids particularly suitable carbon sources for the fungus. Burnett (1968) reports that this fungus is unable to utilize sugars and depends on fatty acids for growth.

Taxonomic details and illustrations are to be found in standard works on phycomycetaceous fungi (Sparrow, 1960). Intermittent constrictions in the hyphae, with shining granules of cellulin (a polysaccharide) sometimes adjacent to the constrictions, give a superficial appearance of septation and make the fungus unmistakable.

Sepedonium sp.

This fungus produces some of the heaviest film growths recorded from percolating filters and was first noted at the Minworth (Birmingham) works of the Upper Tame Main Drainage Authority (Tomlinson, 1941). Despite extensive studies in film and in culture this organism has never produced any spores or reproductive structures other than the characteristic chlamydospores. On the basis of these it was provisionally identified as a new species of *Sepedonium* by Dr Mason of the Commonwealth Mycological Institute. It is sufficiently constant in vegetative growth form both in culture and in film to be readily recognized; four distinct variants are known, separated by the form of the chlamydospores. Comparisons of two of these variants recently in continuous culture has shown that they are similar in growth rate and respond identically to intermittent dosing with nutrient feed (see Fig. 1), while in batch culture in broth they exhibited similar optimal temperature ranges and showed close resemblances in growth response on a variety of solid media (Williams, 1971). Painter (1954) found that three of these variants, including the original isolate, had similar vitamin requirements, thiamine and biotin proving essential for satisfactory growth. In this study the original isolate was shown to need organic sources of nitrogen, to have an optimal pH range of 7·0–8·5, and to be sensitive to zinc at the comparatively low levels of 4–10 mg/l when grown in media partially deficient in other trace metals.

Film dominated by this fungus is olive-green to yellow grey in colour and characteristically forms a thick, tough, leathery mat on the surface, not closely following the contours of the stones or attached to them. Depending on the dosing regime it may occur at greater depth in the filter. In agar plate culture the colony form is always a moderate to scant transparent growth with a slight golden-brown tint in thicker areas. The hyphae are relatively coarse, 9–16 μm in diameter, with widely spaced septa and sparingly branched. Chlamydospores are usually intercalary, sometimes sessile but more often borne on narrow, unicellular stalks or

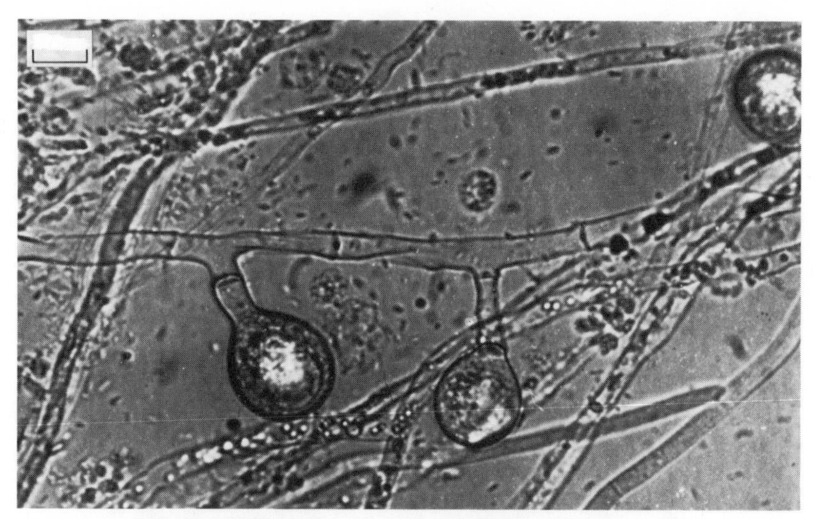

FIG. 11. Chlamydospores of four strains of *Sepedonium* sp. isolated from filter film. (a) Smooth, round type. Scale mark = 20 μm.

recurved on multicellular stalks with enlarged subtending cells; they may be round or oval, smooth, warty or spiny. (Fig. 11a, b, c, d).

Taxonomic information is not included for this fungus as it has only been identified provisionally.

Subbaromyces splendens Hesseltine
(Ascomycotina–Plectomycetes–Microascales–Ophiostomataceae)

First reported and named by Hesseltine (1953) from percolating filters treating the used water from an antibiotic manufacturing plant at Pearl River (N.Y.), this fungus has only occasionally been reported since, always from percolating filters. As it is a slow-growing fungus rarely reaching nuisance proportions it may have often been overlooked. It occurs as discrete masses mainly on the sides and undersides of the stones in the uppermost layers, or as floating clumps in areas of impeded drainage between the stones. At the Langley (Warks) works of the Upper Tame Main Drainage Authority, it occurs regularly in the late summer and autumn in filters dosed continuously, but has not been seen in those subject to periodic dosing at 15 min intervals (Shepherd, 1967). Occasional inspection of canisters of medium sunk in the filters has shown that most of the growth occurs in the upper 30 cm of the filter but smaller amounts are found to a much greater depth. This is the only fungus present in visible amounts in film at this works, where the sewage treated is entirely domestic and notable for its high detergent content (25 to 29 mg/1, Manoxol value, annual average in crude sewage).

This fungus has been shown to grow slowly on all media in culture. On

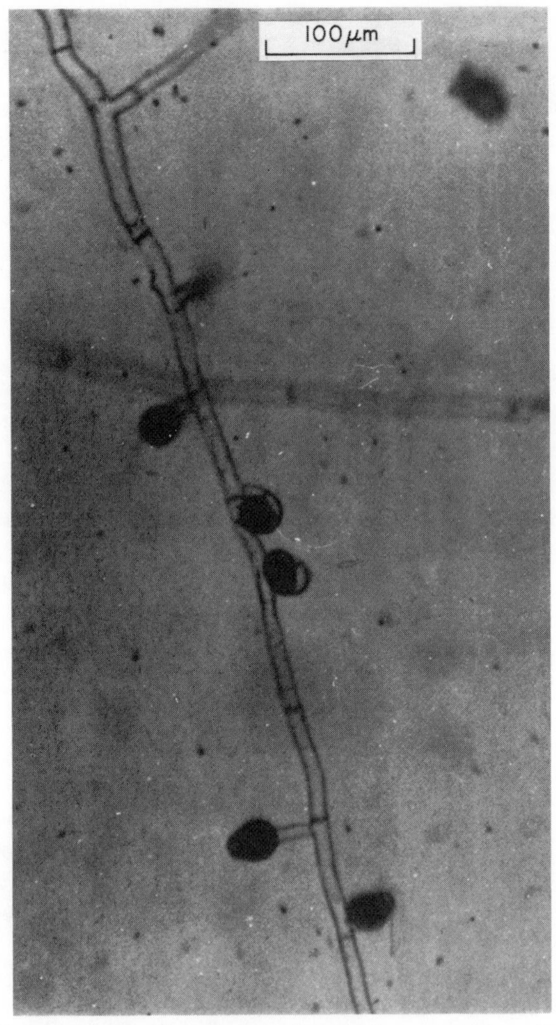

FIG. 11. (*Contd.*) Chlamydospores of *Sepedonium* sp. (b) oval with recurved stalks

agar growth is typically extremely scant and transparent, and aerial mycelium does not normally occur, except where perithecia form. The hyphae are 9 to 12 μm in diameter, extending beneath the agar surface and emerging to bud off rounded masses of conidia at the agar surface. These are single, large, oval, hyaline cells borne on sparsely branched conidiophores (Fig. 12). In film from Langley, hyphae are notably coarser than in culture, septa more frequent, and conidia seen only occasionally and singly. The striking perithecia (Fig. 13) have not so far been produced

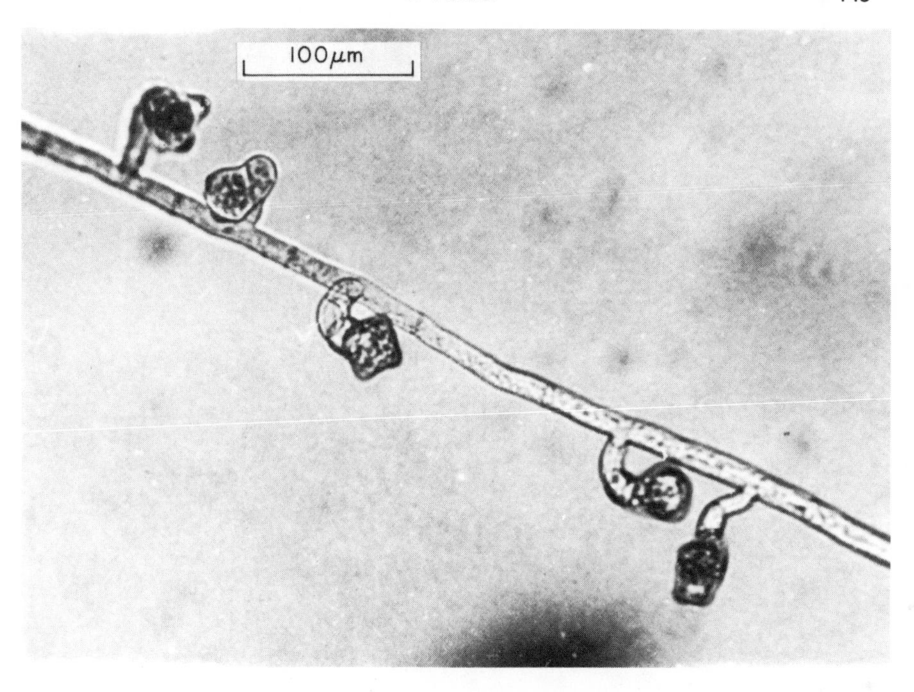

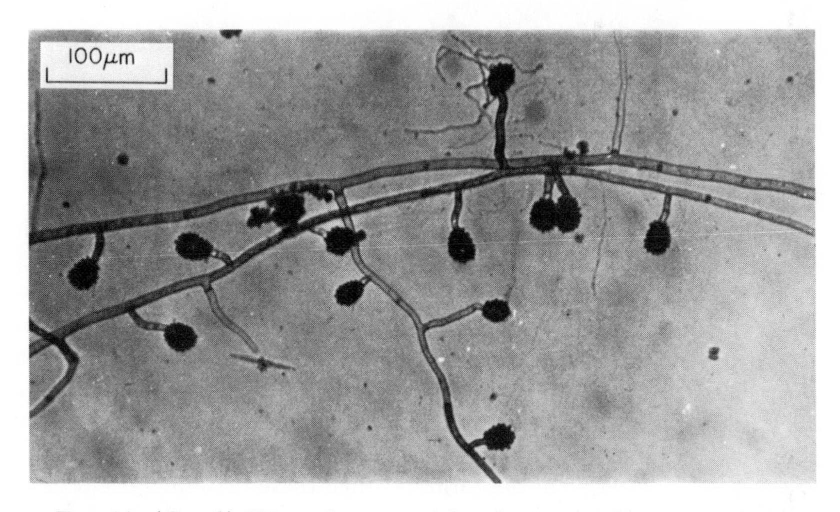

FIG. 11. (*Contd.*) Chlamydospores of *Sepedonium* sp. (c) above, coarsely verrucate; (d) below, finely verrucate

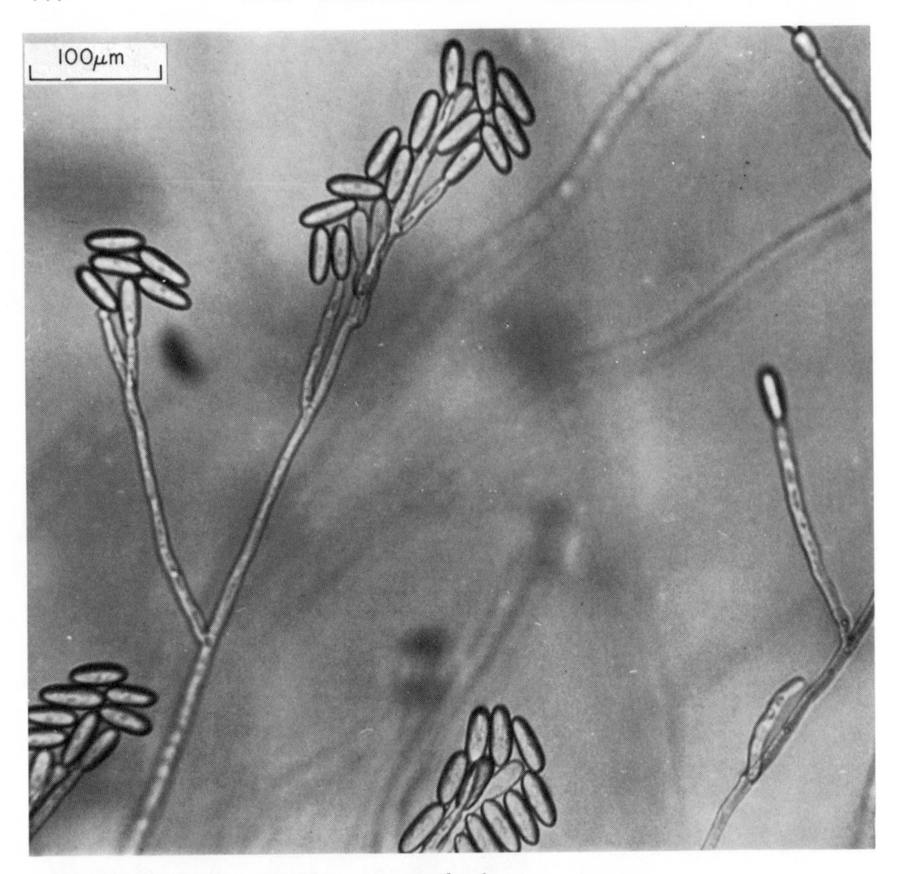

FIG. 12. Conidia of *Subbaromyces splendens*.

in pure culture on agar media but will form freely in fresh isolates in the presence of contaminant bacteria (Hesseltine, 1953; Williams, 1971). They have also been produced on lima beans (Cooke, 1963).

Resting structures considered to be sclerotia are produced in old, drying plate cultures and in younger cultures subjected to certain adverse conditions, for example temperatures above the optimum. These are grey to black bodies of irregularly rounded shape and variable size, and appear to form at the site of large conidial masses. They germinate well and may be used to recover the fungus from old cultures lacking viable conidia. In liquid culture this fungus has not been observed to form surface mats under any conditions, a feature also noted in *Ascoidea rubescens*, and it is interesting to relate this to the form in which these two fungi occur in the percolating filter, where both form discrete lumps or floating masses between the stones.

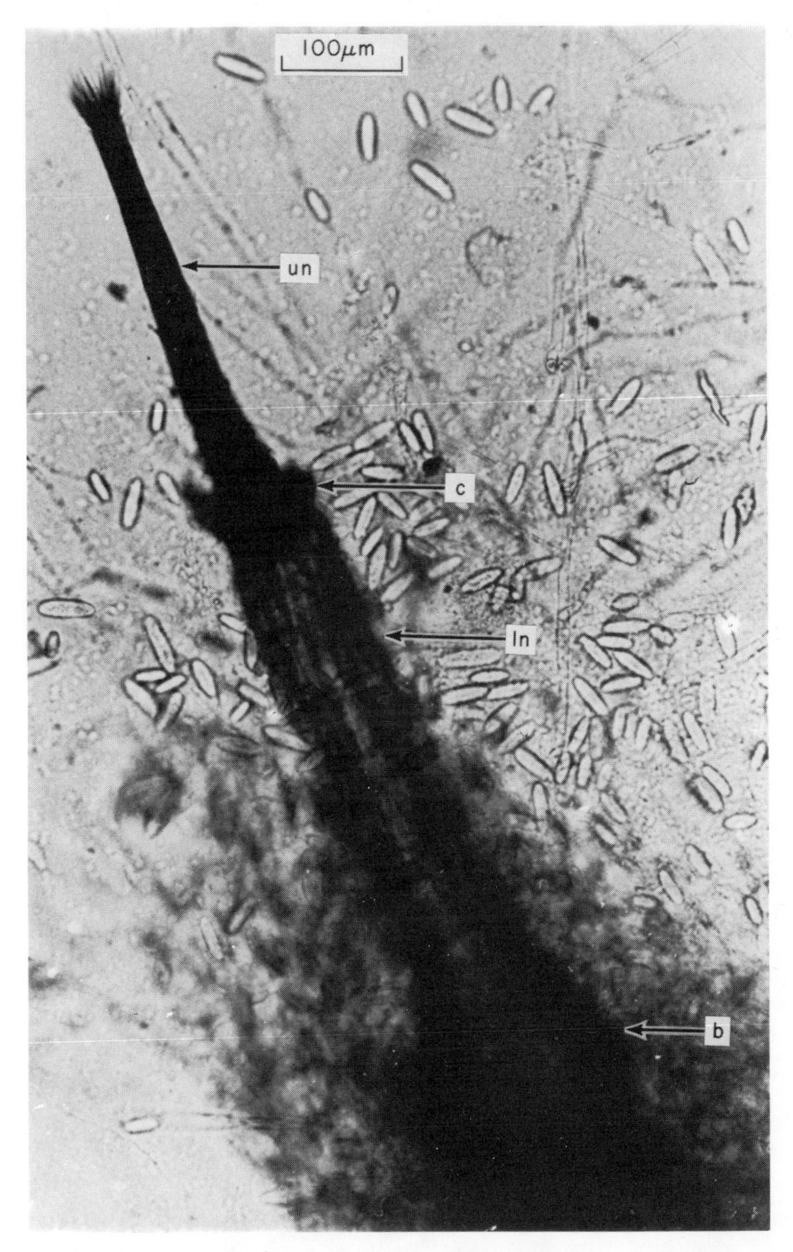

FIG. 13. Mature perithecium of *Subbaromyces splendens* (from culture with bacteria present; the oval bodies are conidia produced by surrounding mycelium). un = upper neck, c = collar, ln = lower neck, b = base.

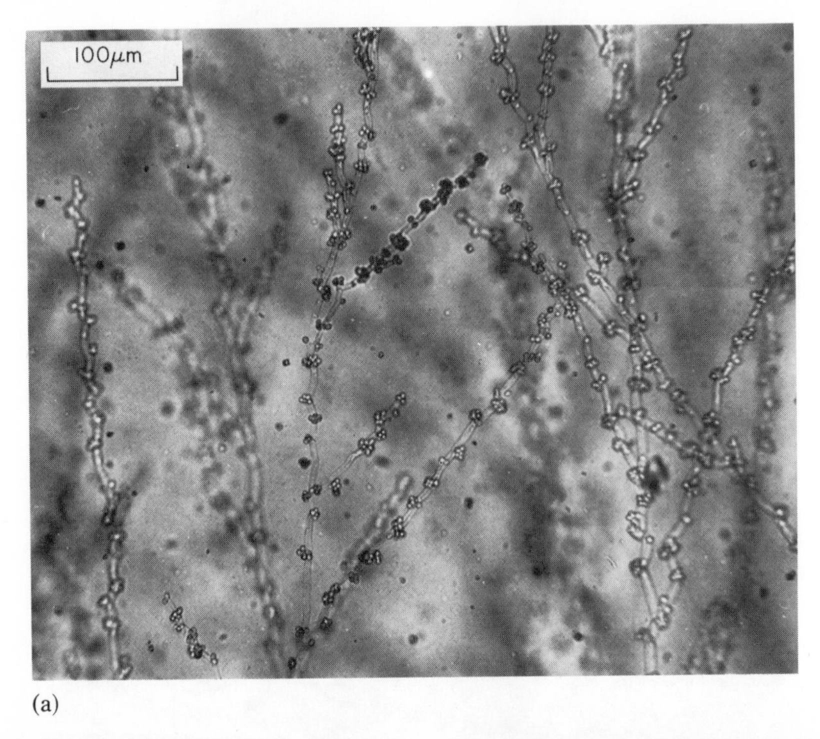

(a)

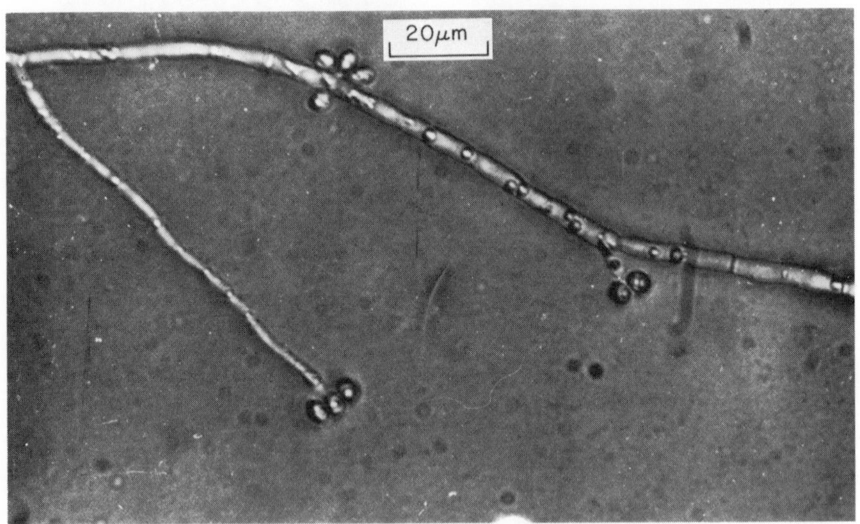

(b)

FIG. 14. *Trichosporon cutaneum* mycelium with blastospores. (a) General appearance. (b) Detail.

Trichosporon cutaneum (de Beurm., Gougeot and Vaucher) Ota
(Deuteromycotina–Hyphomycetes–Moniliales–Cryptococcaceae)

This fungus is not on record as having produced growth in percolating filters in sufficient amount to form recognizable film, but is commonly present (often with *Geotrichum candidum*) and regularly isolated. It resembles *G. candidum* in colony form on agar media and in the production of arthrospores, but is distinguished by the production of blastospores (buds) also, which places it amongst the yeasts. It is therefore fully treated by Lodder (1970). (Fig. 14.)

IV. Present Position and Future Trends

The advancement of knowledge in this field is probably best approached from two angles. The first of these is the follow-up of leads from field work on film accumulation or sludge bulking due to fungi, with further work on the physiology and autecology of the more troublesome species, to allow informed deductions about their role in the treatment process. This has been the approach of British workers in this field to date. Although the view is held in some quarters that any fungus which can be recovered may have a part to play in the ecology of the system, we agree with Sladka and Ottova (1968) that unless a species can make appreciable mycelial growth in the system it cannot be considered to be playing a significant role in purification.

Improvements in design and operation aiming at maximum efficiency from percolating filters have resulted in the widespread adoption of dosing regimes which control most troublesome fungi (Section IIB3(e)). Increasingly strict controls on industrial discharges to sewers, and increased water usage, have tended to cause reduction in sewage strength in recent years. These factors have resulted in reduction of nuisance from fungal growths in works operation, but it is possible that the problem will recur. Restriction on discharge of pollutant material to rivers, economies in water usage as a result of increased charges, and increased use of domestic garbage grinders could all be factors contributing towards stronger sewages. Further, new manufacturing processes may bring new organic compounds of fungus-promoting potential to the sewers, or more toxic compounds to alter the ecological balance in the treatment in favour of the fungi. It is well known that single flushes of acid or toxic waste, or short-term failures in machinery compelling undesirable changes in dosing regimes, can upset the biological treatment plant for prolonged periods, sometimes encouraging fungal growths which persist (Winsor, 1973); the chance of minimizing these effects depends on recognition of the problem and a knowledge of the ecology of the organisms concerned. Continued research is therefore important and advantage should be taken of our present period of respite.

A second approach is through the present trend towards maximum conservation of resources. The ability of fungi to degrade organic compounds in sewage and sludge can be further exploited for the benefit of man. Composting of town refuse with sewage sludge or other organic waste is already widely practised, and the process depends on the degradative action of saprophytic bacteria and fungi to produce a stable end-product valuable as a fertilizer (Gray *et al.*, 1971). Many industrial effluents are purified or rendered inoffensive by the deliberate use of fungi: decolourization (Marton *et al.*, 1969) and clarification of liquors from wood-processing plants (Toni *et al.*, 1968), the use of *Fusarium solani* in cyanide waste treatment (Shimizu *et al.*, 1969) and the removal of organic materials from food processing wastes (Peacock and Page, 1971) are examples. The enormous interest engendered in recent years in the potential of yeasts to solve the world food shortage has led many workers to explore the possibility of using effluents, mainly from industrial processes, for growing yeasts and filamentous fungi for human or animal food. Activity and potential in this field is discussed by Edwards and Finn (1969). A specific study of a large number of fungi in relation to their ability to convert organic material from corn and soy processing wastes to mycelial protein showed that *Trichoderma viride*, *Gliocladium deliquescens* and *Aspergillus oryzae* gave best results. The mycelium could readily be recovered by coarse filtration and had potential as animal feed, and there was over 98% BOD reduction in the waste liquors (North Star Research and Development Institute, 1970). *Candida utilis*, one of the more intensively studied of the food yeasts, has been successfully grown on many industrial effluents including wood hydrolysate, phenolic wastes and effluent from wet carbonization of peat (McLoughlin, 1972).

Fungi have been used for many years to produce a variety of chemical compounds industrially (Prescott and Dunn, 1959), and the possibility that in the future it will be practicable to grow such fungi under controlled conditions at the expense of waste organic material in industrial effluents, or even sewage, so that the waste is purified, the fungus is harvested as protein source and useful products are recovered from the effluent, should prove an attractive prospect for industrialist and conservationist alike.

Acknowledgements

This chapter includes work carried out by us at the Water Pollution Research Laboratory (T.G.T.) and the University of Aston in Birmingham (I.L.W.), to whom we are indebted for facilities. We would like to acknowledge a limited use of data from unpublished work in project reports by the following graduates of the University of Aston in Birmingham: B. K. Manns, J. Battersby and J. Baynes (who also kindly provided Fig. 13).

References

Albert, A. (1960). "Selective Toxicity." Methuen, London.

Alexopolous, C. J. (1962). "Introductory Mycology" (2nd Edn). Wiley, New York.

Becker, J. G. and Shaw, C. G. (1955). *Appl. Microbiol.* **3,** 173–180.

Bell, H. D. (1926). *Survr munic. Cty Engrs* **70,** 561–565.

Brancato, F. P. and Golding, N. S. (1953). *Mycologia* **45,** 848–864.

Bruce, A. M., Merkens, J. C. and Macmillan, S. C. (1970). *J. Inst. Publ. Hlth Engng* **69,** 178.

Burman, N. P., Oliver, C. W. and Stevens, J. K. (1969). *In* "Isolation Methods for Microbiologists" (Eds D. A. Shapton and G. W. Gould). Soc. Appl. Bact. Tech. Ser. No. 3. Academic Press, London and New York.

Burnett, J. H. (1968). "Fundamentals of Mycology." Edward Arnold, London.

Butler, E. E. and Petersen, L. J. (1972). *Mycologia* **64,** 365–374.

Cantino, E. C. (1966). Morphogenesis in aquatic fungi. *In* "The Fungi" (Eds G. C. Ainsworth and A. S. Sussman), Vol. 2, C. 10. Academic Press, New York and London.

Carmichael, J. M. (1957). *Mycologia* **49,** 820–830.

Cochrane, V. (1958). "Physiology of Fungi." Wiley, New York and London.

Cooke, W. B. (1954a). *Sewage ind. Wastes* **26,** 539–549.

Cooke, W. B. (1954b). *Sewage ind. Wastes* **26,** 661–674.

Cooke, W. B. (1957a). *Sydowia, Ann. Mycol.* ser. 2 Beiheft 1, 146–175.

Cooke, W. B. (1957b). *Sewage ind. Wastes* **29,** 1234–1251.

Cooke, W. B. (1958a). *Engng Bull. Ext. Ser., Purdue Univ.* No. 96. *Proc. 13th Ind. Waste Conf.* 26–45.

Cooke, W. B. (1958b). *Sew. ind. Wastes* **30,** 21–27.

Cooke, W. B. (1959). *Ecology* **40,** 273–291.

Cooke, W. B. (1963). "Laboratory Guide to Fungi in Polluted Waters, Sewage and Sewage Treatment Systems." U.S. Dept. of Health Public Health Service Publication No. 999-WP-1.

Cooke, W. B. (1965). *Engng Bull. Ext. Ser., Purdue Univ.: Proc. 20th Ind. Waste Conf.* 6–18.

Cooke, W. B. (1970). *Ohio J. Sci.* **70,** 129–146.

Cooke, W. B. and Hirsch, A. (1958). *Sew. ind. Wastes* **30,** 138–156.

Cooke, W. B. and Ludzack, F. J. (1958). *Sew. ind. Wastes* **30,** 1490–1495.

Cox, C. R. (1921). *Engng News Rec.* **87,** 720–725.

DRESCHLER, C. (1935). *Mycologia* **27,** 6–40.
DRESCHLER, C. (1959). *Mycologia* **51,** 787–823.
DUDDINGTON, C. L. (1956). *Biol. Rev.* **31**(2), 152–193.
DUDDINGTON, C. L. (1957). *Symp. Soc. gen. Microbiol.* **7,** 218–237.
EDWARDS, V. H. and FINN, R. K. (1969). *Process Biochem.* **4**(1), 29.
FARQUHAR, G. J. and BOYLE, W. C. (1971). *J. Wat. Pollut. Control Fed.* **43,** 779–798.
FELDMAN, A. E. (1955). *Sew. ind. Wastes* **27**(11), 1243–1244.
FELDMAN, A. E. (1957). *Sew. ind. Wastes* **29**(5), 538–540.
FINSTEIN, M. S. and HEUKELEKIAN, H. (1967). *J. Wat. Pollut. Control Fed.* **39,** 33–40.
GOTTLIEB, J. (1971). *Mycologia* **63,** 619–629.
GRAY, K. R., SHERMAN, F. and BIDDLESTONE, A. J. (1971). *Process Biochem.* **6**(6), 32–36.
HAENSELER, C. M., MOORE, W. D. and GAINES, J. G. (1923). *Bull. New Jers. agric. Exp. Stn* **390,** 39–48.
HARKNESS, N. and JENKINS, S. H. (1964). *J. Proc. Inst. Sew. Purif.* 1964(3), 280–291.
HARVEY, J. V. (1952). *Sew. ind. Wastes* **24**(9), 1159–1164.
HAWKER, L. E. (1957). *Symp. Soc. gen. Microbiol.* **7,** 238–258.
HAWKES, H. A. (1955). *J. Proc. Inst. Sew. Purif.* 1955(1), 48–59.
HAWKES, H. A. (1957). *J. Proc. Inst. Sew. Purif.* 1957(2), 88–107.
HAWKES, H. A. (1959). *Ann. appl. Biol.* **47**(2), 339–349.
HAWKES, H. A. (1961). *J. Proc. Inst. Sew. Purif.* 1961(2), 105–133.
HAWKES, H. A. (1963). "The Ecology of Waste Water Treatment." Pergamon Press, Oxford.
HAWKES, H. A. (1965a). *Int. J. Air Wat. Pollut.* **9,** 693–714.
HAWKES, H. A. (1965b). *In* "Ecology and the Industrial Society" (5th Symp. British Ecological Society). Blackwell Scientific Publications, Oxford, England.
HAWKES, H. A. and JENKINS, S. H. (1955). *J. Proc. Inst. Sew. Purif.* 1955(4), 352–358.
HAWKINS, H. A. and JENKINS, S. H. (1958). *J. Proc. Inst. Sew. Purif.* 1958(2), 221–226.
HAWKES, H. A. and JENKINS, S. H. (1964). *Sb. vys. Sk. chem. technol. Praze.* **8**(1), 87–122.
HESSELTINE, C. W. (1953) *Bull. Torrey bot. Club.* **80**(6), 507–514.
HOFFMAN, G., SCHWEITZER, T. R. and DALBY, G. (1940). *J. Am. Soc. Chem.* **62,** 988.
HOLMES, P. (1970). M.Sc. Project Thesis, University of Aston in Birmingham (Applied Hydrobiology Section), Birmingham, England.
HOLTJE, R. H. (1943). *Sewage Wks J.* **15,** 14–29.
JENKINS, S. H. and WILKINSON, R. (1940). *J. Soc. chem. Ind., Lond.* **61,** 125–128.
JONES, P. H. (1964). *Engng Bull. Ext. Ser. Purdue Univ.: Proc. 19th Ind. Waste Conf.* **49**(2), 902–914.

Jones, P. H. (1965). *Engng Bull. Ext. Ser. Purdue Univ.: Proc. 20th Ind. Waste Conf.* **49**(4), 297–315.
Lamb, J. C., Westgarth, W. C., Rogers, J. L. and Vernimuran, A. P. (1964). *J. Wat. Pollut. Contr. Fed.* **36,** 1263–1284.
Lodder, J. (Ed) (1970). "The Yeasts" (2nd Edn). North-Holland, Amsterdam, Holland.
Lumb, C. and Barnes, J. P. (1948). *J. Proc. Inst. Sew. Purif.* 1948(1), 83–98.
Lumb, C. and Eastwood, P. K. (1958). *J. Proc. Inst. Sew. Purif.* 1958(4), 380–398.
Martin, J. P. (1950). *Soil Sci.* **69,** 215–232.
Marton, J., Stern, A. M. and Marton, T. (1969). *TAPPI* **52,** 1975–1981.
McLoughlin, A. J. (1972). *Process Biochem.* **7**(1), 27–29.
Mills, E. V. (1945a). *J. Proc. Inst. Sew. Purif.* 1945(1), 35–49.
Mills, E. V. (1945b). *J. Proc. Inst. Sew. Purif.* 1945(2), 94–110.
Monod, J. (1942). "Recherches sur la Croissance des Cultures Bacteriennes." Hermann et Cie, Paris.
Montgomery, H. A. C., Oaten., A. B. and Gardiner, D. K. (1971). *Effl. Wat. Treat. J.* **11,** 23–24.
North Star Research and Development Institute (1970). "Use of Fungi Imperfecti in Waste Control." U.S. Fed. Wat. Qual. Admin., Wat. Pollut. Control Res. Ser. no. 12060EHT 07/70. U.S. Govt. Print. Off., Washington, D.C.
Oxoid Manual (1969). (3rd Edn reprint). Oxoid Ltd., Southwark Bridge Rd, London.
Painter, H. A. (1954). *J. gen. Microbiol.* **10,** 177–190.
Painter, H. A. and Viney, M. (1959). *J. biochem. microbiol. Technol. Engng* **1,** 143–162.
Painter, H. A., Denton, R. S. and Quarmby, G. (1968). *Wat. Res.* **2,** 427–447.
Peacock, G. and Page, O. T. (1971). *Wat. Res.* **5,** 498–502.
Pipes, W. O. (1965). *Engng Bull. Extn Ser. Purdue Univ.* No. 118. *Proc. 20th Ind. Waste Conf.* 647–656.
Pipes, W. O. and Cooke, W. B. (1969). *Engng Bull. Ext. Ser. Purdue Univ.* No. 132. *Proc. 23rd Ind. Waste Conf.* **53**(2), 170–182.
Pipes, W. O. and Jones, P. H. (1963). *Biotechnol. Bioengng* **5,** 287–307.
Pirt, S. J. (1966). *Proc. R. Soc.* **B166,** 369–373.
Prescott, S. C. and Dunn, C. G. (1959). "Industrial Microbiology" (3rd Edn). McGraw-Hill, New York and London.
Rau, W. (1967). *Planta* **72,** 14–28.
Rau, W. and Zehender, C. (1959). *Arch. Mikrobiol.* **32,** 423–428.
Rettger, L. F. (1906). *Engng News, N.Y.* **56**(18), 459–460.
Reynoldson, T. B. (1942). *J. Proc. Inst. Sew. Purif.* 1942, 116.
Robbins, W. J. (1937). *Am. J. Bot.* **24,** 243–250.

SHEPHERD, M. R. N. (1967). M.Sc. Thesis, University of Aston in Birmingham, England.
SHIMIZU, T., TAGUCHI, H. and TERAMOTO, S. (1969). *Chem. Abstr.* **70**(8), 213.
SLADKA, A. and OTTOVA, V. (1968). *Hydrobiologia* **31,** 350–362.
SMITH, G. (1969). "Introduction to Industrial Mycology" (6th Edn). Edward Arnold, London.
SPARROW, F. K. (1960) "Aquatic Phycomycetes." University of Michigan Press, Ann Arbor, Michigan, U.S.A.
TABAK, H. H. and COOKE, W. B. (1968). *Mycologia* **60,** 115–140.
TOMLINSON, T. G. (1941). *J. Proc. Inst. Sew. Purif.* 1941(1), 39–57.
TOMLINSON, T. G. (1942). *J. Soc. chem. Ind., Lond.* **61,** 53–58.
TOMLINSON, T. G. (1946). *J. Proc. Inst. Sew. Purif.* 1946(1), 168–178.
TOMLINSON, T. G. and HALL, H. A. (1955). *J. Proc. Inst. Sew. Purif.* 1955(1), 40–48.
TOMLINSON, T. G. and SNADDON, D. H. M. (1966). *Int. J. Air Wat. Pollut.* **10,** 865–881.
TONI, T., TANI, Y. and ONO, K. (1968). *Chem. Abstr.* 1968 **69,** 7501.
TOUSSOUN, T. A. and NELSON, P. E. (1968). "*Fusarium:* A Pictorial Guide to the Identification of *Fusarium* Species." Pennsylvania State University Press, Pennsylvania, U.S.A.
TRINCI, A. J. P. (1971). *J. gen. Microbiol.* **67,** 325–344.
WALKER, L. B. (1931). *Mycologia* **23,** 51–76.
WALKER, L. B. (1935). *Mycologia* **27,** 102–127.
WATER POLLUTION RESEARCH (1941). "The Treatment and Disposal of Waste Waters from Dairies and Milk Products Factories." W.P.R. Technical Paper No. 8. H.M. Stationery Office, London.
WATER POLLUTION RESEARCH (1955). Pp. 55–57. H.M. Stationery Office, London. 1956.
WATSON, W., HUTTON, D. B. and SMITH, W. S. (1955). *J. Proc. Inst. Sew. Purif.* 1955(1), 73–85.
WILLIAMS, I. L. (1971). M.Sc. Thesis, University of Aston in Birmingham, England.
WINSOR, C. E. (1973). *Wat. Pollut. Control* **72**(5), 560–581.
WISHART, J. M. and WILKINSON, R. (1941). *J. Proc. Inst. Sew. Purif.* 1941, 15–38.
WOLLENWEBER, H. W. and REINKING, O. A. (1935). "Die Fusarien." Paul Parey, Berlin.
ZEHENDER, C. and BOEK, A. (1964). *Zentbl. Bakt. ParasitKde* (Abt. 2) **117,** 399–411.

4

Algae and Bryophytes

Kathryn Benson-Evans
and
Peter F. Williams

Department of Botany
University College
Cathays Park
Cardiff
Wales

I. Introduction

The obvious difference between the algae and other organisms found in used-water treatment processes is the ability of the former to produce oxygen utilizing solar energy in the process of photosynthesis. Light energy is absorbed by chlorophyll and other pigments and converted into chemical energy. The types of pigment present in algae will depend upon their phylogenetic position and upon the ecological conditions to which they have been exposed. Round (1965) gives a good account of algal pigments and their occurrence in the particular algal taxonomic groups. Chemical energy is used to reduce carbon dioxide derived from various forms of oxidized carbon and form sugars; the hydrogen for the reduction is derived from water (photolysis) with the consequent liberation of oxygen. In used-water treatment processes this oxygen is utilized by heterotrophic organisms to oxidize organic matter to carbon dioxide which is again made available to the algae.

In this chapter the authors have endeavoured to summarize a sample of world literature dealing with the occurrence and significance of algae in used-water treatment. After outlining current trends in algal taxonomy and listing the algae found in treatment processes, a more detailed account of the biology and autecology of the more widely quoted genera is given. This is followed by an appraisal of the broader synecological aspects of algae in percolating filters and oxidation ponds. Although reference has been made to laboratory studies in some instances, it cannot be over-emphasized that extreme caution is necessary in applying such results to the field situation.

II. Taxonomic Survey

The algae are a rather heterogeneous group of cryptogamic plants comprising some fourteen or fifteen large classes and several smaller groups that have been incompletely studied. They are common inhabitants of fresh, salt and soil water, the most widely recognized being the seaweeds and, to a lesser extent, large growths of the more obvious green, filamentous forms found in freshwater ponds and streams or the blue-green "water blooms" on lakes or reservoirs in summer. From these observations alone it can be seen that algae differ from other "lower" organisms in that many pigmented forms occur. All contain chlorophyll *a*, but this may be partially masked by other accessory pigments, the importance of which has been recognized since the earliest classification systems of Harvey (1836) and Eichler (1886). This basic system, according to Round (1965), is re-emphasized in a recent text by Christensen (1962).

Fritsch (1935–1945) suggested a rather negative definition of the group: "Unless purely artificial limits are drawn, the designation alga must include all holophytic organisms (as well as their numerous colourless derivatives) that fail to reach the higher level of differentiation characteristic of the archegoniate plants." To decide whether or not a plant may be included in the latter group, an investigation of the structure of the reproductive organs is normally carried out. Where these are multicellular with the outer layer of cells sterile, then the plant is assigned to the archegoniate divisions (Bryophyta—dominant leafy gametophyte, Pteridophyta—dominant leafy sporophyte). In the algae, all sporangia and gametangia are either unicellular or, where multicellular, every cell is fertile. To distinguish the algae from other "lower" organisms, e.g. fungi and bacteria, the algae are regarded as having their chemical economy based on photosynthesis involving chlorophyll *a*, in which the accumulation rather than the breakdown of organic matter predominates. As mentioned by Round (1965), it is considered convenient at present to include all photosynthetic flagellates in the algae, since they all have a photochemical apparatus similar to that of other plants and absent from animals. Round points out that much basic structure and other physiological processes are common to both plant and animal groups, "thus it is better not to be dogmatic and to allow that many of the flagellates are extremely plastic and that convenience is perhaps the only justification for some of the taxonomic positions".

Certain suffixes to be used for the categories of plant classification were laid down by the International Code of Botanical Nomenclature (Utrecht) (Lanjoun *et al.*, 1952). Those which are applicable to the Algae are as follows:

DIVISION (PHYLUM)	—PHYTA
SUB-DIVISION (SUB-PHYLUM)	—PHYTINA
CLASS	—PHYCEAE
SUB-CLASS	—PHYCIDAE
ORDER	—ALES
SUB-ORDER	—INALES
FAMILY	—ACEAE
SUB-FAMILY	—OIDEAE
TRIBE	—EAE

The actual allocation of algal groups to these subdivisions is still open to much discussion and a number of schemes have been put forward by various workers, as may be observed by reference to the many texts dealing with the algae. A list of a representative selection of these is given in "Guide to Taxonomic Works" at the end of this chapter, and many contain good keys for the identification of species.

Characters that are used as a basis for classifications of the algae range from gross morphology and growth form to features of ultrastructure such as plastids, mitochondria and flagella structure. Some of the major characters used in separating the major divisions (phyla) may be biochemical, e.g. pigmentation and food reserves.

Christensen (1962) recognized two major groups of organisms over and above the divisions normally accepted as the largest category. His treatment is basic and reflects a major difference between two groups of the heterogeneous assemblage of organisms referred to as the algae. The two groups are:

I. PROCARYOTA—organisms lacking a definite nucleus and plastids with nuclear substance and photosynthetic structures not delimited by a sharp bounding membrane.

Division (phylum) CYANOPHYTA
Class CYANOPHYCEAE (MYXOPHYCEAE)
(Blue-green algae)

II. EUCARYOTA—algae with a typical nucleus and pigments localized in chromatophores. These may be further sub-divided into the

A. ACONTA—algae with no flagellated stages in their life-cycle.
Division (phylum) RHODOPHYTA
Class RHODOPHYCEAE
(red algae)

B. CONTOPHORA—algae with flagellated stages present in the life-cycle.
1. Division (phylum) EUGLENOPHYTA
(or EUGLENOPHYTINA, a subphylum of the CHLOROPHYTA)
2. Division (phylum) PYRROPHYTA
3. Division (phylum) CRYPTOPHYTA
4. Division (phylum) PHAEOPHYTA
Class PHAEOPHYCEAE
(brown algae)
5. Division (phylum) CHRYSOPHYTA
(a) Class XANTHOPHYCEAE
(Heterokontae)
(b) Class CHRYSOPHYCEAE
(c) Class HAPTOPHYCEAE
(d) Class BACILLARIOPHYCEAE
(diatoms)
6. Division (phylum) CHLOROPHYTA
(Isokontae)
(a) Class CHAROPHYCEAE
(b) Class BRYOPSIDOPHYCEAE
(c) Class CONJUGATOPHYCEAE
(d) Class OEDOGONIOPHYCEAE
(e) Class CHLOROPHYCEAE
(f) Class PRASINOPHYCEAE

The various arguments and opinions for adopting the above system are given fully in Round (1965), who puts forward a similar system based on a combination of the works of Fott (1959) and Christensen (1962).

Only four divisions (phyla) are represented to any extent in used-water treatment processes and plants; the Chlorophyta, Euglenophyta, Cyanophyta and Chrysophyta. In only two of the used-water treatment processes are algae found to any significant extent, namely in percolating filters and in oxidation ponds. Tables I and II list the algae of these two treatment processes. Besides the algae listed in these Tables a few have been quoted in the literature as being associated with used-water treatment without the process being specified; these include *Pseudanabaena catenata* Lauterb. (Cyanophyta); *Vaucheria* sp., *Meridion circulare* Agardh., *Cocconeis* sp. (all in the Chrysophyta); *Enteromorpha* sp., *Oedogonium* sp., *Cladophora* sp., *Spirogyra* sp. (all in the Chlorophyta); and *Batrachospermum* sp. (Rhodophyta).

TABLE I. Algae (and bryophytes) of percolating filters

CHLOROPHYTA	
Volvocales	*Chlamydomonas* spp.
Chlorococcales	*Chlorococcum humicolum* (Naeg.) Rabenh.
	Chlorococcum sp.
	Characium sp.
	Chlorella pyrenoidosa Chick.
	Chlorella vulgaris Beyerinck
	Chlorella sp.
	Oocystis parva West
	Oocystis sp.
Ulotrichales	*Ulothrix tenuissima* Kuetz.
	Ulothrix sp.
	Hormidium flaccidum (Kuetz.) A.Br.
Ulvales	*Monostroma* sp.
Chaetophorales	*Stigeoclonium nanum* Kuetz.
	Stigeoclonium tenue Kuetz.
	Stigeoclonium sp. (coccoid stage)
	Stigeoclonium sp.
	Chaetophora pisiformis (Roth.) Agardh.
EUGLENOPHYTA	
Euglenales	*Euglena* spp.

(*Continued overleaf*)

TABLE I (*Continued*)

CYANOPHYTA	
Chroococcales	*Microcystis aeruginosa* Kuetz.
	(*Anacystis montana* (Lightf.) Drouet and Daily)
Pleurocapsales	"Pleurocapsa-like growth" (*Microcystis* or other?)
Oscillatoriales	*Oscillatoria limosa* (Agardh.) Kuetz.
	Phormidium tenue (Menegh.) Gomont
	Phormidium uncinatum (Agardh.) Gomont
	Phormidium sp.
	Anabaena sp. Bory.
	Amphithrix janthina (Mont.) Bonn. and Flah.
	Amphithrix janthina (var. *torulosa* (Grun.)?)
CHRYSOPHYTA	
Heterococcales	*Bodo caudatus* (Duj.) Stein
	Bodo edax Klebs
	Bodo globosus Stein
	Cercobodo longicauda (Stein) Pascher
Chrysomonadales	*Synura uvella* Ehren.
Centrales	*Melosira* sp.
Pennales	*Diatoma hiemale* (Lyng.) Heiberg
	Amphora ovalis Kuetz.
	Hantzschia amphioxys (Ehren.) Grun.
	Navicula cryptocephala Kuetz.
	Navicula gracilis Ehren.
	Navicula lanceolata (Agardh.) Kuetz.
	Navicula minima Grun.
	Navicula platystoma Ehren.
	Navicula pusilla W. Smith
	Nitzschia palea (Kuetz.) W. Smith
	Nitzschia palea (var. *kuetzingiana* Hilse.?)
BRYOPHYTA	
Hepaticae	*Marchantia polymorpha* L.
Musci	*Ceratodon purpureus* (Hedw.) Brid.
	Bryum argenteum Hedw.
	Bryum bimum (Brid.) Richards and Wallace
	Leptodictyum brevips
	Leptodictyum riparium (Hedw.) Warnst.
	Leptodictyum trichopodium

TABLE II. Algae of oxidation ponds

CYANOPHYTA	
Chroococcales	*Merismopedia minima* Beck.
	Microcystis sp. (*Anacystis* sp.)
	Chroococcus sp.
Oscillatoriales	*Oscillatoria amoena* (Kuetz.) Gomont.
	Oscillatoria chalybea Mertens
	Oscillatoria chlorina Kuetz.
	*Oscillatoria lauterbornii** Schmidle
	*Oscillatoria limosa** (Agardh.) Kuetz.
	Oscillatoria minima Gieklhorn
	Oscillatoria okeni (Agardh.) Gomont.
	Oscillatoria princeps Vaucher
	Oscillatoria putrida Schmidle
	Oscillatoria subtilissima Kuetz.
	Oscillatoria tenuis Agardh.
	Oscillatoria terebriformis (Agardh.) Gomont.
	Oscillatoria sp.*
	Spirulina sp.*
	Arthrospira sp.
	Phormidium foveolarum (Mont.) Gomont.
	Phormidium molle (Kuetz.) Gomont.
	*Phormidium tenue** (Menegh.) Gomont.
	Phormidium sp.
	Anabaena constricta (Szafer) Geit.
	Lyngbya spp.
	Cylindrospermum sp. (*Anabaenopsis* sp.)
	Plectonema nostocorum B. et Thur.
CHRYSOPHYTA	
Heterococcales	*Bodo caudatus** (Duj.) Stein
	Bodo celer Klebs
	Bodo edax Klebs
	Bodo globosus Stein
	Bodo minimus Klebs
	Bodo putrinus (Stokes) Lemm.
	Bodo sp.*
	Cercobodo agilis (Moproff) Lemm.
	Cercobodo sp.
Chrysomonadales	*Oicomonas mutabilis* Scher.
	Anthophysa vegetans (Müller) Bütschli

* Algae also found in anaerobic oxidation ponds.

(*Continued overleaf*)

TABLE II (*Continued*)

Pennales	*Navicula accomoda* Hust.
	Navicula cuspidata Kuetz.
	Navicula cuspidata var. *ambigua* (Ehren.) Cleve
	Navicula excelsa Krasske
	Navicula gregaria Donkin
	Navicula kriegeri Krasske
	Navicula lanceolata (Agardh.) Kuetz.
	Navicula spp.
	Amphora ovalis Kuetz.
	Gomphonema parvulum Kuetz.
	Nitzschia accomadata Hust.
	Nitzschia amphibia Grun.
	Nitzschia communis Rabh.
	Nitzschia diserta Hust.
	Nitzschia fonticola Grun.
	Nitzschia latens Hust.
	Nitzschia palea (Kuetz.) W. Smith
	Nitzschia thermalis Kuetz.
	Nitzschia spp.
	Hantzschia sp.
CHLOROPHYTA	
Volvocales	*Chlamydomonas celerrima* Pasch.
	Chlamydomonas ehrenbergii Gorosch.
	Chlamydomonas reinhardi Dang.
	Chlamydomonas tremulans Skuja
	Chlamydomonas spp.
	Polytoma uvella Ehren.
	Gonium pectorale Müll.
	Gonium sp.
	Pandorina morum (Müll.) Bory
	Pandorina sp.
	Eudorina sp.
	Carteria sp.
	Chlorogonium acus Naijal
	Chlorogonium fusiforme Matw.
	Chlorogonium sp.
	Chlamydobotrys (≡*Pyrobotrys*) *gracilis* Korsch.
	Spondylomorum quaternarium Ehren.
Tetrasporales	*Tetraspora* sp.
Chlorococcales	*Characium* sp.
	Hydrodictyon reticulatum (L.) Lagerh.
	Chlorella ellipsoidea Gerneck

	Chlorella pyrenoidosa Chick
	Chlorella vulgaris Beyerinck
	Chlorella spp.
	Radiococcus sp.
	Golenkinia sp.
	Micractinium pusillum Fresenius
	Micractinium sp.
	Oocystis sp.
	Ankistrodesmus convolutus Corda
	Ankistrodesmus falcatus (Corda) Ralfs
	Ankistrodesmus sp.
	Closteriopsis sp.
	Actinastrum sp.
	Selenastrum capricontium
	Selenastrum minutum (Naeg.) Coll.
	Selenastrum sp.
	Kirchneriella sp.
	Crucigenia irregularis Wille
	Tetrastrum sp.
	Scenedesmus armatus Chodat.
	Scenedesmus bijugatus Kuetz.
	Scenedesmus dimorphus Kg.
	Scenedesmus obliquus (Turp.) Kuetz.
	Scenedesmus ornatus (Lemm.) G. M. Smith
	Scenedesmus quadricauda (Turp.) Bréb.
	Scenedesmus sp.
	Coelastrum sp.
	Sphaerocystis sp.
Chaetophorales	*Stigeoclonium tenue* Kuetz.
	Stigeoclonium sp.
	Chaetopeltis sp.
Desmidiales	*Closterium acerosum* (Shrank) Ehrenb.
	Closterium sp.
	Cosmarium sp.
Zygnemales	*Zygnema* sp.
EUGLENOPHYTA	
Euglenales	*Euglena acus* Ehren.
	Euglena agilis Carter
	*Euglena deses** Ehren.
	Euglena gracilis Klebs
	Euglena granulata (Klebs) Lemm.

(*Continued overleaf*)

TABLE II (*Continued*)

	Euglena intermedia (Klebs) Schmitz
	Euglena klebsii (Lemm.) Mainx
	Euglena minuta Prescott
	Euglena oxyuris Schmarda
	Euglena pisciformis Klebs
	Euglena polymorpha Dang.
	Euglena proxima Dang.
	Euglena rostifera
	Euglena sanguinea Ehren.
	Euglena spathirhyncha Skuja
	Euglena spirogyra Ehren.
	Euglena variabilis Klebs
	Euglena viridis Ehren.
	Euglena sp.*
	Lepocinclis texta (Duj.) Lemm.
	Phacus pyrum (Ehren.) Stein
	Phacus tortus (Lemm.)
	Phacus sp.
	Trachelomonas volvocina Ehren.
CRYPTOPHYTA	
Cryptomonadales	*Cryptomonas erosa* Ehren.

III. Algae, Bryophytes and Their Autecology

While there has been very little ecological work specifically concerning algae and bryophytes in waste-treatment processes there is a wealth of ecological and physiological information available from other habitats and these data are reviewed in this section. The most important algal and bryophyte genera in waste-treatment processes are considered in order of their taxonomic position.

A. Division (Phylum) Chlorophyta

These are green, pigmented algae possessing one or more chromatophores which are often complex and used as diagnostic characters. In siphonaceous genera they are usually discoid. Colourless forms do occur also. The final product of photosynthesis is starch, pyrenoids are usually present and a starch sheath may form around them.

Flagellate cells or reproductive stages usually possess two or four equal flagella. Asexual reproduction is usually by means of zoospores but aplanospores, autospores, akinetes and fragmentation are alternative methods that are frequently reported. Sexual reproduction may be isogamous, anisogamous or oogamous. Mostly, haploid vegetative plants occur, but diploid stages predominate in the coenocytic forms and a well developed alternation of generations occurs in some genera. A morphological range exists from unicellular to colonial, siphonaceous and thalloid plants.

Freshwater, terrestrial, brackish water and marine species occur, as well as symbionts.

1. *Class Chlorophyceae—Volvocales*

This order includes well known genera such as *Chlamydomonas* Ehrenberg, *Gonium* Müll., *Pandorina* (Müll.) Bory, and *Spondylomorum* Ehrenberg.

According to West and Fritsch (1932), *Chlamydomonas* species are usually found in eutrophic conditions in ponds, ditches, rain-pools and slow-flowing water. Prescott (1969) lists this genus among those found in and beneath soil and refers to those species that occur as cryobionts, i.e. in ice and snow at temperatures below freezing, which, although mostly green cells, may become bright red in high light intensities when encysted, due to the production of haematochrome. In reviewing the order, Prescott reports that practically all the Volvocales are freshwater in habitat and many show a preference for water enriched with nitrogen and organic substances, e.g. oxidation ponds and places which receive effluents from slaughterhouses, breweries and dairies. Barnyard pools and drinking troughs are frequently opaquely green because of *Eudorina*, *Chlamydomonas* or *Pandorina* blooms.

Fjerdingstad (1965), in his review, reports that most species of Volvocales occur in localities without organic contamination as well as in heavily polluted localities such as sewage lagoons, and lists them as either saprobionts or saprophilous. Most multiply readily within the pH range of 5·0 to 8·5 and *Pandorina* and *Spondylomorum* are listed as chemobionts, e.g. in wastes from paper factories. Some details of water qualities from different habitats are reviewed.

The Volvocales carry out photosynthesis with light as the energy source but are heterotrophic (i.e. they use organic compounds in the medium) at night (the nutrition is referred to as mixotrophic). Fjerdingstad (1965) states the amino acids and acid amides are the best nitrogen sources for some species, whereas ammonium and nitrate ions may both serve as the only nitrogen source as well as urea and glutamine in others, but none of them serves as the sole carbon source.

Fogg (1953) and Round (1965) indicate that certain species such as

Chalmydomonas reinhardi are "acetate-organisms". These use salts of acetic and butyric acids and therefore occur in quantity where the decomposition of sewage has given rise to these acids and their salts and alcohols in high concentration. According to Wiessner (1962), manganese is necessary for the growth of *Chlamydomonas* at 10^{-7} M, and Syrett (1962) gives more details of specific nitrogen requirements.

Eighteen species of *Chlamydomonas* are known to release extra-metabolites such as carbohydrates like galactose, arabinose and others. There are also records of species being mutually inhibitory even though closely related. Many members of the Volvocales, especially *Chlamydomonas*, have been used for investigations of cellular physiology and of reproductive phenomena in algae. Culture studies have shown that there are homothallic and heterothallic strains of a species. Light, temperature and micro-elements introduced into the culture medium have all demonstrated highly refined adjustments and variabilities, sometimes resulting in gamete fusion, sometimes inhibiting mating of gametes. A reduction in nitrogen, for example, seems to induce sexual reproduction in cultures of *Chlamydomonas*. Because this genus exhibits so many variations in cytological features and in reproductive habits under artificial conditions, it is thought probable that many of the numerous species and varieties of *Chlamydomonas* have been described incorrectly, and that these so-called "species" are incidental variants and not true taxa.

As stated by Hynes (1970), studies on the plankton of larger rivers have been geographically widespread, and Hynes lists a large number of papers describing such studies from various parts of the world. From these he concludes that during the summer or in permanently warm rivers the flora includes a variety of truly planktonic Chlorophyceae which include flagellates such as *Chlamydomonas* spp. Even in fast-flowing, smaller streams, unmodified species, that normally inhabit pools and the edges of ponds, may occur. They are found in quiet areas of water, particularly where sheltered by obstructions, in areas of higher vegetation and in small bays along the banks. Such accumulations of algae occur primarily at times of low water in warm weather and are usually temporary and sporadic, since small changes in water level destroy them. This illustrates the capacity for algae to occupy and exploit favourable situations quickly.

2. Class Chlorophyceae—Chlorococcales

The Order Chlorococcales includes genera such as *Chlorella* Beyerinck, *Chlorococcum* Fries, *Scenedesmus* Meyen, *Oocystis* Naeg., *Ankistrodesmus* Corda and *Selenastrum* Reinsch.

According to Prescott's review (1969), *Chlorella* species are common in moist soil and agricultural loams, being found on the surface and to a depth of many centimetres. Often, they are pioneer species of barren, volcanic or denuded areas, where they may live as saprophytes on organic matter especially in nitrogenous wastes. They form an erosion-preventing crust that retains moisture and provides a substrate for the germination of spores and seeds of other plants. By their own activities, the algae add organic matter to the soil. They are abundant in small ponds and pools especially when these are contaminated with organic matter.

Fogg (1953) records that the cells are extremely resistant to desiccation and can revive after a year in dry conditions. They are dispersed by the wind and so are common contaminants of aquaria and laboratory reagents. Some form symbiotic associations on or in animals (e.g. *Hydra, Stentor, Paramecium*) and beneath the scales of fish and in freshwater sponges, supplying carbohydrates in exchange for nitrogenous materials.

Steemann Nielsen (1955a,b) and Fjerdingstad (1965) have contributed to the understanding of the physiology of *Chlorella pyrenoidosa* Chick. and report it from lagoons and oxidation plants for sewage purification, where it is listed as saprobiontic by the latter author.

Chlorella vulgaris Beyerinck, on the other hand, appears to be more typical of saline localities, tolerating up to 1% of NaCl and also high pH values, e.g. pH 9·0 (Stundl, 1939), and Fjerdingstad (1957) found it in drinking-water works, having a high content of methane. He also claims that it plays an important part in sewage purification lagoons and lists it as a saprophilic organism.

Each species has its own optimum light intensity requirement but this varies with a number of factors, including temperature. For example, *Chlorella* can be grown in light intensities of 3000 f.c. (31 740 lx) when maintained at a high temperature (35–40°C), but at 25°C the same species fails to photosynthesize at 1000 f.c. (10 580 lx). Light intensity has been found to affect the chlorophyll content of the cells, 6·6% of the dry weight being chlorophyll in plants cultured in the shade, but 3·5% of dry weight when grown in the light.

Chlorella has been used widely for research on photosynthesis since an initial series of fundamental experiments by Warburg was published in 1919. Fogg (1953) points out that diffusion of materials between these minute organisms and the surrounding medium is extremely rapid and the whole culture is easily stirred, with the advantage that the concentration of any given substance at the cell surface is the same as in the medium. Under the most favourable conditions the doubling time for *Chlorella* is 15 hours. Similar conditions can obtain in nature in well-lit, mobile water masses. Cytochrome *f* and vitamin K have been found in *Chlorella* and it is supposed that at least this genus makes use of phosphorus in light to bring about chemical energy conversion. Bicarbonate utilization in photosynthesis has been demonstrated by ^{14}C uptake experiments in marine species of *Chlorella* but not in the freshwater species.

Chlorella cells also grow well in the dark if supplied with glucose or other sugars, organic acids and various protein preparations depending on the strain. *Chlorella* can ferment glucose giving rise to acetic acid, lactic acid, hydrogen and carbon dioxide, and Syrett (1958) suggested that the cells must possess hexokinase, the enzyme responsible for the phosphorylation of glucose and fructose. The cells retain chlorophyll although grown in the dark for years and retain the ability to carry out photosynthesis on being returned to the light, with the exception of *C. variegata* which has lost this ability and appears to be a mutation. The latter can be *X*-ray induced and resemble the colourless cells of the genus *Prototheca* sometimes found in exudates from trees and other situations rich in organic matter.

Chlorella will tolerate high concentrations of substances such as carbon dioxide and nutrient salts; indeed, it appears to require relatively large amounts of C, H, O_2, N_2, P, S, K and Mg for growth. According to Fogg (1953), sulphur can be assimilated as sulphate but not as the sulphite or the element. The source of phosphorus is as orthophosphates in solution and as much as 2–3% of their dry weight as phosphorus may be required by the cells. *Chlorella* appears to be able to grow in the apparent absence of Ca, unlike higher plants where substantial amounts are required. According to Wiessner (1962) and Bogorad (1962), iron, manganese, molybdenum and boron are needed for growth and deficiencies give chlorosis, reducing photosynthesis, while the sensitivity of chlorophyll to destruction by light is increased. Optimum levels are given as $1{\cdot}8 \times 10^{-7}$ to $2{\cdot}6 \times 10^{-8}$ M for iron, 10^{-7} M for manganese and 10^{-5} to 10^{-4} M for boron. At higher levels these metals are toxic.

Syrett (1962) states that *Chlorella* can utilize nitrates or ammonium nitrogen and that Mn is essential for the uptake of nitrogen from either source and for the fixation of elemental nitrogen in the absence of bacteria. Glucose or sucrose are necessary as substrates for N_2 fixation in the dark. Therefore *Chlorella* and *Scenedesmus*, which behaves similarly, are often pioneers of soils in waste-ground, endophytes, mud species or are found where sugars are available, such as from the degradation of cellulose below paper-works' effluents, below sewage works' outfalls, etc. Hynes (1970) also cites an example of correlation between nitrate content and numbers of Chlorococcales in a polluted river. *Chlorella* species do not appear to require traces of particular growth substances such as vitamin B_{12} and are apparently completely "self-sufficient", and need no organic materials other than those they can produce themselves from the primary products of photosynthesis (Fogg, 1953). According to Prescott's review (1969), available amounts of several classes of lipids have been studied in *Chlorella*, including phospholipids, sulfoquinovosyl diglyceride and galactolipids; all are highly complex and contain fatty acids. Certain organic substances such as ascorbic acid, glycine and niacin are growth stimulators. Their importance has been indicated by showing the relation of seasonal content in plants to growth periods.

Antibiotics and extra-metabolites are produced. The antibiotic chlorellin apparently is composed of fatty acids and is active against Gram-positive and Gram-negative bacteria, according to Spoehr (1945). It resembles pyocyanase but is not as active against bacteria as is penicillin. However, it can inhibit the growth of other algae, for instance *Nitzschia* and *Chlorella* are found to be mutually inhibitory, and this sort of reaction can help to explain periodicity of algal flora composition. *Chlorella* also exudes glycollic acid and this can be traced in freshwater habitats in amounts as high as 0·2 mg/l. (See also Pratt, 1938, 1942; Pratt and Fong, 1940a,b.)

In addition, the respiration of *Chlorella* is stimulated by concentrations of cyanide which totally inhibit that of most other forms of life. So it appears that *Chlorella* is indeed rather exceptional among the smaller algae in its high rate of growth and its tolerance of a wide range of environmental conditions.

West and Fritsch (1932) describe *Chlorococcum* as a widely distributed terrestrial form which can be conspicuous on stone and brickwork. It has also been obtained regularly from cultures of cultivated soils (Bristol, 1920; Pringsheim, 1946; Bold, 1942; Coyle, 1935), the latter author stating that diatoms, *Chlorococcum humicolum* and some species of *Ulothrix* constitute the flora of acid soils, with blue-green algae exceedingly rare. Tiffany (1951) names this genus as one member of the temporary algal flora that can occur in large numbers in artificial troughs, stagnant pools, small depressions in rocks and rainy-weather puddles. Using the coverglass method, Round, (1963) and Lund (1945) demonstrated the effects of grazing by soil protozoa and rotifers. There is extensive feeding on the smaller coccoid forms including *Chlorococcum*, *Stichococcus*, *Navicula*, *Pinnularia* and *Nitzschia*, and some protozoa show selective grazing.

Less is known of the physiology and biochemistry of this genus than of *Chlorella*, but since it occupies the same ecological niches as the latter and *Stigeoclonium*, and is also frequently confused with stages of these two genera, it may well share some of their characteristics.

The genus *Scenedesmus* was divided by Chodat (1926) into four sections based on differences in cell shape, arrangement and presence or absence of spines or other ornamentation of the membrane. More than 150 species have been described and another monograph of the genus still useful for reference is by Smith (1916). One stage in the life-cycle has been described as another alga, viz. *Dactylococcus infusionum* Naeg., which is a stage found in the natural habitat as well as in cultures. More recently, Trainor (1963a,b,c) observed biflagellate motile stages when cultures of *Scenedesmus obliquus* were maintained under conditions of nitrogen deficiency. Trainor also claims that during culture, *S. obliquus* appeared to pass through stages that resembled not only *Dactylococcus* but also *Oocystis*, *Chlorella* and *Ankistrodesmus*, all of which are described fully in West and Fritsch (1932) and Bourrelly (1966).

Scenedesmus is similar to *Chlorella* in many respects and though a regular constituent of freshwater plankton it occurs most abundantly in stagnant water, particularly in association with *Pediastrum boryanum* and *Coelastrum sphaericum* etc., which are also members of the Chlorococcales. It is equally popular among research workers for studies in genetics, production of antibiotic material, and responses of cells to various chemical substances.

Scenedesmus, like *Chlorella*, can respire anaerobically when provided with carbohydrates such as glucose in a fermentation process, according to Prescott (1969), but unlike *Chlorella*, forms hydrogen and lactic acid. Fogg (1953) describes *Scenedesmus* as one of the "acetate organisms" and indicates that unlike glucose, fructose gives poor growths of the cells.

In many cultures it has been shown that gibberellic acid acts as a growth stimulant as in higher plants, e.g. 20 mg/l in the culture medium increased the reproduction of *Scenedesmus* cells by 30% and the length/width ratio by 32·5/32·7%. Similarly, Algeus (1946) reported that 1 mg/l iodoacetic acid doubles the rate of cell multiplication in *Scenedesmus* and *Chlorella*.

As in *Chlorella*, Syrett (1962) reports that *Scenedesmus* can utilize nitrates, ammonium or elemental nitrogen, traces of molybdenum being necessary for fixation of the latter form, and glucose or sucrose being necessary as substrates for N_2 fixation in the dark. This genus is therefore found in similar habitats (below effluents where sugars are available) as were previously recorded for *Chlorella*.

Since large molecules cannot penetrate algal cell walls readily, extracellular enzymes such as proteinase are produced by *Scenedesmus* and *Nitzschia* cells. Chondrillosterol has been identified in *Scenedesmus*, and an antibiotic called scenedesmine which is inhibitory or lethal to *Pediastrum*, *Cosmarium* and other algae, is secreted by the cells. This is interesting in view of the general increase of these species with increased eutrophication of waters. Both *Scenedesmus* and *Chlorella* have been examined thoroughly as a possible source of food material in animal feed and for humans, and a representative analysis of a mixed culture of these two genera (dried) produced 41·8% protein, 27·4% carbohydrates, 7·2% fat and 19·1% ash. *Scenedesmus*, like *Chlorella*, is tolerant of a wide range of chemical and physical conditions, and is of use neither in any saprobic system of classification of waters nor as an indicator species to utilize in the construction of biotic indices.

Oocystis and *Ankistrodesmus* are reported to be widely distributed in all kinds of aquatic habitats, although most abundant in small ponds, stagnant pools and lakes, some species occurring among submerged *Sphagnum*. According to West and Fritsch (1932), *Selenastrum* was uncommon in the United Kingdom, usually occurring amongst other water plants at the margins of ponds and lakes. Tiffany (1951) quotes Jaag (1945) as reporting growths of algae including *Oocystis* as lithophytes on sandstone, in montane areas where water issues from cracks in rocks or where dripping water is present.

Fritsch (1935–1945) quotes Oettli's work (1927) in which he showed that in media containing glucose, a full development of the colonies of *Ankistrodesmus* only took place if nitrogen was supplied in the form of peptones. Others give accounts of this genus under the name of *Rhaphidium*. Kessler, (1955, 1957a,b, 1959) made a thorough investigation of the effect of light on nitrate and nitrite reduction by *Ankistrodesmus* in the absence of carbon dioxide. Under anaerobic conditions, nitrate was reduced very slowly and probably no direct photochemical reduction of nitrate occurred. Nitrite reduction was rapid, an observation also supported by the work of Bongers (1958). Kessler suggested that it may be stimulated by photophosphorylation.

3. Class Chlorophyceae—Ulotrichales

According to West and Fritsch (1932), *Ulothrix* Kuetzing is generally distributed in stagnant waters of ponds, ditches, troughs etc., where *U. subtilis* var. *variabilis* is probably the most abundant member of the genus. It also occurs in cultivated soils and other species are found especially in base rich loams. *Ulothrix* is prevalent in acid pools with *Sphagnum* and aquatic angiosperms and is usually found with *Mougeotia*, *Cylindrocapsa*, numerous desmids and some diatoms; it is also present in eutrophic meres (Round, 1965).

Ulothrix zonata occurs on rocks and stones in streams and rivers, forming macroscopic growths (Symoens, 1957). This species occurs mainly in early spring and Blum (1956) found it growing in temperatures below 15°C, when it produced zoospores in melting ice water. More recent work (Prance, 1972) indicates that, in cultures under optimum light intensities, at 2°C all filaments grew slowly and gave rise to zoospores, while at 15°C growth was normal and more rapid with some production of zoospores. At 25°C, growth was normal but less than at 15°C. Prance was also able to demonstrate that periods of exposure to low temperatures for up to three weeks did not prevent subsequent growth at normal temperatures. Jaag (1938) suggested that sunlight gives rise to mass development and cloudy days cause the growths of this genus to disappear.

In Lake Windermere (Round, 1965) a rich epilithic flora occurs on the slope of the shore 0–0·5 m below water level, and this includes *Ulothrix* and other filamentous species. In oligotrophic waters, transitory associations of *Ulothrix*, Conjugales and Oedogoniales may be found forming loose, mucilaginous masses on the sediments. Chapman (1940) distinguishes similar communities dominated by *Ulothrix* and blue-green algae such as *Rivularia* and *Lyngbya* on silty shores of salt marshes. *U. pseudoflacca* is often an epiphyte of larger algae of the intertidal zone of the sea-shore, while other species occur in the spray zone of temperate and Arctic coasts.

From an examination of two *Ulothrix* species it appears that the cellulose in the cell walls is not cellulose I as in higher plants (Kreger, 1962), although some algae do possess this. The permeability of the cells to water seems to be very high, and Lenk (1956) tested the permeability of various Ulotrichales to urea and glycerol. There is great variation in the permeability of individual filaments, which may reflect the age of the cells.

Prance (1972) grew *Hormidium flaccidum* (another member of the Ulotrichales) and some *Ulothrix* species in laboratory batch cultures where nitrate and ammonium nitrogen sources were present up to 40 mg N/l in each series individually and with phosphate present up to 50 mg P/l. Polyphosphates have been demonstrated in *Ulothrix* spp. (Kuhl, 1962) and, as in other algae, it is likely that cells in the light convert orthophosphate into intracellular polyphosphates, the biological functions of which involve the storing of both phosphate and energy (Hoffmann-Ostenhof and Weigert, 1952). Polyphosphates should thus act as reservoirs of high-energy phosphate from excess ATP.

Wiessner (1962) mentions that manganese is necessary for growth of *Ulothrix implexa*, and the alteration of even one factor can frequently be sufficient to induce a reproductive stage. Even a change in physical conditions can do this, e.g. Erben (1962) reports that transfer from moving to still water can induce zoospores.

Conrad *et al.* (1959) observed that the net increase in weight of *Ulothrix subtilissima* grown for 15 days in a medium containing 3 μg/l indoleacetic acid (IAA) was 13 times greater than the net increase observed in the same medium in the absence of IAA. Higher concentrations were inhibitory. The same alga gave a seven-fold increase in a medium containing 50 μg/l gibberellic acid (GA) and again, higher concentrations were inhibitory. No ethanol was used in these experiments as this can also be stimulatory. The increase in wet weight was directly correlated with increased cell division. There was no interaction between IAA and GA when tested together. These observations are of interest in view of the reports of auxin-like substances, some similar to IAA, being synthesized by many algae and being detected in the surrounding medium.

Round (1965), in a table of saprobic zones, places the *U. zonata* communities in the mesosaprobic zone. Fjerdingstad (1965) summarized the work of many authors on several *Ulothrix* species. *U. aequalis* is a brackish-water species tolerant of up to 3% NaCl, or is found in other habitats with a sodium chloride content of 36 880 mg/l, but it is only from oligosaprobic localities, and Fjerdingstad classifies it as saproxenous. *U. subtilissima* is a saprophilic species appearing to tolerate a wide range of pH (4·0–8·6), but is most commonly reported from acid waters. *U. tenerrima* is reported from waters ranging in pH from 6·0 to 9·5, but is apparently more common in neutral or alkaline waters than in acid ones. Again, tolerance of sodium chloride is reported (e.g. one record of as much as 59 670 mg/l NaCl content), but morphologically different forms may inhabit fresh and salt water. This is another saprophilous species occurring in α to β mesosaprobic zones. *U. zonata* tolerates high salinity, sulphur and metal wastes in the water, occurs in a wide pH range (2·4–8·5) and is said to be tolerant of a high degree of pollution by organic matter, having been recorded as an epiphyte on *Sphaerotilus natans* (part of the "sewage fungus" complex). Other observers have reported it from cascading water mixed with *Batrachospermum*, *Draparnaldia* and *Chaetophora* and from other oligosaprobic habitats. Highly different views have been voiced concerning the relation of this species to pollution and at least two physiological races have been suggested; one clean water and saproxenous race (cell width 20–30 μm) and one polluted water race (saprobiontic) with cell width 40 μm or more and with longer filaments than the former.

4. Class Chlorophyceae—Chaetophorales

This Order includes the genus *Stigeoclonium* Kuetzing.

According to Prescott (1969), this genus usually grows in flowing water and some species can be used as indices of pollution. They have been recorded from the fringes of stones, others are solitary, and some even grow on snail shells. West and Fritsch (1932) suggest that palmella stages can be induced in sea-water and other solutions of high osmotic pressure, or in dilute solutions to which certain stimulating metallic salts have been added. The species are most commonly found in the Spring in slow-flowing waters or in springs, and are readily recognized as part of the attached community. Round (1965) quotes the work of Behre on eutrophic algal communities and the macroscopic growths of *S. tenue* that can occur in the epilithic community of meres and within the *Schizothrix* dominated community in larger lakes at a depth of more than 25 cm below the mean water level. The genus may also be represented by epiphytic forms attached to higher aquatics by a basal disc or basal system of creeping filaments.

Vertical zonation in running water can be observed on floating objects, e.g. ships' hulls and on woodwork of jetties and boathouses, etc. *Pleurocapsa*, a blue-green alga, grows above the water surface in the region wetted by waves, and reaching to 30–35 cm below the surface is a zone of *Cladophora* and *Stigeoclonium*. This is followed by a diatom zone and below that, bryozoans and mussels. All these are species that otherwise are found in small streams.

Hynes (1970) remarks on the summer scarcity of *S. tenue* together with *Ulothrix zonata* even in areas where trees are absent and no shading occurs—a natural factor found to eliminate other algae at this season. He suggests that it may be a temperature effect.

Current is an important factor and *Stigeoclonium* with other Chaetophorales are confined to rapids in the summer months. Whitford (1960) and Whitford and Schumacher (1961, 1964) demonstrated how the whole metabolism of the plant is speeded up in faster currents.

McLean (1972) demonstrated that *Stigeoclonium tenue* utilized nitrite, nitrate and ammonium salts as sources of nitrogen, the former proving to be the most efficient supply. The alga also utilized urea, certain amino acids and amino sugars as supplies of nitrogen. Growth of this species increased with phosphate phosphorus up to levels of 200 mg/l. Work regarding the effects of zinc, iron, lead, copper and manganese on *Stigeoclonium* growth demonstrated that this species showed tolerance to all these metals. The lead and iron, however, allowed greater growth of the alga than did zinc, copper or manganese.

Hynes (1970) reports *Stigeoclonium* as growing in soft, acid water and in limestone-spring streams. Fjerdingstad (1965) reports that the alga is tolerant of a wide pH range (from 4·3–8·6), though most records indicate its prevalence in acid waters of pools or streams. In reviewing the work of earlier authors on *S. tenue*, Fjerdingstad emphasizes that there are considerable morphological differences that may be linked to nutrition and that the young stages have been described as a separate alga, *S. farctum*.

Butcher (1932, 1946, 1947) showed, by growing algae from rivers on an artificial substrate such as glass slides, that communities change below sewage outfalls and that *Stigeoclonium* often forms a typical community in this area together with *Gomphonema* and *Nitzschia*. The majority of authors state that this alga tolerates high organic pollution, as does Palmer (1969) from his survey of the literature. It is therefore rated after the work of Kolkwitz (1935) as being part of the community of the α and β mesosaprobic zones, and by Butcher (1932) as even in the polysaprobic zone. It has been reported as tolerant to metallic poisons (now substantiated by the work of McLean, 1972), and when toxic ions are present in sufficient concentration to kill or inhibit other algae forming filamentous growths, such as *Cladophora* and *Oedogonium*, the *Stigeoclonium* may survive. Fjerdingstad (1965) also reports tolerance to chromium and phenol. In view of the tolerance shown to high organics and toxicity, it is not surprising that this alga is found below sewage effluents and in sewage purification lagoons.

5. *Class Chlorophyceae—Desmidiales*

Several members of this order have been recorded as casual inhabitants of oxidation ponds and percolating filters. The order is typified by the genera *Cosmarium* Corda and *Closterium* Nitzsch.

Cosmarium is commonly found in ponds and ditches of lowland districts, in bogs of moorland areas, and is abundant in upland *Sphagnum* areas on the leaves of these plants (West and Fritsch, 1932). Some species are usually found on dripping rocks, normally wet, calcareous rocks, whilst Round (1965) reports greenish, mucilaginous masses of desmids from acidic rocks (e.g. *Cylindrocystis* and *Cosmarium*) among similar but discrete masses of blue-green algae and diatoms. He describes *Cosmarium* as being common in acid, low-base status waters of pools in moorland areas and as an epiphyte of the aquatic moss *Fontinalis* in rivers. Seasonal periodicity occurs and some species such as *Cosmarium laeve* may not make regular annual growth but are sometimes absent altogether. Some species of the genus occur regularly as epiphytes on aquatic angiosperms at lake margins. The periodicity must depend on completion of the life-cycle in nature, and the work of Starr (1955, 1959) is of interest as a pointer to the factors that may modify cycles. He effected germination of a considerable number of zygotes of *Cosmarium turpinii* and *C. formosulum*, which had been previously dried and frozen, by immersing them in a fresh medium, i.e. he simulated natural conditions. This was effective even with aged zygotes.

According to Prescott (1969) desmids are abundant in waters where the ratio between calcium and potassium is relatively low. It is not known whether calcium by itself or in combination is an inhibitor or whether calcareous (hard) waters lack certain elements, e.g. trace elements, which are essential for desmids. In high concentrations of bicarbonate as $Ca(HCO_3)_2$ calcification occurs in certain desmids including *Cosmarium quadratum* (Lewin, 1962).

The cells are fairly resistant to radiation, since Godward (1962) reports that *Cosmarium subtumidum* cells could continue to survive with normal nuclear division after dosage with $20–50 \times 10^3$ rad of β rays. Morphological mutants have been induced in *Cosmarium turpinii* apparently by u.v. radiation.

Jacobi (1962) reports the work of Yin (1948) who, by histochemical methods, showed that phosphorylase is localized in the chloroplasts of *Cosmarium*, and Kuhl (1962) who lists this genus among algae in which polyphosphates have been demonstrated. Stadelmann (1962), in his review of work on cell permeability, reports that two groups of desmids can be distinguished on the basis of their behaviour in vital staining. In one group which contains *Cosmarium amoenum* the cells accumulate toluidine blue or brilliant cresyl blue, which in highly alkaline media results in a violet colouration of the cell sap. In the second group, containing *C. cucurbita* and *C. pyramidatum*, the dye accumulates in the form of blue-green spherules in the cytoplasm even at pH values of 7–8. The ion of certain vital dyes may also stain the cell walls of some, but not all, species of *Closterium*. Stadelmann (1962) described the permeability of *Pleurotaenium ehrenbergii* and *Closterium dianae* as normal but other desmids are somewhat more permeable to glycerol.

Jarosch (1962) reports that secretions of mucilage can be made visible in high concentrations of Indian ink or carmine. Desmids, especially *Closterium*, clearly progress over solid substrates by the secretion of mucilage through pores at one cell pole. As the volume of this mucilage increases, apparently by swelling, the cell is pushed along. Speed of movement varies from 200–400 μm per h in *Micrasterias denticulata* up to 112 μm in 30 sec by *Closterium acerosum*. The movement may be light-directed, as for example in studies of the phototactic behaviour of *C. moniliferum*, when Halldal (1962) reported that in low light intensities the cells first become orientated with their long axes in the direction of the light and then moved towards the light source. At intermediate light intensities, the cells were orientated perpendicular to the direction of light without further movement. In high light intensities such as full sunlight there was a negative reaction. Other desmids behave similarly and there is some evidence of chlorophyll *a* involvement in such responses.

Round (1965) says that both iron and manganese compounds are deposited in the cell walls of *Closterium* and some other desmids, e.g. *Penium*. There are some reports of *Closterium* being present in the abundant plankton of a eutrophic pool rich in calcium, and as a common member of the epipelic community of a eutrophic pond (Round, 1965). It does appear to be found in less acid conditions than *Cosmarium*, as is borne out by the notes compiled by Fjerdingstad (1965) on some of the species also found in polluted areas. He describes *Cosmarium botrytis* as occurring in habitats with a pH range of 5–9, from oligotrophic waters as well as below sewage outfalls. He lists it as saprophilous and states that the general consensus of opinion is that it occurs in α and β mesosaprobic zones.

Fjerdingstad (1965) assesses all three *Closterium* species that he considers as saprophilous. *C. acerosum* appears to predominate in alkaline waters (pH 7·2–9·5) and is widely distributed in water polluted with organic matter, and has been described from sewage purification lagoons. It also tolerates chromium. *C. leiblanii* has a wider pH tolerance (6·0–9·0) and occurs in oligo- and eutrophic waters, flowing and static conditions; it tolerates small amounts of iron. *C. moniliferum* has the same pH tolerance but is rather more common in fairly hard waters and, as in the case of the above species, varies from oligo- to β-mesosaprobic zones.

B. Division (Phylum) Euglenophyta

This phylum is represented by a group of unicellular flagellates having a long flagellum and a second, insignificant shorter one in most of the pigmented genera, but with one, two or three in the colourless genera. The flagella arise in an apical invagination, and an eyespot and contractile vacuoles are present. The periplast (pellicle) is soft, rigid, striate or warty, and some genera have an external theca. The chromatophores are green, discoid, rodlike, ribbon or band-shaped or stellate. Assimilatory products are paramylum and oil. Reproduction is by longitudinal fission of the cell with generation of new parts. Sexual reproduction has not been substantiated but cyst formation can occur.

1. Order Euglenales

Members of this order include the well known genera *Euglena* Ehrenberg and *Phacus* Dujardin. There are approximately 40 different euglenoid genera (Huber-Pestalozzi, 1955; Leedale, 1967a,b, 1971) which exhibit a wide range of size (from 10 μm to 500 μm in length) and shape. Details of plastid structure, mitosis, photosynthesis and physiology of *Euglena* are given in Burton (1968).

According to West and Fritsch (1932) and Prescott (1969), in late summer the floras in lower river reaches are usually abundant with *Euglena, Microcystis* and *Aphanizomenon*, which form blooms in those sections of the river which have received an enrichment of nutrients. The blooms may be of short duration but when they occur are indicators of high nutrient content and a source of vitamin B_{12}. Use has been made of this aspect in medical research where *Euglena*, being sensitive to vitamin B_{12}, can be used diagnostically when grown in comparison with a culture in which there is blood of a suspected anaemia patient. It is also of interest that the anti-pernicious anaemia antibiotic, streptomycin, has produced permanently colourless strains of *Euglena*.

Although the euglenoids are predominantly freshwater forms, a few are found in brackish or marine situations and many can adjust to the soil habitat. Agricultural loam soils are said to be much occupied by *Stichococcus, Hormidium, Euglena, Chlorella* and *Navicula*. *Euglena* occurs as solitary cells in the tychoplankton (Prescott, 1969) or as a dense growth that forms a powdery film over the surface of ponds and slow-flowing rivers. At times, a pond may become blood- or brick-red when *Euglena* becomes coloured with haematochrome carotenoids. This pigment seems to form in response to intense light. *Euglena sanguinea* is permanently blood-red but does not occur in dense growths. Many species are abundant in sewage oxidation ponds where they play an active role. Here, they survive well, partly because of their ability to metabolize in anaerobic conditions. Fogg (1953) lists *Euglena* with the "acetate organisms" that use salts of acetic and butyric acids. The genus is therefore to be found together with members of the Cryptophyceae and Volvocales, with *Scenedesmus, Nitzschia* spp. and *Chlamydomonas reinhardi* (carbon source mutants of the latter) in abundance in water containing sewage. The breakdown of the latter gives rise to fatty acids, their salts, and alcohols in high concentration.

The genus is much used in algal culture experiments, both physiological and cytological, and is of special interest for investigators of metabolism in darkness when polysaccharides are used in the medium. A review of this work is given by Danforth (1962), who also suggests that *Euglena gracilis* forms a bridge between the acetate flagellates and the sugar-utilizing algae. However, different strains of *E. gracilis* differ in their ability to utilize different substrates. Syrett (1962) lists *Euglena* as one of the algae that cannot utilize nitrate as a nitrogen source.

Among the characteristics of the cell is the semi-permeability of the periplast. Water can evidently penetrate through the periplast as easily as through many other types of protoplasm which have a low water permeability (Hilmbauer, 1954). The chromosomes may be atypical in centromeric organization in *Euglena* so that radiation-produced fragments are not lost and the nucleus is not thereby inactivated as in other algae such as *Spirogyra* (Godward, 1962). Gliding is found as a mode of movement in the *Euglena intermedia* group especially (Jarosch, 1962). They exhibit a differentiated antero-posterior axis and always glide in one direction unless injured. Bünning and Tazawa (1957) concluded that taxis in *Euglena gracilis* was purely phobic, the maxima for both positive and negative taxis being around 490 μmh, while Brokaw (1962) reports that the speed of forward swimming of *Euglena gracilis* var. *bacillaris* can be correlated with the pH of the medium. Others, e.g. Wolken and Shin (1958) quoted by Prescott (1969), found that the greatest speed of *Euglena* movements occurred when the light intensity was 40 f.c. (433·6 lx). The rhythmic movement of *Euglena* was interestingly demonstrated by a laboratory experiment with *E. deses*. This organism, together with mud from the River Thames in which it lived, was taken into the laboratory, where the cells buried themselves in the mud at the time when high tide would have occurred and reappeared at the surface at times of low tide. Sweeney and Hastings (1962) also described diurnal phototactic rhythms in *Euglena* which may persist for long periods in darkness interrupted by the test light that was shown to be too weak to act as a timing signal. No rhythm was manifest in cultures grown on organically enriched media, a factor that may influence distribution of this genus in oxidation ponds and lagoons.

Fjerdingstad (1965) gives details of the habitats of several species of *Euglena* and their tolerance to pollution. They all appear to tolerate a wide range of pH (4·7–9·6), and Tiffany (1951) indicates the extreme range of pH that can be tolerated in mine waters high in sulphuric acid content and with pH values of 1·8–3·9. The majority serve as indicators of organic pollution particularly where decomposable substances are found, e.g. in sewage or places where decaying vegetation occurs. Many of them are green in colour in clean waters but lose their colour in organically polluted waters. The majority also are listed as saprophilous, and many are chemobionts, resistant to chromium and sulphur, and all tend to occur in the α or β mesosaprobic zones. Many of them are obviously collective species, and this no doubt accounts for the tremendous variation in described habitats.

According to West and Fritsch (1932), *Phacus* is common in habitats frequented by *Euglena* and some can occur in enormous quantities in old cultures in laboratories. Prescott (1969) describes it as occurring as single cells in the tychoplankton and especially in water rich in organic matter.

C. Division (Phylum) Cyanophyta

In this phylum the cells are relatively undifferentiated, DNA and other nuclear materials not being delimited by a sharp, binding membrane.

Pigments are associated in thylakoids situated in the peripheral cytoplasm (chromoplasm), again not delimited from the rest of the cytoplasm (centroplasm) by a bounding membrane. The cell walls are similar to bacterial cell walls in structure and many other features are shared with this group (Echlin and Morris, 1965).

The morphological range includes unicells, colonies, unbranched, falsely branched and truly branching filaments (trichomes with mucilaginous sheaths). Colouration is variable as many genera exhibit the phenomenon of complementary chromatic adaptation. Highly refractive granules of cyanophycin are often present, glycogen is the main food reserve, and pseudo- or gas vacuoles form in some planktonic species.

No flagellate stages have been observed but short lengths of trichomes (hormogonia) move with a gliding motion. No sexual reproduction occurs, though filament anastomosis followed by a possible reduction division has been reported for one species (Lazaroff and Vishniac, 1962). Reproduction is usually by endospores, exospores, encysted fragments of filaments (hormocysts) or unencysted fragments of filaments (hormogonia) that are motile and escape from the parent sheath.

1. Order Oscillatoriales or Hormogonales

This order includes the genera *Oscillatoria* Vaucher, *Phormidium* Kuetzing and *Anabaena* Bory.

Prescott (1969), in his review, records many species of *Oscillatoria* as planktonic, occurring in freshwater or brackish water, and in the latter habitat often densely entangled with another genus, *Spirulina.* These planktonic forms buoyed in the upper water levels often become concentrated at or near the surface because of their pseudovacuoles. The often sudden appearance of these vacuoles has been explained by a change in physiology resulting from oxygen deficiency at lake bottoms, but this is not borne out by observations on the behaviour of plankters. At least in many instances, bloom-producing plankters with pseudovacuoles rise to the surface and then become scattered in relation to light, warmth, and gas expansion. The pseudovacuoles refract light so that cells appear red, purplish, brown or even black under the microscope, and because of their refractibility serve protectively against strong light. *Oscillatoria rubescens, Microcystis aeruginosa, Anabaena flos-aquae, Aphanizomenon flos-aquae and Gloeotrichia echinulata* are examples of surface-loving plankters which contain pseudovacuoles. Pearsall (1922) classified lake waters and pointed out that lakes rich in calcium are also rich in nitrates. Calcium is capable of forming salts of fatty acids derived from sugar of photosynthesis, and in other ways enters into the metabolism of algae. The bottom of a eutrophic lake or lagoon is usually heavily sedimented and there can be profuse development of aquatic plants. The algal flora, especially of cyanophytes such as *Oscillatoria* and diatoms, are rich also, hence these eutrophic bodies of water are often referred to as a "blue-green–diatom" type. Another factor claimed to trigger off the development of blooms of cyanophytes is vitamin B_{12}. The important element in the vitamin is cobalt and it is thought that this element is also involved in the process of nitrogen-fixation by blue-green algae (see later).

Alkaline and calcareous soils (pH 7·2–8·0) are dominated by cyanophytes and diatoms (*Oscillatoria, Phormidium, Porphyrosiphon* and *Nitzschia*). Temperature changes in the terrestrial habitat are more drastic than in the aquatic environment, i.e. changes are more rapid, the fluctuations are greater and the extremes persist longer. Blue-green algae seem to be more adaptable, in general, to temperature changes and extremes than are chlorophytes or diatoms (Marrè, 1962). Hot springs are usually highly mineralized, rich in calcium, silicon, and sulphur, and often contain many gases. Thus, water chemistry as well as temperature plays a role in determining species selectivity. Unicellular, colonial and especially filamentous growth forms occur in hot water, e.g. *Chroococcus, Microcystis, Oscillatoria, Plectonema, Phormidium* and *Scytonema.*

Jarosch (1962) gives an account of gliding movements in *Oscillatoria* and the concept of "contraction waves", which is the most firmly held of all the hypotheses to explain the mechanism. Movement is also associated with the secretion of mucilage against substrates and with the physical changes that occur in the mucilage. A streaming of mucilage up and down the exterior of an *Oscillatoria* trichome has been demonstrated and this induces filament movement in opposite directions. It is interesting that some cyanophytes are responsive to weak light (Halldal, 1962) but not to intense illumination. Different species react to different ranges of wavelengths of light, and by experimentation it has been shown that response to light is to those wavelengths which are absorbed by the carotenoid and the phycobilin pigments. Both phobo- and topophototaxis have been recognized by Drews (1957, 1959).

Stadelmann (1962) reports that *Phormidium* cells show a high permeability to water and that Elo (1937), using non-electrolytes on *Oscillatoria princeps* and *O. limosa,* found that the magnitude of the permeability constant in this case was related to the size of the molecule of the solute, the process of permeation perhaps being one mainly of ultra-filtration through a differentially permeable cell wall. This behaviour is unusual among the algae but is found in *Beggiatoa,* one of the sulphur bacteria, which shares many other characteristics with filamentous cyanophytes. Syrett (1962) reviews nitrogen assimilation and Fogg (1962a,b) nitrogen fixation; both mention species of *Oscillatoria, Phormidium* and *Anabaena*, and other references are given in Fogg (1953).

Nitrate is readily used by all cyanophytes examined in culture, and traces of molybdenum are found to be associated with maximum rates of growth. The first step in the assimilation of nitrate is its reduction to ammonia via nitrite and hydroxylamine. Nitrite is as effective as nitrate as a nitrogen source for *Oscillatoria* and *Microcystis*, although concentrations of over 13·6 mg/l may inhibit growth. It has not been demonstrated that molybdenum is necessary for maximum growth with nitrite as a nitrogen source. All cyanophytes can utilize ammonia and normal growth rates can take place in the absence of the relatively high concentrations of molybdenum which are necessary for growth on nitrate or with free nitrogen. Fogg (1965) writes that three species of planktonic blue-green algae which have been isolated in pure culture, viz. *Microcystis aeruginosa, Aphanizomenon flos-aquae* and *Oscillatoria rubescens*, have proved not to possess the nitrogen-fixing property, but there is indirect evidence that planktonic species of *Anabaena* fix nitrogen and many species of this genus, in culture, have been demonstrated to do so, but have a requirement for molybdenum in the order of 0·1 mg/l.

Fogg (1962a,b) also describes secretion of extracellular products into the medium, e.g. polysaccharides by *Oscillatoria splendida;* organic acids such as oxalic, tartaric, succinic and auxin-like substances were found in lake-water containing a nearly uni-algal growth of *Oscillatoria.* Polyphosphates have been demonstrated in many *Oscillatoria* and *Phormidium* species and one species of *Anabaena.*

Fjerdingstad (1965) describes fourteen *Oscillatoria* species and gives details of their habitats. He lists eleven as saprobiontic; *O. amphibia* and *O. limosa* are saprophilic as probably is *O. curviceps;* and *O. splendida* may have two ecological forms, the second form being saprophobic. The optimum pH range appears to be 7·5–8·0, although the species occur over a much wider range. There are many indications that the genus is tolerant to eutrophication and high organic matter, and there are many reports of its growth in eutrophic streams and rivers as well as in sewage treatment plants, including anaerobic sewage lagoons. Reports of increased growth in media supplemented with sewage are available from laboratory experiments, and some species, e.g. *Oscillatoria* and *Anabaena*, increased the oxygen content and reduced the BOD. Growth was often more rapid and greater at 15°C than at 20°C, and salts of copper, cobalt and zinc could be toxic.

Oscillatoria lauterbornii is considered to be a polysaprobic species and is characteristic of putrefying sludge with a high content of H_2S. *O. limosa* is also an associate species in sulphur-bacteria communities and has been recorded as an epiphyte on *Sphaerotilus natans* in an experimental plant for sewage purification. *O. princeps* is fairly tolerant of iron and zinc, and *O. putrida* is frequent in water with a high H_2S content but differs from the previous sulphur-tolerant species in that the cells become colourless under such conditions. It is also a facultative anaerobe. Other species tend to rise up into the plankton when the H_2S content increases in the lower waters. Fjerdingstad (1965) assigns them to the α mesosaprobic zone.

Species of *Phormidium* are amongst the commonest blue-green algae, occurring on damp earth, wet rocks or being entirely submerged, in waterfalls and swiftly-flowing streams as well as in ditches and lakes. Whilst growing in oligosaprobic conditions, the species of this genus are also frequently associated with polluted water containing *Sphaerotilus* and *Stigeoclonium tenue.* The genus is found over a wide pH range and is recorded from sewage purification plants and is a common constituent of the α mesosaprobic zones. *P. foveolarum* is said to be resistant to toxic wastes, and Fjerdingstad (1965) considers it to be saprophilous and common in α and β mesosaprobic zones.

Anabaena spp., according to West and Fritsch (1932), are relatively abundant in the waters of lakes and ponds. Trichomes may be of very brief duration, and heterocysts with akinetes in groups may be what is mostly to be found of some species. Some are found in cultivated soils and among other algae of still waters. Several species occur in *Sphagnum* bogs or among filaments of aquatic mosses, e.g. *Fontinalis.* A number of planktonic species are of interest because of their toxin-secreting capacity. They share a distinction along with *Microcystis* (*Anacystis*) of being the most poisonous of all blue-green algae (Prescott, 1969). As far as indicating polluted conditions is concerned, Fjerdingstad (1965) suggests they are probably capable of thriving in polluted places as well as oligosaprobic zones.

D. Division (Phylum) Chrysophyta

A group of four classes belong in this phylum, in which the cell wall of the organisms is often composed of two halves (valves); these may be obvious or may not be detectable without chemical treatment of the cells prior to observation. In the Bacillariophyceae (diatoms) the one valve which is the older (epitheca) fits closely over the younger (hypotheca), joined by cementing bands that together form the girdle. The cell walls are frequently impregnated with silica either in the vegetative cells as in the above named class, or in resting stages (cysts). The cells (frustules) are diploid, unicellular, colonial, filamentous or siphonaceous. Some motile cells are found with unequal biflagellate arrangement, others with uniflagellate. Chromatophores are usually two in number in the diatoms but more may occur in the other classes. They are yellow-green in colour since carotenoids predominate. Products of assimilation may be oil, chrysose, chrysolaminarin, leucosin or volutin, but never starch.

During vegetative reproduction in non-flagellate forms, the cells enlarge, and the nuclei, chloroplasts and pyrenoids divide followed by the whole protoplast in a plane parallel to the valves. The two valves separate and a new valve is secreted within the old hypotheca and the epitheca respectively. Thus, during repeated divisions one line of descendents remains the same size, whereas the other line becomes gradually reduced. Having reached a physiologically minimum size, the smaller cells form auxospores by the protoplast escaping from the separated valves and secreting a siliceous wall (perizonium). On germination a cell is formed of normal size. In the flagellate forms, vegetative reproduction by cell division and fragmentation is common. Asexual reproduction is by aplanospores, autospores or zoospores. Sexual reproduction is by isogamy, anisogamy or oogamy. Conjugation occurs in some genera and apogamy in others.

1. Class Bacillariophyceae—Pennales

Members of this order include genera such as *Nitzschia* Hassall (Grunow), *Hantzschia* Grunow, *Navicula* Bory, *Diatoma* de Candolle and *Amphora* Ehrenberg.

Nitzschia spp. are said to be both freshwater and marine and are often exceedingly abundant in ponds and ditches, with some in cultivated soils. Unpolluted streams are also reported to support filamentous algae such as *Cladophora*, upon which epiphytic growths of *Nitzschia linearis* occur (Prescott, 1969). With increased eutrophication downstream, Palmer (1962) points out that *Euglena viridis* and *Nitzschia palea* are predominant with *Stigeoclonium tenue*, *Oscillatoria tenuis* and *O. limosa*. Alkaline and calcareous soils (pH 6·8–8·4) are dominated by cyanophytes and diatoms and the most abundant of the latter group are from the genus *Nitzschia*. Working with *Nitzschia closterium*, Maddux and Jones (1964) found that at high concentrations of phosphate and nitrate, both the light optimum and temperature optimum were higher than with lower concentrations of nutrients. Therefore in oligotrophic situations, optimum conditions for growth would be reached at the depths with reasonably low temperatures and limited light penetration. With an increase in nutrients (eutrophication) light and temperature optima might be raised, with the result that the maximum concentration of the alga would occur closer to the surface. *Nitzschia closterium* has been studied by Hood and Park (1962) and was found to use bicarbonate ions in photosynthesis.

As much as 10·7% of dry weight of *Nitzschia linearis* was found to be fat, of which 0·1% is sterol, 52·6% is ash and 47% organic matter. Oil occurring in droplets is the food reserve, along with chrysolaminarin and volutin. *Nitzschia* like *Chlamydomonas* and *Euglena*, is described as an "acetate" organism (Fogg, 1953), and since large molecules cannot readily penetrate the walls, extracellular enzymes, e.g. proteinase that can liquify gelatine, are produced by the cells, an example being *Nitzschia putrida*. Extrametabolites from the latter genus are known to be inhibitory to the growth of other organisms such as *Chlorella* and *Escherichia coli*, the numbers of which can be greatly reduced in water and filter systems, indicating how such interactions can help to explain periodicity.

Jarosch (1962) describes gliding movements in diatoms; in one active species, *Nitzschia paradoxa*, this movement involves the secretion of mucilage into the raphe. This occurs particularly when cells are in contact with a substratum or with each other. A mucilaginous track is left and cells can move at speeds of 0·2–25 μm per sec. Fjerdingstad (1965) only reviews the habitats for *Nitzschia palea* and agrees with Butcher (1932) that it is typical of α to β mesosaprobic zones. It occurs over a wide pH range (4·2–9·0) and is a chemobiont, being resistant to copper, phenol and chromium. It is characterized by Fjerdingstad as saprophilous.

Hantzschia frequently occurs in prodigious quantities in damp soil, whilst small forms are found regularly in most earthy or silty habitats. Some tidal species show a tidal rhythm in phototaxis (Sweeney and Hastings, 1962). It can tolerate extremes of temperature and pH and withstand high concentrations of H_2S. It is found in blooms together with *Nitzschia*. Fjerdingstad (1965) classifies it as saprophilous.

Navicula species are abundant in freshwater and terrestrial habitats, with the aquatic forms showing a preference for freshwater. Some occur in boggy tracts in elevated regions and many are associated with *Sphagnum* pools. Others are widely distributed in cultivated soils. Several are known for their rapidity of movements (West and Fritsch, 1932). Bearing many of the characteristics of *Nitzschia*, many species are ubiquitous, but Fjerdingstad (1965) describes three in particular as saprophilous. *Navicula accomoda* is said to occur in heavily polluted waters. *N. cryptocephala* is found in abundance below effluents of organic nature from domestic sewage, slaughterhouse effluent, brewery and yeast factory effluents and in sewage purification plants. It is also common in papermill effluents and is tolerant to phenol. Occurring mainly in α to β mesosaprobic zones, the species is also considered to be saprophilic. *Navicula viridula* is tolerant of varying chemical conditions and a wide range of pH. It is copper resistant and thrives in areas of sewage contamination. It is also listed as saprophilous, being found in the same zones as the previous two species.

Diatoma is generally distributed in quiet waters; some are abundant in hilly districts, and the cells often occur in pure masses. *D. elongatum* is found frequently as an epiphyte on sewage fungus growths, as is *D. vulgare*, which thrives in paper-mill effluents and oily sewage and is tolerant of phenols. They are both saprophilous.

2. Class Bacillariophyceae—Centrales

This order includes the genus *Melosira* Agardh., which can be found in oily sewage, but shows a preference for alkaline waters and is tolerant to high temperatures, being found even in hot springs. It may occur in large quantities in ponds, ditches, and slow rivers, on wet rocks, sometimes forming crisp, mat-like masses on dripping sandstone, and is common on carboniferous sandstone. Other species occur in boggy pools and in the plankton or among mosses. Some species growing in caves have a greenish irridescence (West and Fritsch, 1932). It is a genus containing some saprophilous species.

3. Class Chrysophyceae—Chrysomonadales

Synura Ehrenberg is a genus that is said to have an absolute requirement for vitamin B_{12} (Fjerdingstad, 1965). It is euplanktonic and frequently occurs in such numbers that it produces objectionable tastes and odours in reservoir waters. It occurs in a wide range of habitats from soft to hard waters. In northern, temperate latitudes *Synura* appears early in the spring immediately following an intensification of light. Occasionally it appears with diatoms throughout the winter, often with a dominant "bloom" of *Dinobryon* spp. (Prescott, 1969). West and Fritsch (1932) suggest that it is quite rare in lakes but is commonly found in small ditches and pools, particularly if they are of rainwater, and pure gatherings can frequently be obtained in the early summer. It is responsible for a water odour somewhat like that of ripe cucumbers. It is saprophilous and has been reported from a filter in a sewage-treatment plant and from sewage lagoons.

E. Bryophyta

This encompasses a group of plants which may either consist of a thallus showing no differentiation into stem and leaves, as in some of the liverworts (Hepaticae), or may show a definite axis or stem, on which delicate leaves are borne, often only one cell thick. There is never any extensive differentiation in the stem and no vascular tissue, although simple conducting systems are to be found in species growing in dry areas. The leaves, when present, are rarely more than 12 mm in length and may become more than one cell thick in the region of the midrib or at the margin. They are never stalked. Such stem and leaf structure is exhibited in the mosses (Musci). Whether thallose or leafy, the plant is usually attached to the substratum by delicate structures called rhizoids. These are unicellular in the Hepaticae and multicellular and even branched in the Musci.

Apical growth by means of one or more cells is common and the plants show a tremendous capacity for regeneration from stems and leaves and fragments of thalli. Asexual reproduction can occur quite regularly in some forms by means of clusters of small cells or plates of tissue (termed gemmae). These readily germinate to give new plants. The normal sexual reproduction (Watson, 1968) involves special organs, the male antheridium and female archegonium. The contents of the former are antherozooids (male gametes) and these flagellate stages are chemotactically attracted to the female gamete, resulting in a zygote. Various protective outgrowths surround the developing female organs and later may continue to grow to protect the developing zygote. This is a diploid phase termed the sporophyte, which normally consists of a spore-bearing sac which may or may not become elevated on a seta (stalk). At the base there is an absorbing organ (foot) which penetrates the gametophyte (gamete-bearing plant). Various structural features of the capsule (spore-bearing sac) ensure dehiscence and ultimate spread of the haploid spores in either wet or dry conditions according to habitat. On germination a juvenile stage or protonema may develop prior to the mature plant.

The Hepaticae (liverworts) are represented in used-water treatment processes by the single genus *Marchantia* L. This genus occurs near habitation, especially on cinder paths, and is a notable weed of greenhouses. It also occurs on moors and heaths, especially after fires, and on the banks of rivers. *M. polymorpha* var. *aquatica* Nees., a form with more or less erect branches and wavy thallus margins, is found in marshes, streams and rivers. It has been recorded regularly as growing on coal tips of some age and in eutrophic rivers and cement from the side of treatment ponds.

The Musci (mosses) are represented by several genera in waste-treatment. *Leptodictyum* (Hedw.) Warnst. is one genus, and this grows on a variety of substrata including soil, rock and wood, but usually in moist areas at the margin of ponds or on river banks. It is mainly a lowland species and is most common in calcareous districts. It is recorded from rivers containing sewage effluents, and Watson (1968) has found it to be the sole species responsible for extensive growths on the metallurgical coke of the percolating filters of the Reading Sewage Treatment Works.

IV. Ecology of Algae and Bryophytes in the Processes

A. Percolating Filters

The algae most commonly found in percolating filters include *Chlorella, Chlorococcum, Phormidium, Oscillatoria, Stigeoclonium, Ulothrix* and a range of diatoms (Hawkes, 1963, 1965; Cooke and Hirsch, 1958; Fjerdingstad, 1965; Benson-Evans, unpublished), although many others may also occur. An interesting case quoted by Hawkes (1965) describes percolating filters at a sewage works in Warwickshire with extensive foliaceous growths of *Monostroma* sp. The alga was absent from nearby filters at the same works which differed only in the composition of the medium. Fjerdingstad (1965) refers to several species of *Bodo* occurring in the surface layers of percolating filters. This would seem to be an ideal ecological niche since species of this genus are known to be coprophilic. Hawkes (1963, 1965) and Cooke (1954) refer to several bryophytes that grow on filters. A list of algae and bryophytes of percolating filters, based on available literature, is given in Table I. Unfortunately, the literature relating to this topic is not extensive, and it seems reasonable to assume that many algae characteristic of eutrophic or polluted conditions may grow in this situation.

Algae are found in the upper layers of the filter and may form an extensive green covering over the surface, especially in the summer months (Hawkes, 1965; Cooke and Hirsch, 1958). This green layer may be an encrustation of microscopic algae (e.g. *Chlorella, Chlorococcum* and diatoms) or it may be a sheet or a felt of a larger filamentous alga (e.g. *Phormidium* or *Stigeoclonium*). Cooke and Hirsch (1958) have given a detailed account of the algal distribution in the upper layers of filter beds in Ohio, U.S.A. There are three fundamental layers; the main algal layer is situated immediately below a very thin layer of fungal hyphae (0·3 mm thick) and above the basal anchoring layer (Fig. 1). However, algae do occur to some extent in all three layers. It has been calculated that photosynthesis could provide only 5% or less of the oxygen requirements of the micro-organisms of the filter (Bartsch, 1961). Furthermore, photosynthesis could only be an intermittent source of oxygen since it does not take place in the dark and algae are often only present in the summer months. Carbon dioxide produced during the oxidation of waste by the other organisms of the filter might increase the rate of photosynthesis.

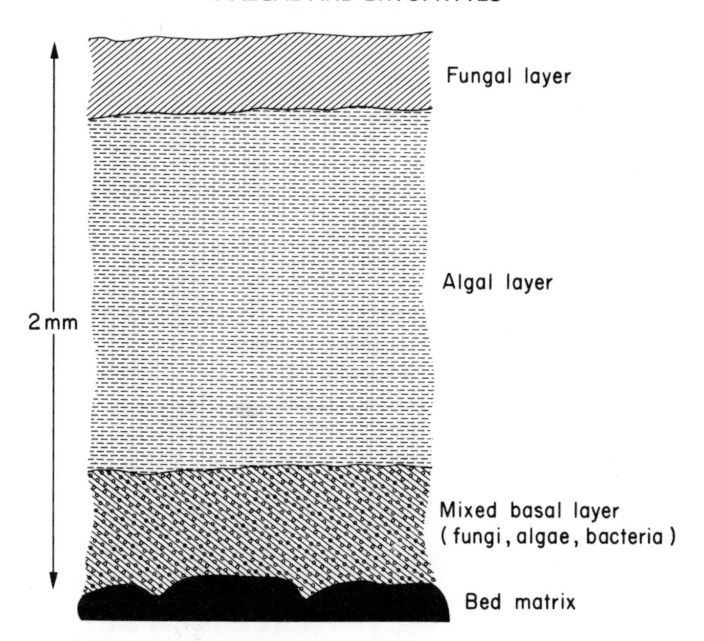

FIG. 1. Diagram of cross-section of surface layers of a percolating filter (modified from Cooke and Hirsch, 1958).

Hawkes (1965) has proposed that algae derive nitrogen and minerals from the waste and that some might be facultative heterotrophs. The widespread occurrence of nitrogen-fixing blue-green algae in percolating filters is also of interest.

One of the major requirements for algal growth is a moist environment, and a continuous flow of effluent gives rise to a more extensive and uniform algal growth than conditions of intermittent flow. This is particularly true of the larger outgrowths of filamentous algae such as *Stigeoclonium* (Cooke and Hirsch, 1958).

Sheets of algal growth on the surface of the filter can impair distribution and decrease ventilation (Hawkes, 1965; Benson-Evans, unpublished). *Phormidium* is probably a major cause of this condition because of its tendency to form leathery mucilaginous patches.

It appears, therefore, that algae play only a minor part in the purification processes of the percolating filter and may even, at times, reduce the efficiency of these processes.

B. Oxidation Ponds

1. *Factors affecting Species Composition*

The algae of oxidation ponds are listed in Table II, which is compiled from the literature of those parts of the world where this process is

TABLE III. Literature surveyed to compile list of algal species in oxidation ponds (Table II)

Locality	Authors
U.S.A.—Texas	Fisher and Gloyna (1965); Dust and Shindala (1970)
U.S.A.—California	Isaac and Lodge (1958)
U.S.A.—Iowa	Rashke (1970)
South Africa	Hodgson (1964); Hemens and Mason (1968); Tschortner (1968)
Middle East	Kott and Ingerman (1966); El-Sharkawi and Moawad (1970)
Scandinavia	Malchow-Møller *et al.* (1955); Fosberg and Fosberg (1971)
New Zealand	Haughey (1970)
U.K.	Stone and Abbot (1950)

employed (Table III). Four algal genera stand out as the most constant and cosmopolitan: *Chlorella, Scenedesmus, Chlamydomonas* and *Euglena.* As well as these four genera, many algae can be considered to be characteristic of the oxidation pond habitat, and the more important of these have been dealt with in the autecological section.

It seems that climatic factors do not play an important part in influencing the algal flora of oxidation ponds (Silva and Papenfuss, 1953). There is little or no evidence for the development of particular local or regional floras, although some of the rarer species that are recorded may reflect the flora of local bodies of water. Steel and Gloyna (1953) found no correlation between the algae of oxidation ponds in the Austin region of Texas and the algae of a variety of aquatic habitats in the area, which included a eutrophic lake receiving varying amounts of sewage overflow.

The diversity of the algal flora in an oxidation pond will depend to a great extent on the physico-chemical conditions prevailing in the pond (Silva and Papenfuss, 1953) and particularly on the rate of breakdown of waste and hence the amount of unoxidized matter present. Isaac and Lodge (1958) state that a pond containing a lot of unoxidized material will have only a few species with one marked dominant, whereas with improved conditions there is a greater variety of species.

There is disagreement in the literature with regard to algal species that occur under different conditions of waste breakdown or stabilization; thus Allen (1955) and Isaac and Lodge (1958) consider the dominance of *Chlorella* to indicate that a pond is working at maximum rate or even over its capacity, while large numbers of *Chlamydomonas* in a mixed flora are associated with complete mineralization or stabilization. In contrast, Steel and Gloyna (1953) refer to the presence of *Chlamydomonas* as a result of overloading, and Kott and Ingerman (1966) regard this alga to be indicative of recurrent anaerobic conditions. Abbot (1960) failed to draw up a system of algal indicators for oxidation ponds, whereby the performance of the pond might be gauged from the algal species present.

The nature of the influent to the pond or particular constituents can influence the flora; thus a high nitrogen level or a low pH, for example, could favour the growth of a particular species or group. El-Sharkawi and Moawad (1970) showed that the flora of an oxidation pond, receiving milk waste from a dairy plant, was dominated by *Selenastrum*. It seems that this alga was able to tolerate the mildly acid conditions (pH 5·8) caused by the fermentation of the milk waste by *Streptococcus lactis* and *Lactobacillus* sp. When this acidity was neutralized the *Selenastrum* gave way to a flora dominated by *Pandorina, Navicula* and *Oocystis* which the authors considered to be adapted to milk waste. It is interesting to note in this connection that Silva and Papenfuss (1953) report the occurrence of *Pandorina* and *Oocystis* in an oxidation pond in California receiving milk processing waste. Huang and Gloyna (1968), as a result of laboratory studies on the toxic effects of certain phenolic compounds, concluded that organic compounds in industrial wastes could destroy chlorophyll in *Chlorella pyrenoidosa* and even cause the death of the alga.

The production by certain algae of various types of growth-promoting and -inhibiting extracellular products could have quite profound effects on the flora of oxidation ponds. This topic has been dealt with to some extent in the autecological section with reference to the particular algae concerned. However, it is a very complex subject and, as yet, many of these compounds have not been identified or their precise interactions worked out. Fogg (1962b, 1965) gives a good basic account of the phenomenon.

It seems, therefore, that the algal species present in an oxidation pond will depend to a large extent on the interaction of all or some of the factors discussed above.

2. *Seasonal Factors*

In addition to the factors considered in the previous section, seasonal environmental factors will influence the numbers and species of algae present in a pond at any given time of year. There is agreement in the literature that there are generally fewer algal numbers, species and genera in winter than in summer; for example, Oswald and Gotaas (1955) quote algal yields of 1 ton per acre (dry weight) in December–January, and yields of up to 5 tons per acre (dry weight) in July–August. Silva and Papenfuss (1953) and Rashke (1970) also present data in agreement with this. Dust and Shindala (1970) quote algal counts ranging from 200 to 5000 cells per litre; although there was no clearly defined seasonal trend, the higher values were attained only during the summer months.

There is little information available about the seasonal incidence of particular algal species in oxidation ponds; nevertheless, several authors comment on the fact that *Chlorella* is able to grow well in the colder winter months when it is often dominant (Golueke, 1960; Dust and Shindala, 1970). The former author states that the shorter days, lower incident light intensities and lower carbon dioxide levels (due to reduced bacterial activity at the colder temperatures) favour the growth of this species. It is not possible to comment more widely on specific seasonal patterns because of inadequate and conflicting information in the literature. Rashke (1970) says that coccoid "greens" and "green flagellates" are the dominant algae for most of the year; however, he did observe dramatic drops in cell numbers ("the water was like pea-soup one day and clear the next"), but he was unable to correlate these sudden changes with any chemical, physical or seasonal factors. Many algae occurring in oxidation ponds have complex life histories that are triggered by critical temperature and light durations (Renn, 1954) and until these are worked out, our knowledge of the cause and duration of algal growths will remain incomplete.

3. Distribution of Algae in the Oxidation Pond

Most of the algae in oxidation ponds are planktonic, especially the coccoid "greens" and "green flagellates" (Rashke, 1970). These algae occur where there is sufficient light for photosynthesis; thus in a shallow pond they may be dispersed throughout, whereas in a deeper pond they will grow in the upper layers only. A vertical migration of *Euglena rostifera* in response to light has been reported by Hartley and Weiss (1970). This alga shows a preference for light intensities not greater than 800 lx. A more detailed account of depth and light penetration is given later.

Some algae may be found at the surface; for example, Oswald and Gotaas (1955) describe a foamy scum of *Chlamydomonas* forming over the surface of an oxidation pond, and Malchow-Møller *et al.* (1955) and Hartley and Weiss (1970) refer to floating mats of the blue-green alga, *Oscillatoria*. These surface growths of algae can impair the efficiency of the pond by preventing light reaching algae in the lower strata and thereby reducing oxygen production.

There are several reports of algae occurring in the benthos particularly of shallow ponds. Rashke (1970) describes a benthic flora of an experimental oxidation pond dominated by pennate diatoms and filamentous blue-green algae. Benthic layers may be attached (Bush *et al.*, 1961) or they may take the form of a loose granular deposition (Hemens and Mason, 1968; Hemens and Stander, 1969). In both examples of benthos referred to above the dominant alga was *Scenedesmus.*

4. The Role of Algae in the Purification Process

In contrast to their role in percolating filters, algae play a very important role in supplying oxygen in the purification process of the oxidation pond. The importance of photosynthetically produced oxygen will depend to a great extent on the area and retention time of the pond. Oswald and Gotaas (1955) consider that in large oxidation ponds (sewage lagoons) with a retention time between three weeks and six months, surface aeration is the most important source of oxygen, while smaller ponds with a retention time of less than one week are highly dependent on oxygen produced during photosynthesis.

A general scheme showing the part played by algae in the biochemical processes of the oxidation pond is shown below in Fig. 2. Organic waste entering the pond is broken down by bacterial decomposition to form simple nutrients which are taken up by the algae and used to synthesize new algal cell material in the presence of light. Carbon dioxide produced

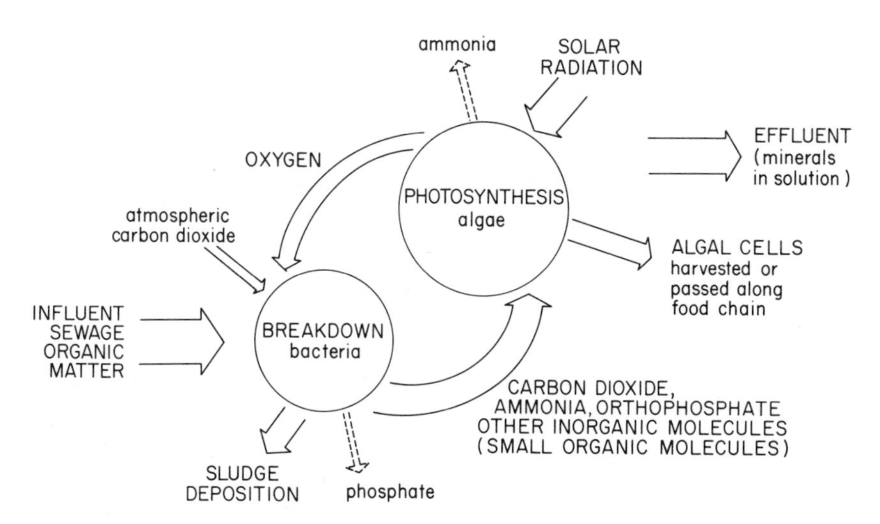

FIG. 2. Schematic representation of the biochemical processes of an oxidation pond.

during the oxidation of the organic matter enhances the photosynthesis of the algae while the oxygen produced by the algae during photosynthesis promotes the aerobic degradation processes of the bacteria and other micro-organisms. The algae produced in the process can be harvested mechanically (Golueke and Oswald, 1965), or they may form the basis of a food chain including invertebrates and, in some cases, fish (Ganapti and Chacko, 1952; Waddington, 1963).

Opinion is divided regarding the extent to which algae assimilate organic molecules directly and the relative importance of this in the overall process. Isaac and Lodge (1958) quote Allen (1955) as saying that algae play no part at all in the breakdown of organic matter and that their sole use is as oxygen suppliers. However, it has been demonstrated that several algae of importance in oxidation ponds are able to take up and utilize simple organic molecules such as amino acids, acid amides, urea, acetate and simple sugars (Fogg, 1953; Syrett, 1958; Pipes, 1961; Fjerdingstad, 1965; Round, 1965). Kott and Ingerman (1966) discuss the non-selective uptake of amino acids in oxidation ponds and present a scheme including algal metabolism of small organic molecules. However, most authors agree that carbon dioxide and/or bicarbonate, nitrate and ammonia nitrogen and phosphate as orthophosphate are of major importance as algal metabolites (Oswald and Gotaas, 1955; Isaac and Lodge, 1958). Renn (1954) refers to the "luxury metabolism" of algae like *Chlorella, Scenedesmus* and *Chlamydomonas* that are able to take up nitrogen and phosphorus in excess when these are present in the medium in high concentrations as ammonium, nitrate and phosphate. This phenomenon has obvious advantages in terms of sewage purification.

In addition to their direct role in the treatment process of the oxidation pond, algae may also play an indirect part by altering the physico-chemical environment. The utilization of carbon dioxide during active photosynthesis causes a shift in the carbonate–bicarbonate equilibrium with a resultant increase in pH due to hydroxyl ion formation; increases up to pH 10·0 and 11·0 are not uncommon (Pipes, 1961). Fitzgerald (1961) and Fitzgerald and Rohlich (1964) have observed the precipitation of phosphate above pH 8·0, and Kreft *et al.* (1958) state that ammonia nitrogen is lost to the atmosphere above pH 8·5 as a result of decreased ionization. Hemens and Mason (1968) and Hemens and Stander (1969) have utilized this algal-induced elevation of pH to bring about effective purification of sewage waste in an experimental stream system. However, other workers dealing with more conventional oxidation pond systems consider pH values in excess of 9·0 to be detrimental to the process because of the resultant decrease in the rate of bacterial breakdown of organic matter (Oswald and Gotaas, 1955; Varma and Talbot, 1965).

5. Factors affecting Photosynthetic Efficiency

This consideration is limited to ponds where photosynthesis is the major source of oxygen.

Several workers recommend the measurement of parameters other than cell numbers when considering the photosynthetic activity of an oxidation pond. Kott and Ingerman (1966) quote an example of a system of ponds where the net weight of dry algae per litre gave a much better index of pond efficiency than cell numbers. Dust and Shindala (1970) and Tschortner (1967, 1968), on the other hand, have used chlorophyll determination as a measurement of pond efficiency.

Retention time is one of the most important considerations in the operation of an oxidation pond. If the retention period is too short there will be insufficient time for the development of an adequate algal population and the breakdown of influent waste will be incomplete (Davis and Wilcomb, 1967). Alternatively, if the retention time is too long, certain nutrients may become limiting, algal growths will become dense with resultant mutual shading and less efficient use of incident light, and finally the algal cells will become older and have a reduced capacity for oxygen production (Golueke, 1960; Meron *et al.*, 1965).

Oswald and Gotaas (1955) have calculated that in latitudes up to 40°N a retention period of three days is adequate to maintain photosynthetic production of oxygen most of the time. The majority of workers agree that retention times should lie between 1 and 6 days depending on season and climate (Oswald and Gotaas, 1955; Oswald *et al.*, 1962; Renn, 1954). However, these figures assume adequate levels of light and temperature, which are not reached in many parts of the world. In parts of the British Isles, for example, conditions may be suitable for algal growth in oxidation ponds for certain periods of the year (Isaac and Lodge, 1958; Stone and Abbot, 1950) but it is unlikely that growths can be sustained in the colder, cloudier months (Forlin and Benson-Evans, 1956). The climate of California with its long hours of sun, low rainfall and warm air temperatures is ideal for this method of sewage treatment. A good ecological review of oxidation ponds in California is given by Silva and Papenfuss (1953).

Light is obviously of fundamental importance in any consideration of oxidation pond efficiency, and much research has been carried out on this subject. It is generally agreed that the minimum light intensity required to maintain photosynthesis lies between 320 and 1070 lx. The intensity of the incident illumination and the density of algal growth will determine the depth at which this minimum intensity is reached. For areas with midday light intensities up to 120×10^3 lx, this depth lies between 60 cm and 10 cm, for algal densities ranging between 50 mg l^{-1} and 400 mg l^{-1}. Depths exceeding 90 cm cannot normally be used (Oswald and Gotaas, 1955; Oswald *et al.*, 1962; Hemens and Stander, 1969). Maximum algal yields occur at light intensities between $4{\cdot}3 \times 10^3$ lx and $13{\cdot}0 \times 10^3$ lx.

Optimum temperatures for algal growth lie between 15°C and 25°C (Isaac and Lodge, 1958); these temperatures could be exceeded in shallow oxidation ponds.

Regular mixing has been proposed as a method of ensuring adequate illumination and uniform distribution of organisms and nutrients (Renn, 1954). Oswald *et al.* (1962) suggest mixing two times a day and point out that too much mixing will result in increased turbidity and reduced light intensity because of the constant cycling of mud and bacterial flocs. It is interesting to note that Klein (1966) draws a comparison between aerated oxidation ponds and activated-sludge plants, since both have no algae because of the lack of light.

An adequate supply of nutrients is essential to maintain algal growth (Steel and Gloyna, 1953) and it is possible for nutrients to become limiting in the oxidation pond environment. Both Oswald and Gotaas (1955) and Meron *et al.* (1965) have found that carbon is a limiting factor during periods of rapid algal growth and that nitrogen, and to a lesser extent phosphorus, could be limiting if levels in the influent were low. However, in recent years with the increased use of artificial fertilizers and detergents, it is unlikely that these latter two would limit growth.

If a pond is overloaded with organic matter and the algae are unable to provide sufficient oxygen to maintain aerobic conditions, then fermentation will be the dominant breakdown process and anaerobic conditions will ensue. Although algae may be present in these anaerobic ponds (for example, Malchow-Møller *et al.* (1955) refer to masses of blue-green algae floating on the surface of such a pond) their presence is incidental and they play little or no part in the purification process.

For mathematical treatments of this subject readers are referred to the work of Oswald and Gotaas (1955) and Tschortner (1968).

Guide to Taxonomic Works

1. English Texts

BLINKS, L. R. (1909). "Autotrophic Micro-organisms." Cambridge University Press, Cambridge, England.

CHAPMAN, V. J. (1964). "The Algae." Macmillan, London.

COLLINS, F. S. (1909). "The Green Algae of North America." Tufts College Studies, U.S.A.

DESIKACHARY, T. V. (1959). "Cyanophyta." Indian Council of Agricultural Research, New Delhi, India.

FOGG, G. E. (1953). "The Metabolism of the Algae." Methuen, London.
FOGG, G. E. (1965). "Algal Cultures and Phytoplankton Ecology." University of Wisconsin Press, London and Milwaukee.
FRITSCH, F. E. (1935–1945). "The Structure and Reproduction of the Algae." Cambridge University Press, Cambridge, England.
GOJDICS, MARY. (1953). "The Genus *Euglena*." University of Wisconsin Press, Madison, Wisconsin, U.S.A.
HUTCHINSON, G. E. (1957). "A Treatise on Limnology." Wiley, New York and London.
MORRIS, I. (1968). "An Introduction to the Algae." Hutchinson (University Library), London.
OKAMURA, K. (1907–1932). "Icones of Japanese Algae." Tokyo, published by the author.
PAPENFUSS, G. F. (1955). Classification of the Algae. *In* "A Century of Progress in the Natural Sciences." California Academy of Sciences, San Francisco.
PATRICK, R. and REIMER, C. W. (1966). "The Diatoms of the United States" Vol. 1. Monographs of the Academy of Natural Science of Philadelphia No. 13.
PRESCOTT, G. W. (1964). "How to know the Freshwater Algae." Wm. C. Brown, Dubuque, Iowa, U.S.A.
PRESCOTT, G. W. (1969). "The Algae: A Review." Nelson, London.
PRINGSHEIM, E. G. (1946). "Pure Cultures of Algae." Cambridge University Press, Cambridge, England.
RAMATHAN, K. R. (1964). "Ulotrichales." Indian Council Agric. Res., New Delhi, India.
ROUND, F. E. (1965). "The Biology of the Algae." Edward Arnold, London.
ROUND, F. E. (1969). "Introduction to the Lower Plants." Butterworth, London.
RUTTNER, F. (1952). "Fundamentals of Limnology." University of Toronto Press, Toronto, Canada.
SETCHELL, W. A. and GARDNER, N. L. (1903). "Algae of North Western America." *Univ. Calif. Publ. Bot.* **1,** 165–418.
SMITH, G. M. (1938). "Cryptogamic Botany" Vol. 1 (latest Edn 1955). McGraw-Hill, New York and Maidenhead.
SMITH, G. M. (1950). "The Freshwater Algae of the United States." McGraw-Hill, New York.
SMITH, G. M. (1951). "Manual of Phycology." Chronica Botanica, Waltham, Mass., U.S.A.
TIFFANY, L. H. and BRITTON, M. E. (1952). "The Algae of Illinois." University of Chicago Press, Chicago, Illinois, U.S.A.
VAN HEURCK, H. (1896). "A Treatise on the Diatomaceae." Wheldon and Wesley, London. (Reprinted 1962.)
WARD, H. B. and WHIPPLE, G. C. (1959). *In* "Freshwater Biology" (Ed. T. Edmonson) (2nd Edn). John Wiley, London and New York.

WEST, G. S. and FRITSCH, F. E. (1932). "A Treatise on the British Freshwater Algae." Cambridge University Press, Cambridge, England.

2. German Texts

EICHLER, A. W. (1886). "Syllabus der Vorlesungen über specielle und medicinischpharmaceutische Botanik" (4th Edn). Berlin.

ENGLER, A. (1954). "Syllabus der Planzenfamilien" (12th Edn), Vol. 1. Gebrüder Borntraeger, Berlin.

ETTL, H. and KOMARCK, J. (1958). *In* "Algolische Studien." Tshechoslowakischen Akademie der Wissenchaften, Prague.

FOTT, B. (1959). "Algenkund." Gustav Fischer, Jena.

GEITLER, L. (1932). Cyanophyceae. *In* "Deutschlands Kryptogamenflora" (Ed. G. L. Rabenhorst), Vol. 14. Akademische Verlagsgesellschaft. m.b.H. Leipzig.

GEITLER, L. (1942). Schizophyta (Cyanophyta). *In* "Die natürliche Pflanzenfamilien" (Eds. H. G. A. Engler and K. A. E. Prantl). Duncker and Humbolt, Berlin.

HUBER-PESTALOZZI, G. (1941, 1950, 1955, 1961). "Die Binnengewasser" (5 vols). Schweizerbart, Stuttgart.

HUSTEDT, F. *et al.* (1930–1959). Die Kieselalgen. *In* "Deutschlands Kryptogamenflora" (Ed. G. L. Rabenhorst). Akademische Verlagsgesellschaft. m.b.H. Leipzig.

KARSTEN, G. (1928). Vol. 2: Diatomeae. *In* "Die natürliche Pflanzenfamilien" (Eds. H. G. A. Engler and K. A. E. Prantl). Duncker and Humbolt, Berlin.

LINDEMANN, E. (1928). Vol. 2: Peredineae. *In* "Die natürliche Pflanzenfamilien" (Eds H. G. A. Engler and K. A. E. Prantl). Duncker and Humbolt, Berlin.

OLTMANNS, F. (1922, 1923). "Morphologie und Biologie der Algen" (Vols 1, 2, 3). Gustav Fischer, Jena.

PASCHER, A. (1939). *In* "Deutschlands Kryptogamenflora" (Ed. G. L. Rabenhorst), Vol. 11. Akademische Verlagsgesellschaft. m.b.H. Leipzig.

PASCHER, A. *et al.* (1913–1930). "Die Süsswasserflora Deutschlands, Oesterreichs und der Schweiz." (Vols 1–12). Gustav Fischer, Jena.

PRINGSHEIM, E. G. (1963). "Farblose Algen." Fischer, Stuttgart.

PRINTZ, H. (1927). Vol. 3: Chlorophyceae. *In* "Die natürliche Pflanzenfamilien" (Eds. H. G. A. Engler and K. A. E. Prantl). Duncker and Humbolt, Berlin.

SCHILLER, J. (1937). Dinoflagellateae. *In* "Deutschlands Kryptogamenflora" (Ed. G. L. Rabenhorst), Vol. 10. Akademische Verlagsgesellschaft. m.b.H. Leipzig.

SCHUSSNIG, B. (1960). "Handbuch der Protophytenkunde, Vol. 11. Gustav Fischer, Jena.

3. French Texts

Bourrelly, P. (1966). "Les Algues d'Eau Douce" (3 vols). Boubée et Cie, Paris.

Chadefaud, M. (1960). Les Vegeteaux non Vasculaire. *In* "Traité de Botanique Systematique" (Eds. M. Chadefaud and L. Emberger). Masson et Cie, Paris.

Dangeard, P. (1933). "Traité d'Algeologie." Lechevalier, Paris.

Frémy, P. (1930). "Les Myxophycées de l'Afrique Équatoriale Française." Arch. de Bot. **3,** No. 2. Caen.

Gomont, M. (1893). "Monographie des Oscillariées (Nostocacées-Homocystées)." *Annls. Sci. nat.* 7 ser. Bot. **15,** 263–368; **16,** 91–264. Paris.

Grassé, P. (1952). "Traité de Zoologie". Masson et Cie, Paris.

Sirodot, S. (1884). "Les Batrachospermes. Organisation, Fonctions, Développement, Classification." G. Masson, Paris.

4. Classical Texts

De-Toni, G. B. (1889–1905). "Sylloge Algarum omnium hucusque cognitarum" (11 vols). Patavii.

Kützing, F. T. (1845–1869). "Tabulae Phycologicae." Nordhausen.

Kützing, F. T. (1849). "Species Algarum" (19 vols). Leipzig.

Nordstedt, C. F. O. (1896). "Index Desmidiacearum citationibus locupletissimus atque bibliographia."310 pp. Berolini.

Nordstedt, C. F. O. (1908). "Supplementum." 149 pp. Berolini.

References

Abbot, A. L. (1960). Unpublished paper presented at C.S.A. Conference, Pretoria.

Algeus, S. (1946). *Bot. Notiser* **9,** 129–278.

Allen, M. B. (1955). California State Water Pollution Control Publication No. 13.

Bartsch, A. F. (1961). *J. Wat. Pollut. Control Fed.* **33,** 239–249.

Bartsch, A. F. (1967). *In* "The Environmental Requirements of Blue-Green Algae." U.S. Dept. of the Interior, Oregon.

Blum, J. L. (1956). *Bot. Rev.* **22,** 291–341.

Bogorad, L. (1962). *In* "Physiology and Biochemistry of Algae" (Ed. R. A. Lewin), pp. 385–404. Academic Press, New York and London.

Bold, H. C. (1942). *Bot. Rev.* **8,** 69–138.

Bongers, L. H. J. (1958). *Neth. J. agric. Sci.* **6,** 70–88.

Bourrelly. P. (1966). "Les Algues d'Eau Douce." Boubée et Cie, Paris.

Bristol, B. M. (1920). *J. Linn. Soc.* (*Bot.*) **44,** 473–482.

Brokaw, C. J. (1962). *In* "Physiology and Biochemistry of Algae" (Ed. R. A. Lewin), pp. 595–600. Academic Press, New York and London.

Bünning, E. and Tazawa, M. (1957). *Arch. Mikrobiol.* **27,** 306–310.

BURTON, D. E., Ed. (1968). "The Biology of *Euglena*." Academic Press, New York and London.

BUSH, A. F., ISHERWOOD, J. D. and RODGI, S. (1961). *Proc. Am. Soc. civ. Engrs* **87,** S.A. 3, 39–57.

BUTCHER, R. W. (1932). *Ann. Bot.* **46,** 813–861.

BUTCHER, R. W. (1946). *J. Ecol.* **33,** 268–283.

BUTCHER, R. W. (1947). *J. Ecol.* **35,** 186–191.

CHAPMAN, V. J. (1940). *J. Ecol.* **28,** 118–152.

CHODAT, R. (1926). *Revue Hydrologie* **3,** 71–258.

CHRISTENSEN, T. (1962). "Botanik. Systematisk Botanik." Bd. II, No. 2. Alger Munksgaard, København, Denmark.

CONRAD, H., SALTMAN, P. and EPPLEY, R. (1959). *Nature, Lond.* **184,** 556–557.

COOKE, W. B. (1954). *Bryologist* **56,** 143.

COOKE, W. B. and HIRSCH, A. (1958). *Sewage ind. Wastes* **30,** 138–156.

COYLE, E. E. (1935). Abstract Doctoral Diss. Ohio State University **38,** 39–45.

DANFORTH, W. F. (1962). *In* "Physiology and Biochemistry of Algae" (Ed. R. A. Lewin), pp. 99–119. Academic Press, New York and London.

DAVIS, E. M. and WILCOMB, M. J. (1967). *Wat. Res.* **1,** 335–343.

DREWS, G. (1957). *Ber. dt. bot. Ges.* **70,** 259–262.

DREWS, G. (1959). *Arch. Protistenk.* **104,** 389–430.

DUST, J. V. and SHINDALA, A. (1970). *J. Wat. Pollut. Control Fed.* **42** (7), 1362–1369.

ECHLIN, P. and MORRIS, I. (1965). *Biol. Rev.* **4o,** 143–187.

EICHLER, A. W. (1886). "Syllabus des Vorlesungen über specielle und medicinischpharmaceutische Botanik" (4th Edn). Berlin.

ELO, J. E. (1937). *Ann. Bot. Soc. Zool. Bot. Fenn. Vanamo* **8** (6), 1–108.

EL-SHARKAWI, F. M. and MOAWAD, S. K. (1970). *J. Wat. Pollut. Control Fed.* **42** (1), 115.

ERBEN, K. (1962). *In* "Physiology and Biochemistry of Algae" (Ed. R. A. Lewin), pp. 701–707. Academic Press, New York and London.

FISHER, C. P. and GLOYNA, E. F. (1965). *J. Wat. Pollut. Control Fed.* **37,** 1511–1520.

FITZGERALD, G. P. (1961). "Algae and Metropolitan Wastes." U.S. Publ. Hlth Service SEC TRW **61-3,** 136–139.

FITZGERALD, G. P. and ROHLICH, G. A. (1964). *Verh. int. Ver. Limnol.* **15,** 597–608.

FJERDINGSTAD, E. (1957). *Arch. Hydrobiol.* **53,** 240–249.

FJERDINGSTAD, E. (1965). *Int. Revue ges. Hydrobiol. Hydrogr.* **44,** 63–132.

FOGG, G. E. (1953). "Metabolism of the Algae." Methuen, London.

FOGG, G. E. (1962a). *In* "Physiology and Biochemistry of Algae" (Ed. R. A. Lewin), pp. 161–168. Academic Press, New York and London.

FOGG, G. E. (1962b). *In* "Physiology and Biochemistry of Algae" (Ed. R. A. Lewin), pp. 475–486. Academic Press, New York and London.

FOGG, G. E. (1965). "Algal Cultures and Phytoplankton Ecology." The University of Wisconsin Press, Milwaulkee and London.

FORLIN, R. and BENSON-EVANS, K. (1956). Report to National Coal Board, S.W. Division Scientific Dept. Ref. DIO/SR/131.

FOSBERG, C. and FOSBERG, A. (1971). *Ambio* **1** (1), 26.

FOTT, B. (1959). "Algenkund." Gustav Fischer, Jena.

FRITSCH, F. E. (1935–1945). "The Structure and Reproduction of the Algae." Cambridge University Press, Cambridge, England.

GANAPTI, S. V. and CHACKO, P. I. (1952). *Indian Geogr. J.* **25,** 35–44. (Also in *Wat. Pollut. Abstr.* **25** (1952), 209.)

GODWARD, M. B. E. (1962). *In* "Physiology and Biochemistry of Algae" (Ed. R. A. Lewin), pp. 551–564. Academic Press, New York and London.

GOLUEKE, C. G. (1960). *Ecology* **41,** 65–75.

GOLUEKE, C. G. and OSWALD, W. J. (1965). *J. Wat. Pollut. Control Fed.* **37,** 471–498.

HALLDAL, P. (1962). *In* "Physiology and Biochemistry of Algae" (Ed. R. A. Lewin), pp. 583–590. Academic Press, New York and London.

HARTLEY, W. R. and WEISS, C. M. (1970). *Wat. Res.* **4,** 751–767.

HARVEY, W. H. (1836). Algae. *In* "Flora Hibernica" (Ed. J. T. Mackay). Dublin.

HAUGHEY, A. (1970). *Br. phycol. J.* **5** (1), 97–102.

HAWKES, H. A. (1963). "The Ecology of Waste Water Treatment." Pergamon Press, Oxford.

HAWKES, H. A. (1965). *In* "Ecology and the Industrial Society" (Eds. G. T. Goodman, R. W. Edwards and J. M. Lambert), pp. 119–148. Blackwell Scientific Publications, Oxford and Edinburgh.

HEMENS, J. and MASON, M. H. (1968). *Wat. Res.* **2** (4), 277–297.

HEMENS, J. and STANDER, G. S. (1969). *In* "Advances in Water Pollution Research" (Ed. S. H. Jenkins). Proc. 4th Int. Conf., Prague, pp. 701–711.

HILMBAUER, K. (1954). *Protoplasma* **43,** 192–227.

HODGSON, H. T. (1964). *J. Wat. Pollut. Control Fed.* **36,** 51–68.

HOFFMANN-OSTENHOF, O. and WIEGERT, W. (1952). *Naturwissenschaften* **39,** 303–304.

HOOD, D. and PARK, K. (1962). *Physiologia Pl.* **15** (2), 273–282.

HUANG, J. and GLOYNA, E. F. (1968). *Wat. Res.* **2** (5), 347–354.

HUBER-PESTALOZZI, G. (1955). "Die Binnengewasser." Schweizerbart, Stuttgart.

HYNES, H. B. N. (1970). "The Ecology of Running Water." Liverpool University Press, Liverpool, England.

ISAAC, P. C. G. and LODGE, M. (1958). *New Biology* **25,** 85–97.

JAAG, O. (1938). *Mitt. naturf. Ges. Schaffhausen* **14,** 1–158.

JAAG, O. (1945). *Beitr. zur Kryptogamenfl. der Schweiz.* **9** (3), 1–560.

Jacobi, G. (1962). *In* "Physiology and Biochemistry of Algae" (Ed. R. A. Lewin), pp. 125–137. Academic Press, New York and London.

Jarosch, R. (1962). *In* "Physiology and Biochemistry of Algae" (Ed. R. A. Lewin), pp. 573–579. Academic Press, New York and London.

Kessler, E. (1955). *Nature, Lond.* **176,** 1069–1070.

Kessler, E. (1957a). *In* "Researches in Photosynthesis" (Ed. H. Gaffron), pp. 250–256. Interscience, New York.

Kessler, E. (1957b). *Planta* **49,** 505–523.

Kessler, E. (1959). *Symp. Soc. exp. Biol.* **13,** 87–105.

Klein, L. (1966). "River Pollution" Vol. 3: "Control". Butterworth, Washington and London.

Kolkwitz, R. (1935). *In* "Planzenphysiologie" (Eds. W. Berecke and L. Jost) (3 Aufl.). 310 pp. Gustav Fischer, Jena.

Kott, Y. and Ingerman, R. (1966). *Air and Wat. Pollut. Int. J.* **10,** 43–54.

Kreft, G., van Eck, H. and Stander, G. J. (1958). *Water and Waste Treatment J.* **7,** 53–60.

Kreger, D. R. (1962). *In* "Physiology and Biochemistry of Algae" (Ed. R. A. Lewin), pp. 315–332. Academic Press, New York and London.

Kuhl, A. (1962). *In* "Physiology and Biochemistry of Algae" (Ed. R. A. Lewin), pp. 211–229. Academic Press, New York and London.

Lanjoun, J., Baehni, C. H., Merrill, E. D., Rickett, H. W., Robyns, W. and Sprague, T. A. (1952). "International Code of Botanical Nomenclature" (Utrecht). Chronica Botanica, Waltham, Mass., U.S.A.

Lazaroff, N. and Vishniac, W. (1962). *J. gen. Microbiol.* **28,** 203–210.

Leedale, G. F. (1967a). "Euglenoid Flagellates." Prentice-Hall, Englewood Cliffs, New Jersey, U.S.A.

Leedale, G. F. (1967b). *A. Rev. Microbiol.* **21,** 31–48.

Leedale, G. F. (1971). "The Euglenoids." *In* "Oxford Biology Readers—5" (Eds J. J. Head and O. E. Lowenstein). Oxford University Press.

Lenk, I. (1956). *Öst. Akad. Wiss. math-naturw. Kl. Sitzber. Abt. I* **165,** 173–279.

Lewin, J. C. (1962). *In* "Physiology and Biochemistry of Algae" (Ed. R. A. Lewin), pp. 457–466. Academic Press, New York and London.

Lund, J. W. G. (1945). *New Phytol.* **44,** 196–219.

Maddux, W. S. and Jones, R. F. (1964). *Limnol. Oceanogr.* **9** (1), 79–86.

Malchow-Møller, O., Bonde, G. J. and Fjerdingstad, E. (1955). *Schweiz. Z. Hydrol.* **17,** 98–122.

Marre, E. (1962). *In* "Physiology and Biochemistry of Algae" (Ed. R. A. Lewin), pp. 541–549. Academic Press, New York and London.

McLean, R. O. (1972). Personal communication. Ph.D. Thesis, University of Wales, Cardiff.

Meron, A., Rebhum, M. and Sless, B. (1965). *J. Wat. Pollut. Control Fed.* **37** (12), 1657–1670.

Oettli, M. (1927). *Bull. Soc. bot. Genève* **19,** 1–91.

Oswald, W. J., Golueke, C. G., Cooper, H. K., Gee, H. K. and Bronson, J. C. (1962). Int. Conf. Wat. Pollut. Res. Sec. 2, Paper No. 25., 1–22.

Oswald, W. J. and Gotaas, H. B. (1955). *Proc. Soc. civ. Engrs* **81,** separate 686, 29 pp.

Palmer, M. C. (1962). "Algae in Water Supplies." U.S. Department of Health, Education and Welfare. Public Health Service Publication No. 657.

Palmer, M. C. (1969). *J. Phycol.* **5,** 78–82.

Pearsall, W. H. (1922). *J. Ecol.* **9,** 241–253.

Pipes, W. C. (1961). *Water and Sewage Works* **108,** 176–181.

Prance, N. B. (1972). Personal communication: Ph.D. Thesis, University of Wales, Cardiff.

Pratt, R. (1938). *Am. J. Bot.* **25,** 498–501.

Pratt, R. (1942). *Am. J. Bot.* **29,** 142–148.

Pratt, R. and Fong, J. (1940a). *Am. J. Bot.* **27,** 431–436.

Pratt, R. and Fong, J. (1940b). *Am. J. Bot.* **27,** 735–743.

Prescott, G. W. (1969). "The Algae—A Review." Nelson, London.

Pringsheim, E. C. (1946). "Pure Cultures of Algae." Cambridge University Press, Cambridge, England.

Rashke, R. L. (1970). *J. Wat. Pollut. Control Fed.* **42** (4), 518–530.

Renn, C. E. (1954). *Am. J. publ. Hlth* **44,** 631–634.

Round, F. E. (1963). *Br. phycol. Bull.* **2** (4), 224–246.

Round, F. E. (1965). "The Biology of the Algae." Arnold, London.

Silva, P. C. and Papenfuss, F. (1953). "Report on a Systematic Study of the Algae of Sewage Oxidation Ponds." State Water Pollution Control Board, Sacramento, California. Publication No. 7, 34 pp.

Smith, G. M. (1916). *Trans. Wis. Acad. Sci. Arts Lett.* **18,** 422–530.

Spoehr, H. A. (1945). *Biochemical Investigation, Carnegie Inst. Washington Year Book* **42,** 83–87.

Stadelmann, E. J. (1962). *In* "Physiology and Biochemistry of Algae" (Ed. R. A. Lewin), pp. 493–523. Academic Press, New York and London.

Starr, R. C. (1955). *Am. J. Bot.* **42,** 577–581.

Starr, R. C. (1959). *Arch. Protistenk.* **104,** 155–164.

Steel, E. N. and Gloyna, E. F. (1953). "Oxidation Ponds—Radioactivity Uptake and Algae Concentration." University of Texas, Technical Report No. 1.

Steemann Nielsen, E. (1955a). *Physiologia Pl.* **8,** 106–115.

Steemann Nielsen, E. (1955b). *Physiologia Pl.* **8,** 317–335.

Stone, A. R. and Abbot, W. E. (1950). *J. Inst. Sew. Purif.* **2,** 116–124.

Stundl, K. (1939). *Arch. Hydrobiol.* **34,** 81–102.

Sweeney, B. M. and Hastings, J. W. (1962). *In* "Physiology and Biochemistry of Algae" (Ed. R. A. Lewin), pp. 687–698. Academic Press, New York and London.

Symoens, J. J. (1957). *Bull. Soc. bot. Belg.* **89,** 111.

SYRETT, P. J. (1958). *Nature, Lond.* **158,** 1734.
SYRETT, P. J. (1962). *In* "Physiology and Biochemistry of Algae" (Ed. R. A. Lewin), pp. 171–183. Academic Press, New York and London.
TIFFANY, L. H. (1951). *In* "Manual of Phycology" (Ed. G. M. Smith), pp. 293–310. Chronica Botanica, Waltham, Mass., U.S.A.
TRAINOR, F. R. (1963a). *Bull. Torrey bot. Club* **90** (2), 137–138.
TRAINOR, F. R. (1963b). *Science, N.Y.* **142,** 1673–1674.
TRAINOR, F. R. (1963c). *Can. J. Bot.* **41,** 967–968.
TSCHORTNER, V. S. (1967). *Wat. Res.* **1,** 785–793.
TSCHORTNER, V. S. (1968). *Wat. Res.* **2,** 327–346.
VARMA, M. M. and TALBOT, R. S. (1965). "Reaction Rates of Photosynthesis." Proc. 20th Ind. Waste Conf. Engineering Extension Series No. 118, p. 146.
WADDINGTON, I. (1963). *J. Inst. Sew. Purif.* **3,** 214–215.
WATSON, E. V. (1968). "British Mosses and Liverworts." Cambridge University Press, Cambridge, England.
WEST, G. S. and FRITSCH, F. E. (1932). "A Treatise on the British Freshwater Algae." Cambridge University Press, Cambridge, England.
WHITFORD, L. A. (1960). *Trans. Am. microsc. Soc.* **79,** 302–309.
WHITFORD, L. A. and SCHUMACHER, G. J. (1961). *Limnol. Oceanogr.* **6,** 423–425.
WHITFORD, L. A. and SCHUMACHER, G. J. (1964). *Ecology* **45,** 168–170.
WIESSNER, W. (1962). *In* "Physiology and Biochemistry of Algae" (Ed. R. A. Lewin), pp. 267–279. Academic Press, New York and London.
WOLKEN, J. J. and SHIN, E. (1958). *J. Protozool.* **5** (1), 39–46.
YIN, H. C. (1948). *Nature, Lond.* **162,** 928.

5

Protozoa

C. R. Curds

Department of Zoology
British Museum (Natural History)
London SW7
England

I. Introduction

The presence of protozoa in biological used-water treatment processes was noted almost as soon as each process was introduced, but it is only in recent years that the significance of these organisms has begun to emerge. Much of the early work was concerned with the description of the types of protozoa (generally genera) found in these habitats, but this was quickly followed by a period when several workers attempted to relate the types, and sometimes species, of protozoa to effluent quality and plant performance. Much of the work has been concerned with the activated-sludge process since this habitat is far more amenable to meaningful sampling routines than are percolating filters. However, it seems likely that many of the conclusions reached are equally applicable to both of these aerobic processes. Inevitably there are far more ecological data available in the literature concerning protozoa than data directly pertaining to used-water treatment processes, and the ecological data are mentioned when they are likely to be of relevance.

II. Taxonomy

In view of the various taxonomic advances and subsequent changes that have been and are being made in protozoological nomenclature for the sake of simplicity, all methods of classification used in original published work have been standardized in this chapter to conform with the most recent scheme of classification of the phylum Protozoa suggested by the Committee on Taxonomy and Taxonomic Problems of the Society of Protozoologists (Honigberg *et al.*, 1964).

Some members of the class Phytomastigophorea are regarded both as protozoa and as unicellular algae; those organisms sometimes regarded as protozoa have been included in this chapter even though they may have been mentioned in the previous chapter (ch. 4) concerning phytoflagellates and algae.

A. Taxonomic Keys

There are three types of key available for the identification of protozoa in used-water treatment processes; those originally written for the specialist in protozoan taxonomy, those written for the general biologist, the undergraduate and amateur naturalist (usually limited to the identification of common genera), and those written specifically for the identification of protozoa found in treatment processes. The three types of key have their advantages and disadvantages but all are likely to be used by the serious research worker.

Recently three keys have been written specifically about the protozoa found in treatment processes. Calaway and Lackey (1962) deal exclusively with the flagellated protozoa, Curds (1969) exclusively with the ciliated protozoa, while Martin (1968) includes all forms of protozoa. These three are written in simple language and together provide an excellent introductory basis for the non-specialist. For those who read Japanese a key similar to those mentioned above is available and deals with all forms of protozoa (Morishita, 1968). A key, in English, prepared by Bick (1972) for ciliates used as biological indicators in freshwater biology generally, is also useful since many of the species included are also found in used-water treatment processes.

One of the better generalized keys is contained within the invaluable textbook by Kudo (1932) which gives a considerable amount of broad background information concerning the various taxonomic groups of protozoa and contains workable keys to the family level. Furthermore, the principal morphological characteristics of all major genera are included and many specific examples are given. It is sometimes obvious from published lists of protozoa that some authors have relied solely upon this book without reference to the specialist literature, but every research worker should also refer to such specialist literature and at the end of this chapter there is a reference section devoted to lists of some of the more useful keys and taxonomic works.

B. Survey of Species

Five classes of protozoa are found in used-water treatment processes; the Phytomastigophorea, the Zoomastigophorea, the Rhizopodea, the Actinopodea and the Ciliatea. Although many lists of protozoa which inhabit treatment processes have been reported, they are often given as secondary information. Table I lists the protozoa found in the various processes compiled from the major lists published by Barker (1943), Calaway and Lackey (1962), Clay (1964), Curds (1969), Curds and Cockburn (1970a), Lackey (1925), Martin (1968) and Morishita (1968, 1970). Only those organisms named specifically in the above works have been included here, though other genera have been recorded. A very large number of genera were included in the list compiled by Clay (1964), who covered a very wide field of literature. She expressed concern that many authors neither mention species nor quote the authority for the species when given, so that sometimes it is difficult to determine precisely which species was intended. Perhaps it should be mentioned here that the valuable lists compiled by Clay (1964) include all organisms, both plant and animal, associated with sewage and used-water treatment processes.

1. Protozoa in Aerobic Processes

(*a*) *Percolating filters.* Protozoa are plentiful in percolating filters and there is as great a variety of species as in any of the other aerobic biological treatment processes. It has been noted that the fauna of filters does not conform to the usually accepted ecological principle that the size

TABLE I. A checklist of protozoa reported to inhabit used-water treatment processes (PF—percolating filters, AS—activated sludge, IT—Imhoff tanks)

	PF	AS	IT
Phylum **PROTOZOA**			
Sub-phylum I. **SARCOMASTIGOPHORA**			
Superclass 1. **MASTIGOPHORA**			
Class 1. PHYTOMASTIGOPHOREA			
Order 1. Chrysomonadida			
Anthophysa vegetans (Müller)	+	+	+
Monas amoebina Meyer	+		+
Monas fluida Dujardin	+	+	
Monas minima Meyer	+	+	+
Monas obliqua Schewiakoff	+	+	
Monas vestita Kent		+	
Monas vivipara Ehrenberg	+		
Monas vulgaris (Cienkowsky)	+		
Oicomonas mutabilis Kent	+		
Oicomonas ocellata Scherfel		+	
Oicomonas socialis Moroff	+	+	+
Oicomonas steinii, Kent	+	+	
Oicomonas termo (Ehrenberg)	+	+	+
Stylobryon petiolatum Dujardin	+		
Synura uvella Ehrenberg	+	+	
Order 2. Cryptomonadida			
Chilomonas paramecium Ehrenberg	+		+
Cryptomonas erosa Ehrenberg		+	
Cryptomonas ovata Ehrenberg		+	
Cyathomonas truncata Ehrenberg	+	+	+
Order 3. Volvocida			
Carteria globosa Korschikoff	+		
Polytoma uvella Ehrenberg	+	+	+
Order 4. Euglenida			
Anisonema emarginatum Stein	+		
Anisonema grande Ehrenberg	+	+	
Anisonema ovale Klebs	+		
Anisonema truncatum Stein	+		
Astasia dangeardii Lemmerman	+		
Astasia inflata Dujardin	+		
Astasia quartana Moroff	+		
Distigma proteus Ehrenberg	+		+
Entosiphon sulcatum (Dujardin)	+		+
Euglena agilis Carter	+		

TABLE I (*Continued*)

	PF	AS	IT
Euglena gracilis Klebs		+	
Euglena intermedia Klebs		+	
Euglena polymorpha Dangeard		+	
Euglena viridis Ehrenberg	+		
Heteronema acus (Ehrenberg)	+	+	+
Notosolenus orbicularis Stokes	+		+
Peranema trichophorum Ehrenberg	+	+	+
Petalomonas carinata France	+		+
Petalomonas mediocanellata Stein	+		+
Petalomonas stenii Klebs	+		
Rhabdomonas incurva Fresnel	+		
Class 2. ZOOMASTIGOPHOREA			
Order 1. Rhizomastigida			
Mastigamoeba radiosa Lackey			+
Mastigamoeba longifilum Stokes	+		+
Mastigamoeba reptans Stokes	+		+
Mastigamoeba viridis Lackey			+
Order 2. Choanoflagellida			
Codosiga botrytis (Ehrenberg)	+	+	
Monosiga ovata Kent	+		
Monosiga steinii Kent	+	+	
Order 3. Bicoecida			
Poteriodendron petiolatum Stein	+		
Salpingoeca amphoridium Clark	+		
Salpingoeca marsonii Lemmerman			+
Order 4. Kinetoplastida			
Bodo alexeieffii Lemmerman	+		+
Bodo angustus (Dujardin)	+	+	+
Bodo caudatus Dujardin	+	+	+
Bodo edax Klebs	+	+	
Bodo glissans Lackey	+		+
Bodo globosus Stein	+	+	
Bodo lens (Müller)	+	+	+
Bodo mutabilis Klebs	+		+
Bodo putrinus Stokes	+		+
Bodo saltans Ehrenberg	+	+	
Cercobodo crassicauda (Alexeieff)	+	+	+
Cercobodo longicauda (Stein)	+	+	+
Cercobodo ovatus Lemmerman	+	+	+
Cercobodo radiatus (Klebs)	+		

TABLE I (*Continued*)

	PF	AS	IT
Pleuromonas jaculans Perty	+	+	+
Rhynchomonas nasuta (Stokes)	+		
Order 5. Diplomonadida			
Collodictyon triciliatum Carter	+		
Hexamitus crassus Klebs	+		
Hexamitus fusiformis Klebs	+		
Hexamitus inflata Dujardin	+		+
Tetramitus decissus Perty	+	+	+
Tetramitus pyriformis (Klebs)		+	+
Tetramitus sulcatus Klebs		+	
Trepomonas agilis Dujardin	+		+
Trepomonas rotans Klebs	+	+	
Superclass 2. **SARCODINA**			
Class 1. RHIZOPODEA			
Order 1. Amoebida			
Amoeba actinophora Auerbach	+	+	
Amoeba guttula Dujardin	+		+
Amoeba limicola Rhumbler		+	
Amoeba proteus Leidy	+	+	+
Amoeba radiosa Ehrenberg	+	+	
Amoeba striata Penard	+	+	
Amoeba verrucosa Ehrenberg	+	+	
Amoeba vespertilio Penard		+	
Amoeba villosa Wallick	+		
Hartmanella hyalina (Dangeard)	+	+	+
Naegleria gruberi (Schardinger)	+		+
Vahlkampfia albida (Dujardin)	+	+	+
Vahlkampfia fragilis Lackey			+
Vahlkampfia limax (Dujardin)	+	+	+
Vahlkampfia minuta	+		
Order 2. Arcellinida			
Arcella discoides Ehrenberg	+	+	
Arcella dentata (Leidy)	+		
Arcella hemisphaerica Perty	+		
Arcella vulgaris Ehrenberg	+	+	
Centropyxis aculeata Stein	+	+	
Chlamydophrys minor Belar	+	+	+
Chlamydophrys stercorea Cienkowski	+		+
Cochliopodium bilimbosum (Averbach)	+	+	
Cyphoderia ampulla (Ehrenberg)	+		

TABLE I (*Continued*)

	PF	AS	IT
Difflugia acuminata Ehrenberg	+		
Difflugia constricta (Ehrenberg)		+	
Difflugia corona Wallick	+		
Difflugia oblonga Ehrenberg	+	+	
Difflugia urceolata Carter		+	
Euglypha alveolata Dujardin	+		+
Euglypha tuberculata Dujardin	+	+	
Hyalosphenia papilio Leidy		+	
Pamphagus mutabilis Bailey	+		
Pseudodifflugia gracilis Schlumberger	+	+	
Trinema enchelys Ehrenberg		+	
Trinema lineare Penard	+	+	+
Class 2. ACTINOPODEA			
Order 1. Actinophryida			
Actinophrys sol Ehrenberg	+	+	
Actinophrys vesiculata Penard		+	
Actinosphaerium arachnoideum Penard		+	
Actinosphaerium eichhorni Penard	+	+	
Heterophrys glabrescens Penard		+	
Heterophrys myriopoda Archer	+	+	
Raphidiophrys elegans Hertwig & Lesser	+		
Order 2. Proteomyxida			
Hyalodiscus rubicundus Hertwig & Lesser	+		
Nuclearia simplex Cienkowski	+		
Vampyrella lateritia (Fresnius)	+		
Sub-phylum II. **CILIOPHORA**			
Class 1. CILIATEA			
Sub-class (a). HOLOTRICHIA			
Order 1. Gymnostomatida			
Sub-order (a). Rhabdophorina			
Amphileptus claparedei Stein	+	+	
Coleps bicuspis Noland		+	
Coleps elongatus Ehrenberg		+	
Coleps hirtus Nitzsch	+	+	
Chaenea teres Dujardin	+	+	
Didinium nasutum Müller	+	+	
Dileptus anser (Müller)	+		
Enchelys curvilata Smith		+	
Enchelys simplex Kahl	+		
Enchelyomorpha vermicularis (Smith)	+	+	

TABLE I (*Continued*)

	PF	AS	IT
Hemiophrys branchiarum (Wenrich)		+	
Hemiophrys fusidens Kahl	+	+	
Hemiophrys pleurosigma Stokes	+	+	
Litonotus anguilla Kahl		+	
Litonotus carinatus Stokes	+	+	
Litonotus cygnus (Müller)	+	+	
Litonotus fasciola Ehr.-Wrzesniowski	+	+	+
Litonotus lamella Schewiakoff	+		
Loxodes rostrum (Müller)	+		
Loxophyllum helus (Stokes)		+	
Loxophyllum meleagris Dujardin	+		
Loxophyllum multinucleatum Kahl		+	
Prorodon armatus Clap. & Lachmann	+		
Prorodon discolor Ehr.–Blochm.–Schewiakoff		+	
Prorodon griseus Clap. & Lachmann	+		
Porodon teres Ehrenberg	+		
Rhopalophrya crassa Kahl	+		
Spathidium hyalinum Dujardin		+	
Spathidium spathula Müller	+	+	
Trachelophyllum clavatum Stokes		+	
Trachelophyllum pusillum Perty–Clap. & Lach.	+	+	
Urotricha farcta Clap. & Lachmann	+		
Sub-order (b). Cyrtophorina			
Chilodonella cucullulus (Müller)	+	+	
Chilodonella fluviatilis Stokes		+	
Chilodonella uncinata Ehrenberg	+	+	
Dysteria fluviatilis Stein		+	
Orthodonella gutta Cohn	+	+	
Trochilia minuta Roux	+	+	
Order 2. Trichostomatida			
Colpoda aspera Kahl	+	+	
Colpoda cucullus Müller	+	+	
Colpoda inflata (Stokes)	+	+	+
Colpoda steinii Maupas		+	
Drepanomonas revoluta Penard		+	
Plagiopyla nasuta Stein	+	+	+
Trichopelma opaca (Penard)	+		
Trimyema compressum Lackey	+		+
Trimyema pura (Ehrenberg)	+	+	
Order 3. Hymenostomatida			

TABLE I (*Continued*)

	PF	AS	IT
Sub-order (a). Tetrahymenina			
Cohnilembus pusillus Kahl	+		
Cohnilembus verminus (Müller)	+		
Colpidium campylum (Stokes)	+	+	
Colpidium colpoda Stein	+	+	
Colpidium striatum Stokes	+		
Glaucoma scintillans Ehrenberg	+	+	+
Loxocephalus granulosus Kent	+	+	
Pseudoglaucoma muscorum Kahl		+	
Sathrophilus oviformis Kahl		+	
Tetrahymena pyriformis (Ehrenberg)	+	+	
Tetrahymena rostrata (Kahl)		+	
Tetrahymena vorax (Kidder, Lilly & Claff)		+	
Uronema nigricans (Müller)	+	+	
Sub-order (b). Peniculina			
Cinetochilum margaritaceum Perty	+	+	+
Paramecium aurelia Ehrenberg	+	+	
Paramecium bursaria (Ehrenberg)	+	+	
Paramecium caudatum Ehrenberg	+	+	
Paramecium putrinum Clap. & Lachmann	+	+	+
Paramecium trichium Stokes	+	+	
Urocentrum turbo (Müller)	+		
Sub-order (c). Pleuronematina			
Cyclidium citrullus Cohn		+	
Cyclidium glaucoma Müller	+	+	+
Cyclidium lanuginosum Penard	+		
Cyclidium oblongum Kahl	+		
Pleuronema crassum Dujardin	+		
Sub-class (b) PERITRICHIA			
Order 1. Peritrichida			
Carchesium polypinum Linnaeus	+	+	
Epistylis anastatica Linné		+	
Epistylis articulata Fromentel		+	
Epistylis cambari Kellicott		+	
Epistylis elongata Stokes		+	
Epistylis lacustris Imhoff		+	
Epistylis plicatilis Ehrenberg	+	+	
Epistylis pyriformis d'Udekem		+	
Epistylis rotans Svec	+	+	
Epistylis umbilicata Clap. & Lachmann		+	
Opercularia asellicola Kahl		+	

TABLE 1 (*Continued*)

	PF	AS	IT
Opercularia berberina Linné	+	+	+
Opercularia coarctata Clap. & Lachmann	+	+	
Opercularia curvicaula (Penard)	+	+	
Opercularia glomerata Roux		+	
Opercularia microdiscum Fauré-Fremiet	+	+	
Opercularia minima Kahl	+	+	
Opercularia phryganeae Kahl	+	+	
Opercularia plicatilis Stokes	+		
Platycola decumbens Ehrenberg	+		
Pyxicola socialis Gruber	+		
Pyxidiella collare Kahl		+	
Pyxidiella henneguyi Fauré-Fremiet		+	
Pyxidiella invaginatum Stokes		+	
Pyxidiella urceolatum Stokes		+	
Pyxidiella vernale Stokes		+	
Rhabdostyla conipes Kahl		+	
Rhabdostyla inclinans (Müller)	+		
Rhabdostyla pyriformis Perty		+	
Rhabdostyla vernalis Stokes		+	
Telotrochidium henneguyi (Fauré-Fremiet)		+	
Vaginicola crystallina Ehrenberg		+	
Vaginicola striata Fromentel	+	+	
Vaginicola tincta Ehrenberg	+		
Vorticella aequilata Kahl	+	+	
Vorticella alba Fromentel	+	+	
Vorticella campanula Ehrenberg	+	+	
Vorticella communis Fromentel	+	+	
Vorticella convallaria Linnaeus	+	+	+
Vorticella elongata Fromentel	+	+	
Vorticella extensa Kahl		+	
Vorticella fromenteli Kahl	+	+	
Vorticella hamata Ehrenberg	+		
Vorticella longifilum Kent		+	
Vorticella microstoma Ehrenberg	+	+	+
Vorticella monilata (Tatem)		+	
Vorticella muralis Penard		+	
Vorticella nebulifera var. *similis* Stokes	+	+	
Vorticella nutans Müller		+	
Vorticella picta (Ehrenberg)		+	
Vorticella punctata Dons		+	
Vorticella pusilla Müller		+	
Vorticella putrina Müller–Kent		+	

TABLE I (*Continued*)

	PF	AS	IT
Vorticella pyrum Mereschkowsky		+	
Vorticella striata var. *octava* (Stokes)	+	+	
Vorticella submicrostoma Ghosh		+	
Vorticella vernalis Stokes		+	
Zoothamnium aselli Clap. & Lachmann		+	
Zoothamnium mucedo Entz	+	+	
Zoothamnium pygmaeum d'Udekem		+	
Sub-class (c). SPIROTRICHIA			
Order 1. Heterotrichida			
Blepharisma persicinum Perty		+	
Blepharisma undulans Stein	+	+	
Climacostomum virens (Ehrenberg)		+	
Metopus es Müller	+	+	
Metopus sigmoides Clap. & Lachmann			+
Metopus setosus Kahl		+	
Spirostomum ambiguum Müller–Ehrenberg	+	+	
Spirostomum ambiguum var. *minor* Roux	+		
Spirostomum intermedium Kahl	+		
Spirostomum teres Clap. & Lachmann	+		
Stentor igneus Ehrenberg		+	
Stentor multiformis Müller		+	
Stentor polymorphus (Müller)	+	+	
Stentor roeseli Ehrenberg	+	+	
Order 2. Oligotrichida			
Halteria grandinella (Müller)	+		
Order 3. Odontostomatida			
Saprodinium putrinum Lackey	+		
Order 4. Hypotrichida			
Aspidisca costata (Dujardin)	+	+	+
Aspidisca lynceus Ehrenberg	+	+	
Aspidisca polystyla Stein		+	
Aspidisca sulcata Kahl		+	
Aspidisca turrita Ehrenberg	+	+	
Euplotes aediculatus Pierson		+	
Euplotes affinis f. *typica* Dujardin	+	+	
Euplotes carinatus Stokes		+	
Euplotes charon (Müller)	+	+	
Euplotes eurystomus Wrzesniowski	+	+	
Euplotes harpa Stein	+		
Euplotes moebiusi f. *quadricirratus* Kahl	+	+	
Euplotes patella (Müller)	+	+	

TABLE 1 (*Continued*)

	PF	AS	IT
Euplotes patella f. *latus* Kahl		+	
Euplotes patella f. *variabilis* Stokes		+	
Histriculus similis Corliss	+	+	
Histriculus vorax Corliss	+	+	
Holosticha hymenophora Stokes		+	
Holosticha vernalis Stokes		+	
Opisthotricha similis Engelmann	+	+	
Oxytricha fallax Stein	+	+	
Oxytricha ludibunda Stokes	+	+	
Stylonychia mytilus Ehrenberg	+	+	
Stylonychia pustulata Ehrenberg	+	+	
Stylonychia putrina Stokes		+	
Tachysoma parvistyla Stokes		+	
Tachysoma pellionella (Müller–Stein)	+	+	
Uroleptus limnetis Stokes		+	
Uroleptus longicaudatus Stokes		+	
Uroleptus musculus (Müller)	+		
Uroleptus piscis (Müller)	+		
Urostyla weissei Stein	+		
Sub-class (d). SUCTORIA			
Acineta cuspidata Stokes		+	
Acineta grandis Kent		+	
Acineta foetida Maupas	+	+	
Acineta tuberosa Ehrenberg		+	
Discophrya elongata (Clap. & Lachmann)		+	
Podophrya carchesii (Clap. & Lachmann)		+	
Podophrya elongata Wailes		+	
Podophrya fixa (Quennerstedt)	+	+	
Podophrya maupasi Bütschli	+	+	
Sphaerophrya magna Maupas	+	+	
Sphaerophrya pusilla Clap. & Lachmann		+	
Thecacineta cothurnoides Collin		+	
Tokophrya cyclopum (Clap. & Lachmann)			
Tokophrya infusionem (Stein)		+	
Tokophrya mollis Bütschli	+	+	
Tokophrya quadripartita Clap. & Lachmann		+	

of a population is inversely proportional to the diversity of species (Barker, 1943, 1946; Crozier, 1923; Crozier and Harris, 1923a, b).

Lists of protozoa found in percolating filters have been given by many authors, including Agersborg (1929), Agersborg and Hatfield (1929), Barker (1942, 1943), Brink (1967), Cooke (1959), Crozier (1923), Curds and Cockburn (1970a), Frye and Becker (1929), Hausman (1923), Lackey (1924), Liebmann (1949), Pattamapirat (1963), Tomlinson (1946) and others. It may be seen from the list of organisms in Table I that 218 species of protozoa have been identified in percolating filters and these species are distributed throughout the classes following the scheme; 35 species of Phytomastigophorea, 30 species of Zoomastigophorea, 31 species of Rhizopodea, 7 species in the Actinopodea, and 116 ciliate species. The distribution of species between the five protozoan classes found in percolating filters is shown in Fig. 1, where a percentage scale has been used so that the species distribution of the different treatment processes can be compared. It can be seen from the figure that by far the largest proportion of protozoan species that have been identified in filters

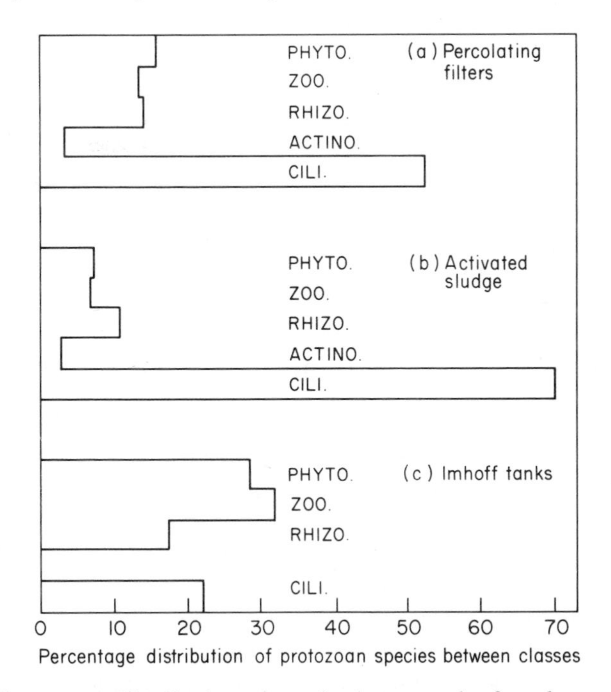

FIG. 1. Percentage distribution of species between the five classes of protozoa represented in used-water treatment processes. Phyto = Phytomastigophorea, Zoo = Zoomastigophorea, Rhizo = Rhizopodea, Actino = Actinopodea, Cili = Ciliatea.

belong to the class Ciliatea, although the amoebae and flagellated protozoa are also well represented.

The class of protozoa which contributes the greatest numbers of individuals to the protozoan population of a filter has been discussed by a number of authors, and it is generally agreed that although the ciliates are normally numerically dominant, this need not necessarily be the case. Barker (1942) listed the numbers of protozoa found in percolating filters to be in the following ranges; rhizopods 100–4600/ml, flagellates 200–13 000/ml, and ciliates 500–10 000/ml of liquor, and he states that ciliates are generally the most abundant. It is unfortunate that no data, however imprecise, exist on biomass estimations; obviously it is not meaningful to compare numbers of organisms when their sizes are significantly different. Barker's (1942) statement concerning ciliate dominance would have been far more convincing if biomass data had been given, since ciliates in general are much larger than flagellates. Frye and Becker (1929) stated that although small amoebae were present in virtually all the samples examined during their investigation, they were generally present only in small numbers; similarly they usually found flagellates present in small numbers although on occasion they were dominant. Frye and Becker (1929) concluded from their observations that ciliates were the most abundant protozoa present in their filter treating milk-waste, and found very large numbers of *Colpoda* sp. and *Uronema* sp.

The most comprehensive survey to date is that of Curds and Cockburn (1970a); they examined effluent samples from 52 percolating filters situated in 46 sewage works all over England, Scotland and Wales. All the filters they examined, including all those treating industrial wastes, contained protozoa, and they concluded that ciliates were generally dominant. Brink (1967) similarly observed that ciliates were generally the most common protozoa in his experimental sand filters, though occasionally amoebae reached similar proportions and sometimes even surpassed them.

Figure 2a shows the distribution of the protozoan species between the orders of the classes identified in percolating filters. The histogram shows that in the Phytomastigophorea the two dominant orders are the Euglenida and the Chrysomonadida; in the Zoomastigophorea the Kinetoplastida and the Diplomonadida are the two dominant orders. In the Rhizopodea and the Actinopodea the percentage distribution of species between their orders is approximately equal. Table II summarizes the observations of Barker (1942, 1943), Curds and Cockburn (1970a) and Frye and Becker (1929), on the most frequently seen species of protozoa in percolating filters. In the Phytomastigophora the species *Oicomonas termo*, *Notosolenus orbicularis* and *Peranema trichophorum* appear to be

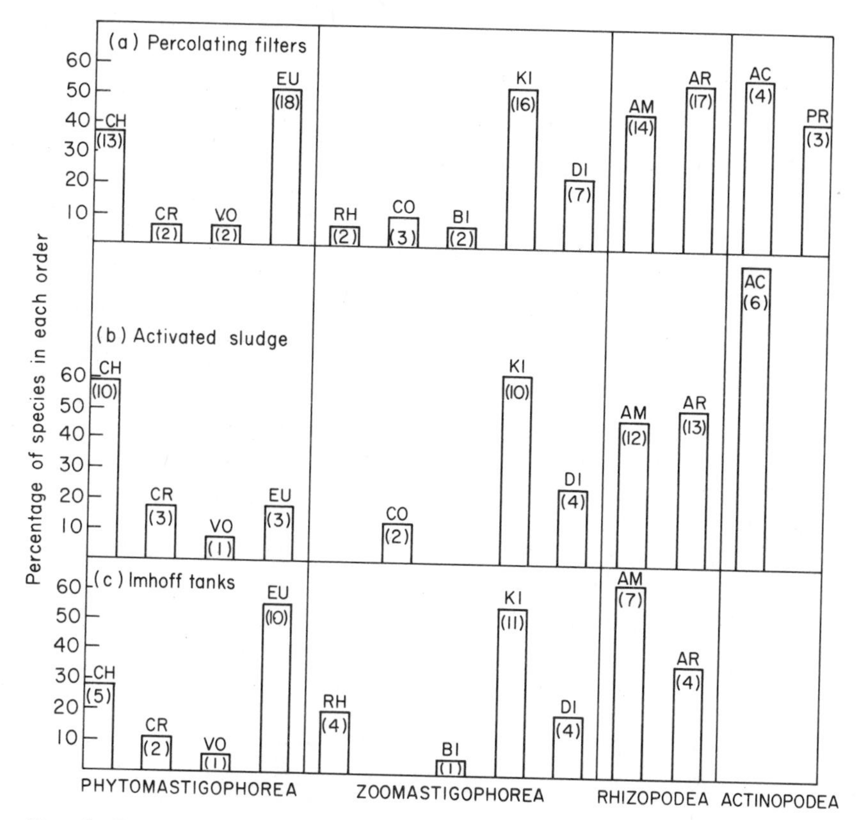

FIG. 2. Percentage distribution of species between some of the Orders of protozoa represented in used-water treatment processes. CH = Chrysomonadida, CR = Cryptomonadida, VO = Volvocida, EU = Euglenida, RH = Rhizomastigida, CO = Choanoflagellida, BI = Bicoecida, KI = Kinetoplastida, DI = Diplomonadida, AM = Amoebida, AR = Arcellinida, AC = Actinophryida, PR = Proteomyxida. Numbers in parenthesis indicate actual numbers of species represented in each order.

the most commonly encountered species, and these are illustrated in Fig. 3. Curds and Cockburn (1970a) stated that 29% of the effluent samples they examined contained *P. trichophorum*. Frye and Becker (1929) did not mention any member of this class as being common. In the Zoomastigophorea (Table II) there was general agreement that *Bodo caudatus*, *Cercobodo* sp. (usually *C. crassicauda*), *Trepanomonas agilis* and *Pleuromonas jaculans* are the most frequently encountered species, and these are illustrated in Fig. 3. In the Rhizopodea there is less agreement, but both naked and testate amoebae seem equally well represented. The testate amoeba *Arcella vulgaris* is the only species common to all lists,

TABLE II. The most frequently observed species of protozoa in percolating filters

Class	Barker (1942)	Barker (1943)	Curds and Cockburn (1970a)	Frye and Becker (1929)
Phytomastigophorea	*Monas* sp. *Oicomonas* sp.	*Oicomonas termo* *Notosolenus orbicularis*	*Peranema trichophorum*	
Zoomastigophorea	*Bodo* sp. *Cercobodo* sp. *Trepomonas* sp. *Pleuromonas* sp.	*Pleuromonas jaculans*	*Bodo caudatus* *Trepomonas agilis*	*Bodo* sp. *Cercobodo* sp.
Rhizopodea	*Amoeba* sp. *Vahlkampfia* sp. *Arcella* sp. *Euglypha* sp.	*Amoeba guttula* *Vahlkampfia limax* *Arcella vulgaris* *Cochliopodium bilimbosum* *Trinema lineare*	Small amoebae *Arcella vulgaris*	*Naegleria gruberi* *Vahlkampfia albida* *Vahlkampfia limax* *Arcella vulgaris* *Chlomdophrys* sp. *Euglypha alveolata*
Ciliatea	*Carchesium* sp. *Chilodonella* sp. *Cinetochilum* sp. *Cyclidium* sp. *Opercularia* sp. *Urotricha* sp.	*Carchesium polypinum* *Chilodonella uncinata* *Cinetochilum margaritaceum*	*Aspidisca costata* *Carchesium polypinum* *Chilodonella uncinata* *Cinetochilum margaritaceum* *Opercularia coarctata* *Opercularia microdiscum* *Trachelophyllum pusillum* *Vorticella convallaria* *Vorticella striata* var. *octava*	*Colpoda* sp. *Uronema* sp.

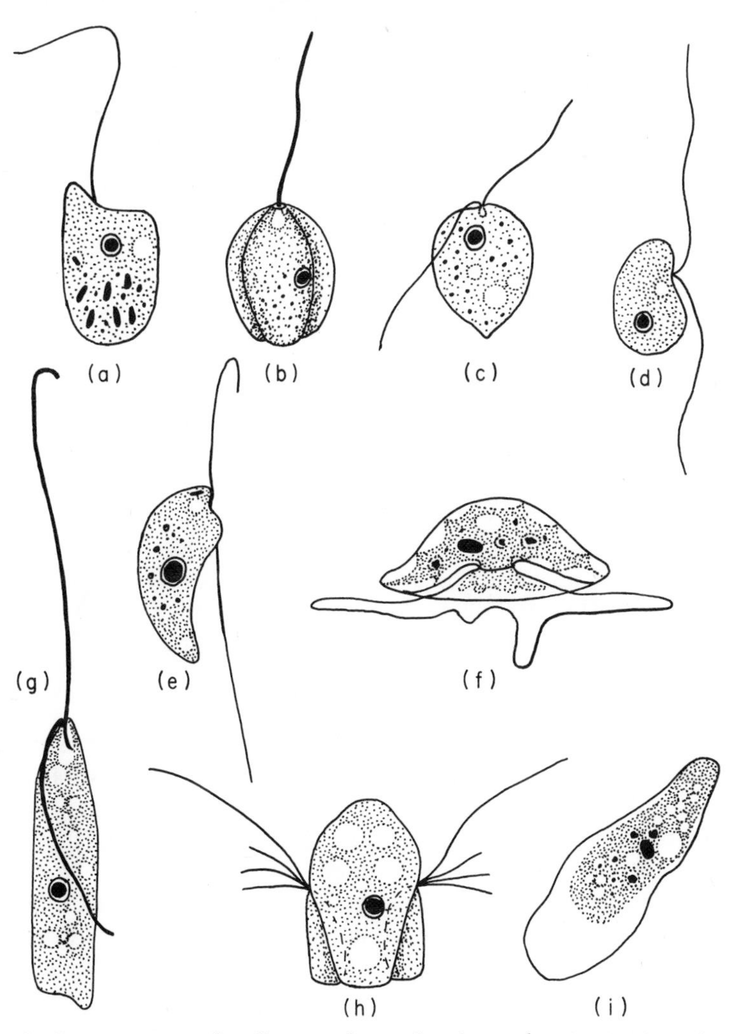

FIG. 3. Some common flagellates and amoebae in used-water treatment processes. (a) *Oicomonas termo*, (b) *Notosolenus orbicularis*, (c) *Cercobodo crassicauda*, (d) *Pleuromonas jaculans*, (e) *Bodo caudatus*, (f) *Arcella vulgaris*, (g) *Peranema trichophorum*, (h) *Trepomonas agilis*, (i) *Vahlkampfia limax*.

and Curds and Cockburn (1970a) found it in 46% of their samples. *Vahlkampfia limax* and *Arcella vulgaris* are the examples of this class illustrated in Fig. 3.

Nine of the orders of the Ciliatea are represented in the percolating-filter fauna, and Fig. 4 illustrates the distribution of the species between these orders. Members of the orders Peritrichida and Gymnostomatida

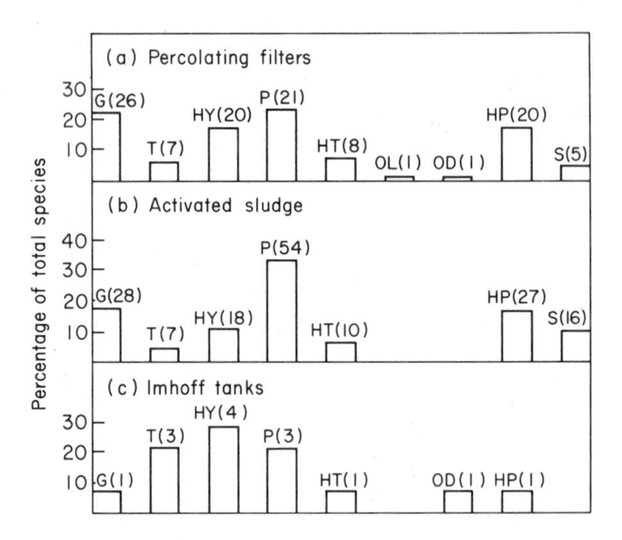

FIG. 4. Percentage distribution of species between the orders of the class Ciliatea found in used-water treatment processes. G = Gymnostomatida, T = Trichostomatida, HY = Hymenostomatida, P = Peritrichida, HT = Heterotrichida, OL = Oligotrichida, OD = Odontostomatida, HP = Hypotrichida, S = Suctorida. Numbers in parenthesis indicate actual numbers of species represented in each order.

are the most well-represented, but a large variety of hymenostome and hypotrich species have also been identified, and Lloyd (1945) was the first to suggest that the peritrichs and *Chilodonella* sp. were of major importance. Curds and Cockburn (1970a) identified 19 holotrichs, 20 peritrichs, 11 spirotrichs and 3 suctorians in their survey, but found that only 8 ciliate species were ever found in large numbers; these were *Chilodonella uncinata* (found in large numbers in 4% of the samples), *Carchesium polypinum* (15%), *Vorticella alba* (2%), *Vorticella convallaria* (10%), *Vorticella striata* var. *octava* (2%), *Opercularia coarctata* (2%), *Opercularia microdiscum* (44%) and *Opercularia phryganeae* (4%). Approximately 50% of the species they identified were found in moderate numbers and these included 10 holotrichs, 11 peritrichs, 5 spirotrichs and one suctorian. Curds and Cockburn tentatively suggested that primary filters contain larger numbers of ciliate individuals than secondary filters, and found that certain species, e.g. *Colpidium colpoda*, *Colpidium campylum*, *Chilodonella cucullulus*, *Paramecium caudatum*, *Epistylis plicatilis*, *Tachysoma pellionella* and *Hemiophrys fusidens*, were limited to primary filters. Curds and Cockburn (1970a) found sedentary ciliates attached to a variety of substrata, e.g. bacterial-film surface and insect debris, but the peritrich *Rhabdostyla inclinans* was restricted to the ends of the setae of the worm *Nais variabilis*. The two loricate peritrichs *Vaginicola striata* and *Pyxicola socialis* were generally attached to insect debris, particularly to the wings of dead flies (*Psychoda* sp.).

Table II lists those species of ciliate most commonly identified in percolating-filter samples, and with the exception of Frye and Becker (1929), whose experimental filters were treating milk-wastes, there is reasonable agreement between the other authors. Those ciliates in Table II taken from Curds and Cockburn (1970a) were found in at least 50% of the samples examined. These two workers found that certain species were found in a very large proportion of the samples, e.g. *Chilodonella uncinata* was found in 90% of the samples, *Vorticella convallaria* in 83%, *Opercularia microdiscum* in 81% and *Carchesium polypinum* in 62%. However, they also found that the frequency of occurrence was not necessarily associated with a dominant numerical position in the populations. For example, although *Cinetochilum margaritaceum* is a common ciliate found in 54% of the samples, it was never recorded in large numbers and was found in moderate numbers in only 2% of the samples. Other species such as *Trachelophyllum pusillum*, *Hemiophrys pleurosigma* and *Vorticella microstoma*, although frequently found, were usually present in small numbers. Some of the most commonly observed ciliates in percolating filters are illustrated in Figs 5 and 7.

(*b*) *Activated sludge.* Protozoa are commonly observed in the mixed liquor of activated-sludge plants and numbers in the order of 50 000 cells/ml are often reported. Calculations based on numerical concentrations such as these (Ministry of Technology, 1968) indicate that the protozoan population could constitute approximately 5% of the dry weight of the mixed liquor-suspended solids. Since Johnson (1914) observed the presence of large numbers of protozoa in activated sludge, they have been reported by a number of workers. Lists have been provided by Agersborg and Hatfield (1929), Ardern and Lockett (1936), Bark (1971), Barker (1942, 1949), Brown (1965), Buswell and Long (1923), Clay (1964), Curds (1963a, 1966, 1969), Curds and Cockburn (1970a), Hawkes (1963), Kolkwitz (1926), Morishita (1970), Read (1966), Richards and Sawyer (1922), Schofield (1971), Taylor (1930) and others. Table I lists 228 species of protozoa that have been identified in activated sludge and the species are distributed amongst the classes as follows; 17 species of Phytomastigophorea, 16 species of Zoomastigophorea, 25 species of Rhizopodea, 6 species of Actinopodea and 160 species of Ciliatea. If this distribution is compared with that of the protozoan fauna of percolating filters (Fig. 1), it can be seen that 70% of the species belong to the class Ciliatea, which is a considerably higher proportion than in the fauna of filters. This means that although the total

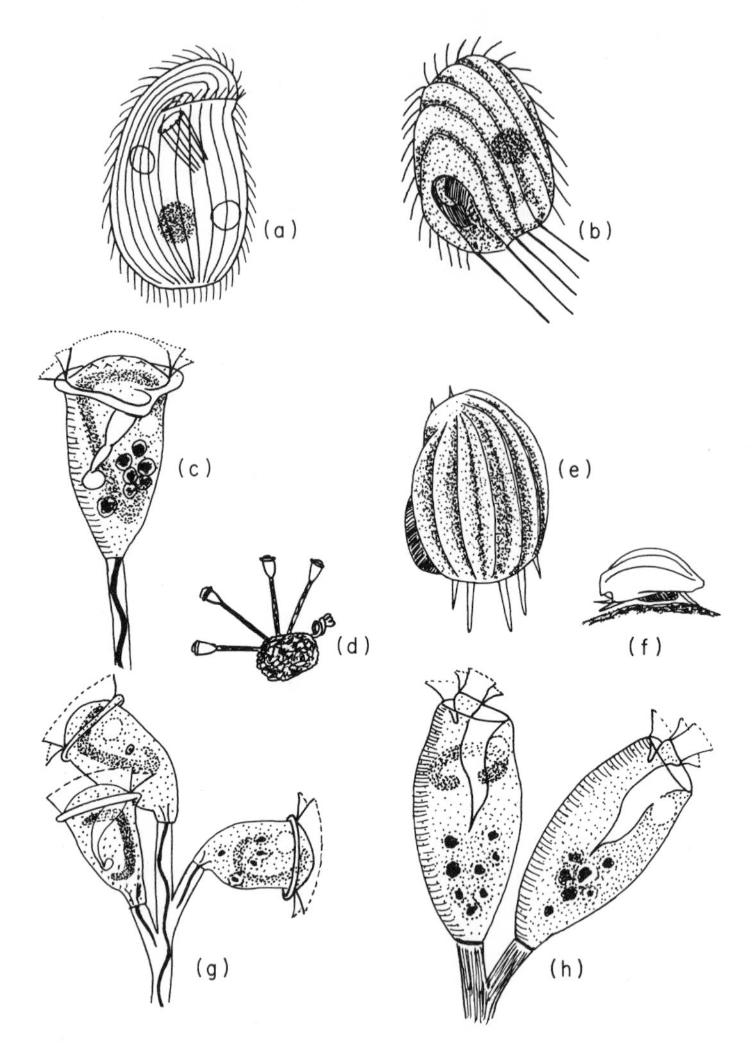

FIG. 5. Some common ciliates in percolating filters. (a) *Chilodonella uninata*, (b) *Cinetochilum margaritaceum*, (c) *Vorticella convallaria*, (d) *V. convallaria* group on sludge floc, (e) *Aspidisca costata*, (f) *A. costata*, lateral view showing crawling habit, (g) *Carchesium polypinum*, (h) *Opercularia microdiscum*.

number of species is greater in activated sludge they are mainly ciliates, and consequently the variety of the protozoan fauna is rather limited at the class level in comparison with the fauna of percolating filters.

It is quite clear from the literature that although the protozoa from the two flagellate classes (the Phyto- and Zoomastigophorea) are found in activated-sludge samples, generally they are not present and are usually associated with inferior operational factors (overloaded plants, etc.). Flagellates appear to occur less frequently, are in smaller numbers, and there are fewer species than in percolating filters.

For many years the general opinion expressed in the literature has been that the ciliated protozoa are more often than not the class which contributes the largest numbers both of species and of individuals to the protozoan fauna of activated sludge. It is clear from Table I and Fig. 1 that more ciliate species have been identified in activated sludge than in any of the other processes. However, some recent authors (Bark, 1971; Schofield, 1971; Sydenham, 1968) suggest that various amoebae, both testate (e.g. *Cochliopodium* spp.) and small naked forms, are just as frequent and are often the dominant organisms both in terms of numbers (Schofield, 1971) and in terms of estimated biomass (Sydenham, 1968). For example, Scofield (1971) stated that on one occasion *Cochliopodium* sp. was found in numbers in the order of 100 000/ml in the sludge at the Leicester Sewage Treatment Works, whereas the largest number of any one ciliate species was only 20 000/ml (*Trachelophyllum* sp.). The observations concerning amoebae have so far been limited to a few plants in the London area and one at Leicester, whereas Curds and Cockburn (1970a) examined mixed liquor samples from 56 plants situated in works all over England, Scotland and Wales; they concluded that ciliates were normally the dominant class of protozoa in activated sludge, although on occasion amoebae and flagellates were seen in moderate numbers. Furthermore Brown (1965), in his survey of the activated-sludge plant at Hogsmill, Surrey, noted that over a complete year ciliate species were dominant, with one exception when *Euglypha* sp., a testate amoeba, was the dominant protozoan species for a short time. Barker (1949) found that rhizopods, with the exception of *Amoeba proteus*, and flagellates were favoured in activated sludge immediately after a plant was started, but after 20 days these were replaced by ciliates which usually remained dominant. It seems likely, therefore, that although the rhizopods can dominate the protozoan population of activated sludge, as is the case with percolating filters, this is infrequent and usually the ciliates hold the dominant position.

Figure 2b shows that the majority of the flagellate species which have been identified in activated sludge belong to the orders Chrysomonadida and Kinetoplastida. No species belonging to the orders Rhizomastigida and Bicoecida have yet been reported in activated sludge although a few species are known to occur in filters. Both the naked and the testate amoebae are well represented but no proteomyxid actinopods have been observed. Table III summarizes the observations of five authors, Barker (1949), Brown (1965), Curds and Cockburn (1970a) and Schofield

TABLE III. The most frequently observed species of protozoa in activated sludge

Class	Barker (1949)	Brown (1965)	Curds and Cockburn (1970a)	Schofield (1971)
Phytomastigophorea	*Anthophysa vegetans* *Oicomonas steinii* *Oicomonas termo*		*Peranema trichophorum*	
Zoomastigophorea	*Bodo caudatus* *Bodo lens* *Cercobodo crassicauda* *Cercobodo longicauda* *Cercobodo ovatus* *Pleuromonas jaculans*			
Rhizopodea	*Amoeba actinophora* *Amoeba guttula* *Vahlkampfia limax*	*Euglypha alveolata*	Small amoebae *Arcella vulgaris* *Euglypha* sp.	*Amoeba* sp. *Cochliopodium* sp.
Ciliatea	*Carchesium polypinum* *Opercularia* sp. *Oxytricha fallax* *Paramecium* sp. *Vorticella alba* *Vorticella campanula* *Vorticella communis*	*Aspidisca costata* *Aspidisca lynceus* *Aspidisca robusta* *Chilodonella cucullulus* *Epistylis plicatilis* *Euplotes aediculatus* *Litonotus fasciola* *Vorticella campanula* *Vorticella convallaria* *Vorticella microstoma* *Vorticella nebulifera* var. *similis* *Vorticella striata* var. *octava*	*Aspidisca costata* *Carchesium polypinum* *Euplotes moebiusi* *Opercularia coarctata* *Trachelophyllum pusillum* *Vorticella alba* *Vorticella convallaria* *Vorticella fromenteli* *Vorticella microstoma*	*Aspidisca costata* *Chilodonella uncinata* *Drepanomonas* sp. *Epistylis* sp. *Hemiophrys* sp. *Vorticella convallaria* *Vorticella microstoma*

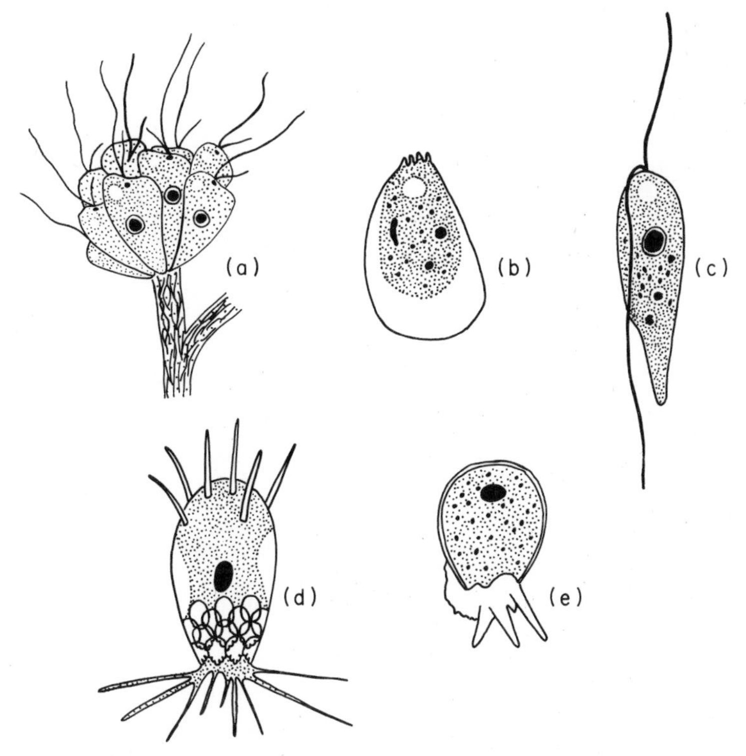

FIG. 6. Some common flagellates and amoebae in activated sludge. (a) *Anthophysa vegetans*, (b) *Amoeba guttula*, (c) *Cercobodo longicauda*, (d) *Euglypha alveolata*, (e) *Cochliopodium bilimbosum*.

(1971), on the most frequently seen species of protozoa in activated-sludge plants, and some are illustrated in Figs 6 and 7. It is apparent from Table III that little mention has been made of flagellates in activated sludge; however, those listed are in general very similar to those listed (Table II) for percolating filters.

Only seven orders of the class Ciliatea are represented in activated sludge, and Fig. 4 illustrates the distribution of the reported species between these orders. It is quite clear that far more species of the order Peritrichida have been identified than any of the other ciliate orders, and this is in agreement with the observations of Morishita (1970) and Curds and Cockburn (1970a). The latter authors found that 23 of the ciliate species they identified were found in large numbers on at least one occasion; these species included 8 holotrichs, 10 peritrichs and 5 hypotrichs. Furthermore they found that approximately half of the species were present in moderate numbers and these included 13 holotrichs, 12

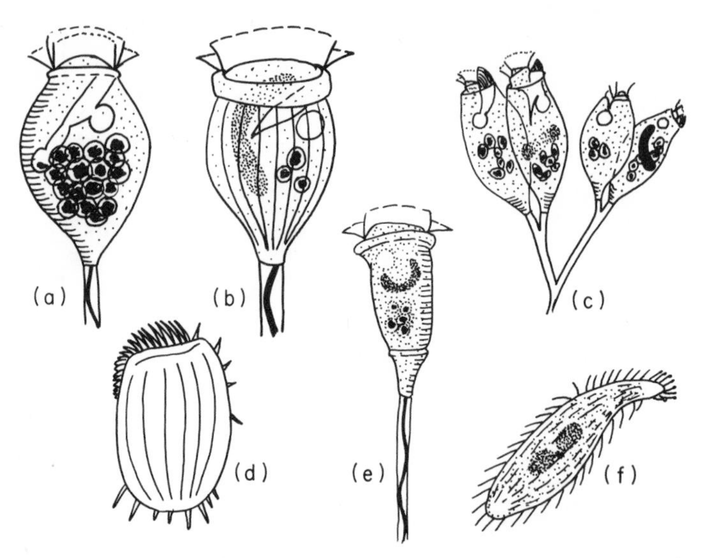

FIG. 7. Some common ciliates of percolating filters and activated sludge. (a) *Vorticella microstoma*, (b) *Vorticella alba*, (c) *Opercularia coarctata*, (d) *Euplotes moebiusi*, (e) *Vorticella fromenteli*, (f) *Trachelophyllum pusillum*.

peritrichs, 9 hypotrichs and 2 suctorians. The most frequently encountered ciliate species in activated sludge are slightly different from those of percolating filters, and Curds and Cockburn (1970a) found that *Trachelophyllum pusillum* was present in 64% of the samples, *Vorticella convallaria* in 58%, *Vorticella microstoma* in 75%, *Vorticella alba* in 38%, *Opercularia coarctata* in 54%, *Euplotes moebiusi* in 35% and *Aspidisca costata* in 69%. Some of these species are illustrated in Fig. 7.

Brown (1964, 1965) first divided the protozoa of activated sludge into two groups: (i) those protozoa which were found in the sludge for long periods throughout the year and appear to be well adapted to life in that habitat, and (ii) those protozoa which were found in the sludge only at irregular intervals and soon disappeared. Brown (1965) found that *Aspidisca costata, Vorticella convallaria, V. similis, V. microstoma* and *V. campanula* could be included in the first group, and *Euplotes aediculatus, Chilodonella cucullulus, Aspidisca robusta, A. lynceus, Epistylis plicatilis, Euglypha alveolata* and *Litonotus fasciola* could be included in the second group. Schofield (1971) also used these groupings and termed those species in the first group as being indigenous; he found that *V. convallaria, V. microstoma, Epistylis* sp., *A. costata, Hemiophrys* sp. and *Cochliopodium* sp. were indigenous to the activated sludge at Leicester. These six species had an incidence of 97% and above, although the actual numbers varied considerably.

(*c*) *Other aerobic processes.* Protozoa are also present in oxidation ponds, but they are not as frequently mentioned in the literature as are the phytoflagellates and algae which are of more importance to the process. Bick and Scholtyseck (1960), however, mentioned protozoa in their study on the ecology of a sewage lagoon. They found that the ciliate population counts were considerably lower than in other aerobic sewage-treatment processes, and the dominant ciliate *Cyclidium citrullus* reached a concentration of only about 1000/ml. The commonest ciliate species were *C. citrullus*, *Phascolodon vorticella*, *Vorticella* sp. and *Strombilidium humile*.

Lackey and Dixon (1943) noted the presence of protozoa in the Hay's process of contact aeration. These workers identified 16 species of flagellate but the most common were the diplomonads *Hexamitus inflata*, *Tetramitus pyriformis*, *T. decissus*, *Trepomonas agilis* and *Trigomonas compressa;* these were present in vast nunbers, but no explanation for their abundance was given. Other flagellates, such as *Bodo caudatus*, *Oicomonas* sp. and *Monas* sp., were also found but less frequently. Lackey and Dixon further reported that amoebae were virtually absent. The most common ciliates were *Cyclidium ecaudatum*, *Enchelys vermicularis*, *Metopus* sp., *Paramecium caudatum*, *P. putrinum* and *Trimyema compressa*. These ciliates, with the exception of *P. caudatum*, are characteristic of Imhoff tanks (see later in this chapter) and of foul waters containing very low concentrations of oxygen and some hydrogen sulphide. The most abundant ciliates were *P. caudatum*, *C. ecaudatum* and *Metopus* sp., in that order. Most of the other ciliates were peritrichs, dominated by species of the genera *Epistylis* and *Vorticella;* these, however, appeared on the surface of the aerators and wherever some dissolved oxygen was present.

2. *Protozoa in Anaerobic Processes*

Most of the work concerning protozoa in anaerobic digestion tanks is confined to Imhoff tanks. The bulk of the liquor in these tanks is anaerobic; however, since it is directly exposed to the atmosphere, some oxygen is available at the surface. It follows from the above considerations that the fauna which survive must be able to live on minimal amounts of oxygen and are either facultative or obligatory anaerobes; this results in a specialized fauna, as compared with the aerobic processes, although some of the facultative forms are found in both aerobic and anaerobic processes (see Table I). Some of the species found are known to be coprozoic (Lackey, 1925) but the source of others is unknown. The ciliates which Noland (1925) found living in natural waters with low oxygen contents (below 45% saturation) have been observed by Lackey

(1932a, b) in the sludge of digestion tanks. The protozoan fauna of anaerobic processes seems to be ubiquitous (Lackey, 1925, 1926), as is the fauna of aerobic processes, although the information available is far more limited than for aerobic processes.

Lists of protozoa present in digestion tanks (Imhoff tanks) have been published by Agersborg (1929), Hausman (1923), Kshirsagar (1962), Lackey (1924, 1925, 1926, 1932a, b), Liebmann (1936) and Rudolfs *et al.* (1924). A list of protozoa found in such tanks is included in Table I and has been compiled from Lackey (1925, 1932a, b), Kshirsagar (1962) and others. The histogram in Fig. 1 shows that only four of the classes of protozoa are present and the majority which have been identified are flagellates. Table I and Fig. 2c show that both the number of orders is smaller and the total number of species is considerably reduced, to less than one third of the total species count of the aerobic processes, so that only 63 species of protozoa have been identified in anaerobic processes. This is a reflection of the far more exacting environmental conditions which prevail. The numbers of ciliate orders and species are drastically reduced (Fig. 4c) and this class is the one which is affected the most, with the exception of the Actinopodea which is completely absent. Some of the more common species are illustrated in Fig. 8. Lackey (1925) observed that *Trepomonas agilis*, *Hexamitus inflata*, *Pleuromonas jaculans*

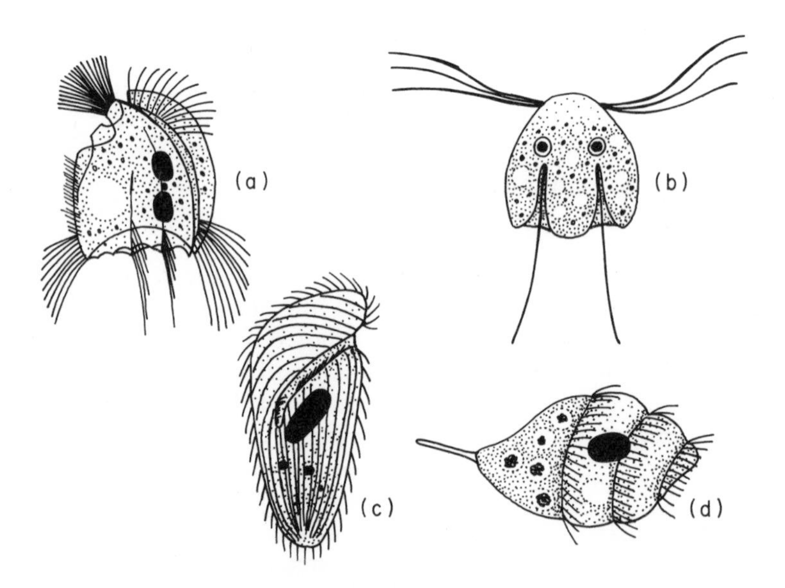

FIG. 8. Some ciliates and flagellates found in anaerobic used-water treatment processes. (a) *Saprodinium putrinum*, (b) *Hexamitus inflata*, (c) *Metopus sigmoides*, (d) *Trimyema compressa*.

and *Oicomonas* sp. were the most common flagellates; none of the larger flagellates are common but a colourless variety of *Euglena gracilis* is sometimes observed.

The rhizopod group is small in numbers of genera and in individuals, although most samples contain them; Lackey (1925) found that *Amoeba gruberi*, *A. guttula* and *Hartmanella hyalina* are the most numerous. Among the ciliates there are only four common species and these are *Holophrya* sp., *Metopus sigmoides*, *Saprodinium putrinum* and *Trimyema compressa*. All of these are strict anaerobes, and repeated attempts by Lackey (1925) to observe them in the inflowing sewage failed.

Lackey (1925) summarized the effects of the environmental conditions in the chamber of an Imhoff tank to be the selection of organisms which are facultative or strict anaerobes, which do not contain chlorophyll unless they are able to survive without it, and the encouragement of sapropelic organisms. Lackey (1925) considers that there are three groups of organisms present in Imhoff tanks, first a small group which are fully adapted to the conditions which prevail and so thrive; second, a group of organisms which may survive for short periods of time but rarely thrive; and third, a group of "chance stragglers" which enter in the sewage and live only for a few hours. The first two groups include amoebae, flagellates and ciliates, although the flagellate population is by far the most important, both in terms of numbers of species and in numbers of individuals.

III. Distribution

A. Factors Determining the Presence, Abundance and Absence of Protozoa in Used-Water Treatment Processes

Noland and Gojdics (1967) observed that when dealing with the general ecology of protozoa (or indeed any other organisms), some factors are largely of a permissive nature in that they set the outer limits beyond which the organisms cannot live. These factors allow growth and reproduction to take place within these physical and chemical boundaries, whereas other factors exert a much more positive action on protozoan populations in that they may, for example, directly supply the necessary materials for metabolic growth and reproduction or in some way directly determine the abundance of the organisms. Yet other factors are purely biological and these might aptly be included under the general title of biological interactions.

1. Permissive Factors

(*a*) *Dissolved oxygen.* Most protozoa require free oxygen for survival. There are few anaerobes, a larger number of facultative anaerobes, but the numbers of obligate aerobic species is enormous in comparison. The bulk of the literature available concerns the effect of low dissolved oxygen concentrations or complete lack of oxygen on protozoa, although Noland (1925) lists those ciliates associated with a wide range of dissolved oxygen concentrations in natural freshwater environments. Dissolved oxygen is recognized as one of the most important ecological factors which determines the presence or absence of particular protozoan species, which may be divided into those which must have free oxygen, those which will use it when it is present, and those which do not use it and are often unable to grow in its presence.

The two major aerobic processes are usually operated in such a way as to keep the dissolved oxygen concentration of the liquor at about 1–2 mg/l. However, even in aerobic processes anaerobic conditions may prevail at certain stages, e.g. the sedimentation tank of an activated-sludge unit is often anaerobic at the bottom. Many protozoa, however, can survive without oxygen for a limited period (see review by Stout, 1956), but this usually involves a reduction in activity (Pantin, 1930; Kitching, 1939; Stout, 1956) and therefore metabolism, The present author does not know of any work concerning the activity of protozoa sampled at depth from a sedimentation tank.

There can be little doubt that the absence of oxygen from digesters is the largest single factor which contributes to the relative paucity of the fauna. Lackey (1938a) listed a variety of protozoa which were able to survive under the anaerobic conditions of an Imhoff tank, and he lists the following as being obligate anaerobes; *Trepomonas agilis*, *Metopus es*, *Saprodinium putrinum* and *Trimyema compressa*. *Metopus es*, however, has been shown by others to be a facultative anaerobe which tolerates moderate concentrations of hydrogen sulphide (Bick, 1958; Liebmann, 1951; Noland, 1925). Lackey (1932a) showed that the flagellate *T. agilis* would survive in stagnant sewage but died in aerated sewage. On the other hand the ciliate *Opercularia* sp. died in the stagnant sewage but lived when aerated. One behavioural characteristic shown by *M. es* is that under conditions of low dissolved-oxygen concentrations it demonstrates a positive geotactic response and would in its natural habitat and in digesters migrate to a position of even lower dissolved oxygen; in contrast, *Paramecium caudatum* under similar conditions demonstrates a marked negative geotropic response (Curds, 1963a).

(*b*) *Temperature*. The temperature in used-water treatment plants is generally a few degrees centigrade below the ambient temperature in the summer and a few degrees above ambient in the winter. Brown (1965) found that an activated-sludge plant in the London area was 9°C in winter and 19·5°C in summer. Percolating filters are sometimes covered in the more northerly areas, where winter temperatures cause ice to accumulate on the surface with subsequent ponding; anaerobic digesters are sometimes heated.

It is a general biological principle that the activity of an organism is affected by temperature, and protozoa are no exception; the growth rate increases with rises in temperature to a pronounced optimum which is usually near the upper limit of the range over which division is possible (Mitchell, 1929; Phelps, 1946; Schoenborn, 1947; Smith, 1940; Stout, 1955). Issel (1906) reported that the following species were capable of survival in hot springs at the indicated temperatures; *Actinophrys sol* (38°C), *Amoeba proteus* (35–45° C), *Enchelys* sp. (42–43°C), *Chilodonella cucullulus* (35–39°C), *Colpoda cucullus* (35–39° C), *Paramecium aurelia* (38–40°C), *Metopus sigmoides* (35–45°C), *Oxytricha fallax* (32–45°C) and *Euplotes patella* (32–35°C), all of which are inhabitants of used-water treatment processes. Phelps (1961) suggested that thermophilic strains were selected in these habitats, since he found that whereas a strain of *Tetrahymena* sp. isolated from a hot spring would grow readily at 41·2°C, strains from cooler sources all died at 40°C. At the other end of the temperature scale it is common for protozoa to be recovered from soil samples that have been exposed to temperatures well below zero. It would appear from the above data that the temperatures likely to be encountered in treatment processes are unlikely to be limiting; however, temperature will affect the metabolic and growth rates, and hence the efficiency of the organisms in carrying out their various roles in the purification processes.

Some work has been carried out in an attempt to correlate the seasonal fluctuations in temperature with the protozoan populations; however, a correlation of this sort in the field does not necessarily mean that temperature is directly influencing the species in question—it might act indirectly through food chains or through the predators of the species concerned. For example, Gray (1952) found that numbers of ciliates in a brook were at their minimum in July when the temperature was at its maximum, while at the same time predatory dipteran larvae were at their maximum. Barker (1946) claimed a negative correlation between the numbers of protozoa in percolating filters and the numbers of flies emerging when temperature was not involved at all. Moreover, other factors beside temperature vary seasonally, and some of these will be considered later in this chapter. Brown (1965) could find no evidence to suggest that there was any relation between the temperature of an activated sludge and its constituent protozoan community.

(*c*) *Hydrogen-ion and carbon dioxide concentrations.* Anyone who has cultured protozoa knows the importance of controlling the pH of the medium. A considerable amount of information has been published concerning the outer limits of pH for growth and survival of particular species in culture (see Hall, 1953, for review) but there is little information available concerning the reasons why certain species are more tolerant than others to acidity or alkalinity. The effects of pH may be very complex, and Schoenborn (1949) showed that the flagellate *Astasia longa* had different pH ranges over which it would grow, depending on whether the basal medium was inorganic or organic.

Viehl (1932) studied the relation of pH to protozoa in activated sludge and found a parallel between the activity of the sludge and the activity of the protozoa. A pH of lower than 5·0 was inimical to the protozoa and Viehl found that a neutral activated sludge contained more species than either an acid or an alkaline one. Cramer and Wilson (1928) gave the optimum pH range of activated sludge to be 7·2–7·4. However, Noland (1925) found no evidence to suggest that pH either promoted or inhibited the occurrence of freshwater ciliates. Low pH values have been found to promote encystment of certain protozoa (Darby, 1929; Koffmann 1924). The buffering capacity of activated sludge is well established (Ingols and Heukelekian, 1940); generally the pH of an activated sludge or percolating filter is unlikely to vary much out of the pH range 6·0–8·0 and many species of protozoa have an optimum pH value within this range. It seems likely, therefore, that pH will not usually be inhibitory to the exclusion of protozoa but will play a part in the selection of species.

In nature pH is intimately linked with both the concentration of carbon dioxide and the alkali reserve of the water, and therefore has the property that it can be modified by the presence of living organisms. Gradients may be set up around the sludge or filter film, and Picken (1937) suggested that such chemical gradients and thigmotaxis could play a part in holding a community of free-living protozoa together when in association with "sewage fungus" and filamentous algae.

Carbon dioxide is the main carbon source for the photosynthetic flagellates which may be found at the top of a filter or in an oxidation pond, but it is very toxic in high concentrations. Stout (1955) regards carbon dioxide as the primary adverse factor for limiting the distribution of ciliates, and the ability to resist it, is the distinct features of species associated with organically polluted environments. Nikitinsky and Mudrezowa-Wyss (1930) introduced protozoa into water which had previously been bubbled with carbon dioxide and showed that protozoa were able to survive for varying lengths of time, for example *Paramecium putrinum*, *Vorticella microstoma* and *Polytoma uvella* lived for 6 h, *Paramecium caudatum* and *Spirostomum ambiguum* survived for 55 min, *Glaucoma scintillans* for 28 min, *Colpidium colpoda* for 25 min and *Euglena viridis* for only 20 min. The above authors concluded that the ability to resist carbon dioxide was a better index of the species' ability to live in polluted waters than their ability to survive a lack of oxygen.

(*d*) *Light.* Light is the energy source for photosynthetic organisms and is very limited in some used-water treatment processes. Light does not penetrate below a few inches in a percolating filter, and the suspended solids in the mixed liquor of an activated-sludge plant prevent its penetration. The lack of light is probably the primary reason why so few photosynthetic protozoa are found in activated sludge and why they are confined to the extreme upper layers in a percolating filter. Light, on the other hand, is essential for the proper operation of an oxidation pond, but this will be dealt with in detail in other chapters. It is known that some normally green photosynthetic species can occur as colourless forms in the absence of light, when they then utilize carbon sources such as acetate; Lackey (1925), for example, found a colourless variety of *Euglena gracilis* in the depths of an Imhoff tank.

(*e*) *Salinity.* In certain arid regions of the world where freshwater is of great value, seawater has been used for flushing purposes. Pillai and Rajagopalan (1948) reported that an activated-sludge plant operated on seawater in India developed a ciliate population comprised of a number of marine species closely related to those found in neighbouring freshwater plants.

(*f*) *Current.* The rate of flow of used water to a plant affects both the dilution rate (the reciprocal of the sewage retention time) of the plant and the velocity of the liquid passing over the surfaces. The effects of the latter only are to be included here; the effects of dilution rate will be considered in later parts of this chapter. English writers ascribe the control of filter film to the grazing macro-invertebrates (Reynoldson, 1939; Lloyd, 1945; Tomlinson, 1946), but many American workers consider the scouring action of the liquid and the microbiological action to be of greater importance (Lackey, 1926; Holtje, 1943; Heukelekian, 1945; Cooke and Hirsch, 1958). Hawkes (1963) suggested that the difference of opinion might be explained by the greater use, in America, of high-rate filters, with attendant greater scouring action of the liquid than in Britain. It is therefore possible that current could indirectly affect the protozoan populations by sloughing off the microbial film from the filters. To the author's knowledge the only work carried out to assess directly the effect of current upon protozoa was reported by Zimmerman (1961). He studied the effects of three velocities of flow, on ciliated protozoa, of various dilutions of sewage in a concrete channel over several months. The only protozoa regularly observed, along with many macro-invertebrates, were *Vorticella* sp. and *Carchesium polypinum*. *Vorticella* was found to grow best in rapidly flowing sewage (15–25 cm/sec) whereas *Carchesium* preferred the lower flow rates (5–15 cm/sec).

(g) *Toxic wastes*. Trade wastes are often combined with domestic sewage, and some of these wastes are toxic. Although there is a considerable quantity of literature concerning the inhibition of various used-water treatment processes by industrial wastes, there is virtually no work at all on their action or effects upon protozoa. It is not possible to state accurately the concentrations at which various substances seriously inhibit the overall effectiveness of a biological-treatment plant, since toxicity is, in many cases, dependent upon other environmental factors such as pH, dissolved oxygen, other toxic substances and temperature etc. Certain wastes are biodegradable, e.g. phenol, while others such as heavy metals are not. Jones (unpublished data) found that the ciliate *Tetrahymena pyriformis* would grow in a small-scale activated-sludge plant treating a flow of medium containing 500 mg/l of phenol; it survived because the bacterium present reduced the phenol concentration to below levels of detection (below 1 mg/l). If a toxic waste is biodegradable it is theoretically advisable to use a completely-mixed reactor which enables considerable dilution of the inflowing toxic material.

In the case of heavy metals, Curds (unpublished data) found that copper and nickel lowered the growth rate of *Tetrahymena pyriformis* at concentrations of below 1 mg/l, and observed them to be lethal at concentrations above 5 mg/l in pure culture. However, when these elements were added to samples of activated sludge, even in concentrations of 50 mg/l, no effect was detectable on the protozoan populations. It is clear, therefore, that data obtained from pure cultures are of little practical value, since activated sludge appears to have considerable powers of chelation or adsorption.

Recently, Cairns and Dickson (1970) investigated the effects of a shock exposure of zinc and copper on free-living protozoa contained in experimental troughs with a continuous supply of unfiltered lake water. In that experiment, the most dramatic effect of 24-h exposure to 24 mg/l of zinc or copper was a great reduction in the numbers of species present. Results of one-h exposure to 24 mg/l of zinc had a negligible effect upon the number of species present.

Schofield (1971) operated two small-scale plants, one of which was receiving 0·5 mg/l cyanide in the sewage. The protozoan population in the plant treating cyanide was reduced in numbers, although the species structure did not change. Later, another batch of sewage also containing 0·5 mg/l cyanide completely killed the protozoan population. This illustrates the complexity of toxic reactions.

Cairns *et al.* (1971) studied the response of protozoa to one of the "enzyme-active" detergents which have been introduced in recent years. The studies were in two parts. First the "time until death" curve of *Paramecium caudatum* was measured; in these tests the ciliate was exposed to concentrations of 30–100 mg/l of the detergent-enzyme Axion. At 30 mg/l a 100% kill was observed in 200 h, whereas at 100 mg/l the same kill was noted in 20–30 min. The second part of the studies involved the exposure of freshwater protozoan communities to selected (20–200 mg/l) concentrations of the detergent-enzyme for 3 h in a continuous flow of lake water. When 56 mg/l detergent was used there was a 35% reduction in the number of species present noted in 5 h, whereas a 78% reduction was obtained at a detergent concentration of 200 mg/l. In both cases the original number of species returned within 150 h.

Curds and Cockburn (1970a) found that all the percolating filters they surveyed contained ciliated protozoa, and found no evidence to suggest that the inclusion of industrial wastes in the sewages had any harmful effect upon the protozoan populations, even though substances, including heavy metals, known to be toxic to protozoa were present in some cases. In fact they found that the average number of ciliate species at works treating sewage containing industrial waste in excess of 10% of the flow tended to be greater than in plants treating sewage containing less than 10% industrial waste. In contrast, Curds and Cockburn (1970a) found three activated-sludge plants which did not contain ciliated protozoa; two contained flagellates alone whilst the other contained no protozoa of any description. It seems likely that several factors were involved; in the case of the plant operating without protozoa, a small-scale plant treating the same sewage contained a variety of protozoan forms. The major difference, other than size, was the retention time of the sewage; in the full-scale plant it was 3 h, but 24 h in the pilot plant. It is known that protozoa can survive in plants operating at retention times of below 3 h, so it is possible that both inhibition by a toxic waste and retention time were involved. The two plants containing flagellates were also treating trade wastes at relatively short retention times.

2. Positive Factors

(*a*) *Nutrition.* Settled sewage is often described as a weak solution of organic and inorganic materials; it is weak in comparison with conventional microbiological culture media, but this is a misleading concept, since it is very rich in comparison with unpolluted natural waters. Painter and Viney (1959) analysed many of the components of sewage. In addition to soluble components, sewage has considerable quantities of colloidal materials and bacteria suspended in it. The bacterial populations of sewage have been studied by Harkness (1966), and a reasonable estimate for the dry weight concentration of bacteria in sewage may be calculated to be about 30 mg/l.

The continual flow of sewage with its wide and relatively concentrated variety of dissolved and suspended substances is one of the primary positive factors which determines why used-water treatment processes contain such large populations of organisms. The soluble organic compounds provide a constant food supply for the saprozoic organisms and the suspended bacterial populations of the sewage supply the holozoic protozoa. Noland (1925) and Sandon (1932) regard food as the most important ecological factor which determines the abundance and distribution of protozoa in nature.

The phylum Protozoa includes a very wide range of nutritional types and the following four types are represented by the used-water treatment protozoa.

(i) *Autotrophs.* These are all members of the class Phytomastigophorea and their role in ecology is as primary producers capable of absorbing sunlight and fixing carbon dioxide; the resultant organic compounds are stored and these become available to herbivores. In addition to light they require a number of inorganic ions and trace elements, all of which are present in sewage. Hutner *et al.* (1949) have shown that *Euglena gracilis* requires vitamin B_{12} (cobalamin); this could be limiting in natural waters but is plentiful in sewage. Light is the most likely limiting factor for the autotrophs.

(ii) *Saprozoic forms.* A number of flagellates in the class Zoomastigophorea neither contain chlorophyll nor require light, but rely upon the dissolved organic complexes of sewage. Many of these types, belonging to such genera as *Polytoma, Chilomonas* and *Polytomella*, require vitamins such as thiamin, and it is well known that many bacteria are able to produce vitamins. The saprozoic protozoa compete with the heterotrophic bacteria for the organic complexes in sewage, but for reasons to be given in the mathematical section of this chapter the flocculating bacteria become dominant.

(iii) *Phagotrophs.* The nutritional demands of the phagotrophic protozoa are principally supplied by bacteria but are highly complex, and are known completely only in a few species (Kidder and Dewey, 1951; Wagtendonk, 1955). There are numerous reports which show that not all bacteria serve as a complete diet when supplied alone; Hetherington (1933), Burbanck (1942), Curds and Vandyke (1966) and Leslie (1940) have all demonstrated this to be true for a variety of freshwater ciliates, and Singh (1941, 1945) has shown it to be so in the case of soil amoebae and ciliates. Curds and Vandyke (1966) presented 19 strains of bacteria to five species of ciliate which had been isolated from activated sludge; they found that whereas certain strains of bacteria would support the growth of ciliates for long periods, others proved to be inadequate, and after a short time the ciliates showed signs of starvation. Furthermore, the results demonstrated that although one bacterial strain was favourable for one ciliate species, it was not necessarily favourable for another ciliate species. Bick (1958) has presented ecological evidence which suggests that anaerobic ciliates such as *Metopus* sp. and *Caenomorpha* sp. live mainly on sulphur bacteria. A number of workers, including Burbanck (1942), Curds and Vandyke (1966), Groscop and Brent (1964), Kidder and Stuart (1939), Singh (1941, 1945) and Strong (1944), have shown that certain bacterial strains are toxic to amoebae and ciliates; generally the pigments from chromogenic bacteria have been shown to be the toxic agents. One favourable bacterium may be more favourable than another to a ciliate, since Burbanck (1942), Curds and Vandyke (1966) and Hetherington (1933) have all shown that different bacterial strains promote different maximum growth rates in a single ciliate species. Hardin (1944) made the interesting observation that *Oicomonas termo* would grow on 83% of the bacteria that could be isolated from its original culture fluid, but not on the remaining 17%. It is unfortunate that all of the above work has been confined to the presenting of a single bacterial strain to protozoa, whereas in nature a mixed population is commonly available. It is possible, for example, that two bacteria, inadequate as diets when presented alone, might prove to be ideal when supplied together.

In recent years a number of advances have been made concerning the quantitative aspects of protozoa feeding on bacteria. Cutler and Crump (1924) were among the first to show that the division rate of a ciliate is dependent upon the concentration of its available bacterial food supply, although later Harding (1937) could find no evidence to suggest that the fission rate of *Tetrahymena pyriformis* was influenced by concentrations of bacteria in the range of 0·6–7·0 million bacteria/ml.

More recent work of Proper and Garver (1966), who studied the feeding and growth of the ciliate *Colpoda steinii* on the bacterium *Escherichia coli*, suggested that the growth rate of this ciliate was directly dependent upon the concentration of bacteria, and the relation between the specific growth rate, μ, of the ciliate and the bacterial concentration, b, could be described adequately by the Michaelis–Menten kinetic equation

$$\mu = \mu_m b/(K_b + b) \tag{1}$$

where μ_m is the maximum specific growth rate and K_b the saturation constant which is numerically equal to the bacterial concentration when $\mu = \mu_m/2$.

Later, Curds and Cockburn (1968) studied the growth and feeding rates of the ciliate *Tetrahymena pyriformis* feeding on the bacterium *Klebsiella aerogenes*, and found that in batch culture the Michaelis–Menten had to be amended to take account of a population factor, so that

$$\mu = \mu_m b/(Cn + b) \tag{2}$$

where C "is a growth parameter which is constant under defined conditions" (Contois, 1959) and n is the number of ciliates present. Curds and Cockburn (1971) later found that this equation did not apply in continuous culture. The self-regulating mechanisms of continuous cultures of the "chemostat" type ensure that the population is in dynamic equilibrium, and it is not possible to reach a steady-state condition at which the protozoan population is too great for the bacterial food supply. Although Curds and Cockburn (1971) found that the mass of the individual cell had to be included in the equation to obtain a precise relationship between the specific growth rate of the organisms and the concentration of available bacteria, Curds (1971 b,c) found that the normal Michaelis–Menten equation (Eqn 1) was usually an adequate description. This conclusion was also reached by Hamilton and Preslan (1969, 1970), who worked on the marine ciliate *Uronema* sp., and by Proper and Garver (1968). Similar quantitative batch experiments have been carried out using the soil amoeba *Acanthamoeba* sp. feeding on the yeast *Saccharomyces cerevisiae.*

The yield constant for protozoa feeding upon bacteria may be defined as

$$Y = \frac{\text{weight of protozoa formed}}{\text{weight of bacteria ingested}} \tag{3}$$

and the values of those estimated for protozoa are remarkably high for animal cells where yields of about 0·1 are typical for invertebrates. Curds and Cockburn (1968, 1971) have reported values of 0·5–0·54 for the ciliate *Tetrahymena pyriformis;* Coleman (1964) estimated the yield of the anaerobic ciliate *Entodinium caudatum* to be 0·5; and Heal (1967a) found the yield of *Acanthamoeba* sp. to be 0·37 when fed upon yeast. Proper and Garver (1966) gave a value of 0·78, which is particularly high. Yields of 0·37–0·54 are fairly common for aerobic bacteria (Hadjipetrou *et al.*, 1964), and it was suggested by Heal (1967b) that the difference between the micro-organisms and the metazoan invertebrates is that the micro-organisms do not possess a stage in their life-cycle equivalent to the adult metazoan; they generally grow to a certain size, divide, and proceed to grow again, whereas macro-invertebrates generally pass through a series of larval stages before becoming adults. The adult invertebrates only grow in the sense that tissues such as reproductive cells, skin etc. are produced.

(iv) *Predators.* Sewage-treatment processes often contain predatory ciliated protozoa, belonging to such genera as *Hemiophrys, Amphileptus, Litonotus* and *Trachelophyllum,* which generally prey upon sessile peritrichs. Suctoria are also found and these feed upon other protozoa. Lilly (1953) found that the hypotrichs *Stylonychia pustulata* and *Euplotes patella* and the suctorian *Tokophrya infusionum* were not able to grow on the ciliate *Tetrahymena pyriformis* alone but required yeast extract in order to divide. Apart from the true carnivorous ciliates it is well known that many can become cannibals, for example the genera *Blepharisma* (Dawson, 1929; Giese, 1938), *Stentor* (Gelei, 1925; Tartar, 1961) and *Stylonychia* (Giese and Alden, 1938) all have species which are known to behave as cannibals. The present author has noted *Trachelophyllum pusillum* attacking its own species in an activated-sludge sample.

Predation is not limited to ciliate eating ciliate; many of the hypotrichs ingest flagellates, and *Stentor* has even been reported (Schaeffer, 1910) to be able to distinguish between two species of the alga *Phacus.* Little work has been carried out on the quantitative and kinetic aspects of protozoan predation, although in recent years Salt (1967, 1968, 1969) has studied amoebae feeding upon ciliates and ciliates upon ciliates.

(*b*) *Source of organisms.* If a habitat is continually inoculated with a supply of organisms from outside, provided the habitat is not immediately toxic then these organisms will have an advantage over others not supplied from the outside. There are two obvious sources of inoculation, the sewage and the atmosphere. Organisms may be introduced into sewage with the faeces; Alexeieff (1917, 1926), Dobell and O'Conner (1921) and Hoare (1927) have all noted the presence of various protozoa in faeces. Others may be picked up during its passage to the sewage works. Land drainage and run-off presumably provide a rich source of soil protozoa, particularly after heavy rain (Brown, 1965; Gray, 1952). However, there is no quantitative information available on the numbers and species of protozoa borne in sewage.

Many protozoa are able to encyst, and these remain potentially viable for many years; *Oicomonas termo* (Noc, 1914), *Naegleria* sp. and *Tetramitus* sp. (Noland and Gojdics, 1967), *Colpoda cucullus* (Bodine, 1923; Dawson and Hewitt, 1931) and *Oxytricha* sp. and *Spathidium* sp. (Dawson and Mitchell, 1929) are just a few examples of protozoa which are known to be able to survive in the encysted state for many years. Cysts may be transmitted in water, in the atmosphere or even on the appendages of birds and flying insects (Maguire, 1959, 1963).

(*c*) *Operation of plant.* All used-water treatment processes depend upon the intimate contact of sufficient organisms with the waste for a sufficient length of time. In aerobic processes, it is desirable to keep oxygen in excess, and this is frequently the case, since only 1–2 mg/l dissolved oxygen is considered to be sufficient to be not limiting. Sewage retention time is probably the most important variable under the control of the operator. In percolating filters and activated-sludge plants the retention time of sewage is determined for a given size of plant by the sewage flow rate. In percolating filters the concentration of organisms is beyond the direct control of the operator, provided that sufficient surfaces are present for the attachment of the organisms, whereas it is possible to select predetermined concentrations of activated sludge which the operator knows by past experience will produce an effluent of the required standard. In order to keep the solids at the desired level the controller wastes sludge at a particular rate, and it should be emphasized that the specific growth rate of the sludge is the primary factor which determines effluent quality and plays an important part in the control of population size and species structure. This will be developed more fully in the mathematical part of this chapter (Section V).

A number of workers have suggested that the protozoa found in aerobic used-water treatment processes may be conveniently divided into three groups by their habit, but Curds (1971a, 1973a) was the first to demonstrate the significance of these groups. The groups suggested are those which swim freely in the liquid phase of the mixed liquor or filter film, those which crawl over the surfaces of sludge and film, and those which attach themselves directly to the sludge or film. One can immediately recognize that an organism which is able to attach itself to a substratum has a great advantage over an organism which remains suspended in the liquid phase where it is more susceptible to washout in the effluent flow. An organism which settles upon the floc in an activated-sludge plant will only be able to leave the system when some of the sludge is wasted, and those attached to filter film will only be removed when the film sloughs off the filter medium. Crawling ciliates are intermediary forms, and these may be considered to be inferior settling forms which on occasion are washed out in the effluent. It will be shown later that the habit of an organism is a key factor for its survival in the activated-sludge process.

3. *Biological Interactions*

Bungay and Bungay (1968) list seven different known types of microbial interactions, and some of these are likely to exist in the protozoan populations of used-water treatment processes.

(*a*) *Competition.* Bungay and Bungay (1968) defined competition as "a race for nutrients and space". Many of the protozoa in used-water treatment processes compete with one another for bacteria as a food supply, while others compete with bacteria and other saprozoic protozoas for soluble substrates, and this has already been mentioned under the earlier heading "Nutrition" (Section III A2(a)).

The fate of two micro-organisms competing with each other for the same substrate in a continuous-culture plant has been treated theoretically by Powell (1958) and Moser (1958); from their work it is clear that provided the newly introduced organism is not washed out before its first few divisions, then the organism which is able to grow the faster under the prevailing conditions (i.e. the organism which has the highest maximum specific growth rate and the lowest saturation constant) will become dominant. However, if this were strictly true, then used-water treatment plants, for example, would theoretically contain cultures of few organisms. Bungay and Bungay (1968) point out in their paper that other factors can override these theoretical considerations, and they give the example, highly pertinent to treatment processes, that organisms capable of attaching themselves to the walls of the culture vessel survive and repeatedly re-inoculate the culture medium.

(*b*) *Predation and stimulation.* Although these two interactions are normally regarded as distinct, they have over some years become inseparable when considering economic protozoology. Several investigators have noted that the purification of sewage, as measured by the rate of oxygen consumption accompanying its degradation, occurs more rapidly and proceeds further in the presence of ciliates and bacteria than when bacteria alone are present (Butterfield *et al.*, 1931; Javornický and Prokesova, 1963). Furthermore, Nasir (1923), Cutler and Bal (1926) and Hervey and Greaves (1941) have stated that rhizopods, flagellates and ciliates are all, to varying degrees, capable of stimulating the fixation of atmospheric nitrogen by *Azotobacter chroococcum*. Similarly Meiklejohn (1932) found that *Colpidium colpoda* could increase ammonia production by soil bacteria. All of these authors have stated that the increased activity noted when protozoa were present was achieved indirectly by their predatory activities, and Johannes (1965) has summarized the effects of predation: "Grazing ciliates prevent bacteria from reaching self-limiting numbers; the bacterial populations are thus kept in a prolonged state of physiological youth, and their rate of assimilation of organic materials is greatly increased." It is evident that these authors believe that simple predation stimulates bacterial activity. Such a possibility is conceivable if the substrate being measured is not the limiting factor—in this case some other factor such as dissolved oxygen might be limiting; under these circumstances the reduction of the bacterial population by predation would be beneficial, and this could explain some of the observations. However, if it is assumed that the nutrient being measured is the limiting factor, then it is quite clear (Curds, 1971c) that the reduction of the bacterial population by predation will, in continuous culture, increase the concentration of substrate, and it is this increase in substrate concentration that raises the specific growth rate of the bacteria (see Eqn 1). However, in a continuous culture more substrate will emerge in the effluent when predators are present, and in batch culture it will take a longer time to reach the same end-point for substrate when predators are present than when bacteria are grown alone.

Another explanation for these phenomena has emerged more recently. Nikoljuk (1969) confirmed that amoebae and ciliates were capable of increasing the nitrogen fixation capabilities of *Azotobacter chroococcum* and stated that this was due to a secretion by the protozoa. The substance secreted was identified to be 3-indolylacetic acid, a heteroauxine. Nikoljuk states that the positive action of soil protozoa upon the bacterial flora, on the development of plants and on soil processes in general, may be accounted for by this capability to produce powerful activators such as heteroauxine. There are no indications as yet whether or not such activities take place in used-water treatment processes. Lilly *et al.* (1960) noted that certain ciliates could stimulate the growth rates of each other. They found that the growth rates of both *Paramecium caudatum* and *Stylonychia pustulata* were greater when grown together in axenic culture than when grown apart.

(*c*) *Self-inhibition.* This is a self-inhibiting interaction which allows other organisms to take the place of the initial organisms. Examples are well known in protozoan batch cultures but not specifically in used-water treatment processes. Woodruff (1913) showed that certain waste products excreted by ciliates were self-inhibitory but had little direct effect upon other species; he suggested that waste products of this nature might help to explain the succession of protozoa commonly found in laboratory hay infusions. Darby (1929) suggested later that these successions might result from changes in pH of the medium brought about by the primary organisms which favoured secondary ones more tolerant to pH. Used-water treatment processes are continuous, not batch, processes and for this reason Curds (1963a) suggested that this type of interaction was unlikely to be of significance in the activated-sludge process. It is possible, however, that such interactions could play a role in the vertical distribution of protozoa in a percolating filter.

B. Spatial Distribution, Successions and Indicator Organisms

It is evident from the literature that spatial distribution, successions and the use of protozoa as indicator organisms are all closely linked and associated with the degree of saprobity of the immediate environment. For this reason the subject of saprobity will be briefly mentioned here.

The saprobic system is a method (or group of methods) proposed for the evaluation of the overall degree of organic pollution on a single scale, using the presence or absence of biological species or communities as the parameters. Many such schemes have been introduced since Kolkwitz and Marsson (1908, 1909) first suggested that rivers or sections of rivers could be classified according to their degree of organic pollution into the four grades; the polysaprobic or highly polluted zone, the α and β mesosaprobic zones and the relatively clean or unpolluted oligosaprobic zone. A full account of the modern methods of zone description and the measurement of saprobity are given by Sladecek (1969). Numerous authors, including Kolkwitz (1950), Lackey (1938a, 1941) and Mohr (1952), have investigated the presence of micro-organisms in rivers with special reference to their degree of organic pollution, and Liebmann (1951) has described a large number of species as being characteristic of a particular degree of sewage pollution. Protozoa vary considerably in their abilities to withstand the effects of pollution (Bick, 1957; Ganczarcyzk and Domanski, 1953; Šrámek-Hušek, 1954, 1956; Turoboyski, 1953), and this influences their distribution.

1. Distribution within Percolating Filters

It is well known that different organisms tend to predominate at certain levels in a percolating filter, e.g. Frye and Becker (1929) noted that metazoa seemed to favour the lower levels while protozoa favoured the upper ones. Holtje (1943) also referred to protozoa as being organisms typically found in the upper regions of a filter. Johnson (1914) was probably one of the first to note the transition of certain protozoan species throughout the depth of a filter; he found a restricted protozoan fauna at the surface which changed further down to a community of greater variety, and finally in the lower regions the community was dominated by *Carchesium polypinum*. Since that time other workers including Barker (1946), Cutler *et al.* (1932), Frye and Becker (1929), Ingram and Edwards (1960) and Lackey (1924, 1925) have reported upon the distribution of protozoa throughout the depth of percolating filters. Brink (1967), Frye and Becker (1929), Hausman (1923) and Crozier and Harris (1923a, b) have all noted that amoebae are particularly prevalent in the surface layers of filters; flagellates are also said to be found in those regions, but Frye and Becker (1929) found *Bodo* sp. throughout all depths. These latter authors list the genera found at one-foot intervals throughout the depth of their experimental filters.

The stratification of organisms in a filter is, of course, dependent upon a great many environmental factors and interactions between organisms from many phyla. Liefmann (1951) stated that nutrition is responsible for the holozoic protozoa being found mainly in the upper layers while the carnivorous forms are usually found further down. Those organisms requiring light will of course be limited to the extreme upper layers. Probably the most common explanation for the vertical distribution of protozoa and other organisms in filters is the association of the organisms with the different levels of saprobity found in that habitat. Gradual purification of the sewage takes place during its passage through the filter so that polysaprobic conditions occur at the top, followed by mesosaprobic zones of diminishing intensities occurring towards the base (Yasuko, 1960). Lymaxenes within the filter vary in their degree of tolerance to sewage (these have been listed by Liebmann, 1949) and their zonation in bacteria beds related to their position in the saprobic system. Liebmann (1949) defined the following protozoa as being polysaprobic species and therefore prevalent in the surface layers: *Trepomonas agilis, Bodo putrinus, Trimyema compressa, Enchelys vermicularis, Colpidium colpoda, Glaucoma scintillans* and *Vorticella microstoma.* He designated the following species as being α mesosaprobic, which would be found in the middle regions of a filter: *Urostyla weissei, Oxytricha fallax, Paramecium caudatum, Chilodonella uncinata, C. cucullulus, Uronema marinum, Amphileptus claparedei, Litonotus fasciola, Opercularia coarctata* and *Podophrya fixa.* Liebmann (1949) finally lists the following as being β mesosaprobic, and these would be expected to inhabit the lower regions of the filter: *Amoeba verrucosa, Arcella vulgaris, Aspidisca costata* and *Euplotes charon.* It should be noted, however, that the position of an organism in a filter is purely relative, since changing conditions in flow or sewage strength can change the precise depth of a particular saprobic zone. For example, an increase in sewage strength usually results in a downward shift in the saprobic grades, i.e. the polysaprobic and α mesosaprobic zones would extend down further and it could result in the extinction of the β mesosaprobic zone. The change in zone positions would then be followed by the pertinent animals (Johnson, 1914; Liebmann, 1949).

In the opinion of the present author the application of saprobic systems to percolating filter distributions is unlikely to supply the reasons why certain organisms are found at particular levels. Only detailed research, initially on specific organisms and later upon communities of organisms, is likely to supply the answers.

2. *Activated-sludge Successions and Indicator Species*

Although the protozoa of percolating filters and activated sludge are similar, the spatial zoning of organisms, characteristic of a filter, is not found in activated-sludge plants owing to the continual mixing and recirculation of the sludge (Hawkes, 1963). In nature, during the self-purification of streams and in the development of mixed populations in sewage or hay infusions, successive types of protozoan populations are dominant (Barritt, 1940; Bick, 1958, 1960; Rudolfs, *et al.* 1926; Woodruff, 1912). In the activated-sludge process there is also a similar marked temporal-zonation or succession of protozoan types during the development and establishment of a mature activated sludge.

Buswell and Long (1923) were possibly the first authors to suggest that there might be a distinct succession of protozoa; they recorded that holotrichous ciliates were the first to appear during the establishment of a sludge but these, in time, gave way to other forms, particularly the peritrich ciliates *Carchesium* sp. and *Vorticella* sp. which appeared after several days' aeration. Kolkwitz (1926) observed a retrogression from hypotrichs to peritrichs in an "over-ripe" sludge. Agersborg and Hatfield (1929) noted that aeration favoured a succession of flagellates, hypotrichs and peritrichs. Similarly, Horosawa (1950) traced three stages in the development of a sludge during which time the flagellates decreased and the ciliates increased until the mature activated-sludge types (i.e. peritrichs) predominated. McKinney and Gram (1956) suggested that flagellates predominate in the system only at the early stages because of their lower energy requirements, and they are succeeded by ciliates since the latter are able to obtain their food more rapidly.

Figure 9 shows the succession of protozoa which Curds (1966) observed in a small-scale pilot plant. This illustrates a typical succession; flagellates were the first dominant protozoan types and included such forms as *Oicomonas socialis, Heteronema acus, Peranema trichophorum* and *Bodo* sp., which were also noted in the sewage reservoir. As the flagellates decreased they were slowly replaced by free-swimming ciliates such as *Uronema nigricans* and *Paramecium caudatum.* The free-swimming forms reached a peak after 20 days but were replaced by crawling hypotrichs such as *Aspidisca costata* and *Aspidisca lynceus* and finally by attached peritrichs such as *Vorticella nebulifera* var. *similis, Epistylis rotans* and *Vorticella campanula.* Both Curds (1966) and Brown (1965) noted an inverse relation between the peritrich and *Aspidisca* spp. populations in a mature sludge. Curds (1966) listed the ciliate species in order of appearance in the developing activated sludge, together with their positions in the saprobic systems of Kolkwitz and Marsson (1908, 1909) Kolkwitz (1950) and Liebmann (1951); Curds (1966) found that the lists suggested that there was a tendency for polysaprobic species to occur in the early stages of development and for the β meso- and β oligosaprobic ciliate species to occur in the later stages when the sludge had matured. In the opinion of Curds (1966) nutrition was a major factor in promoting

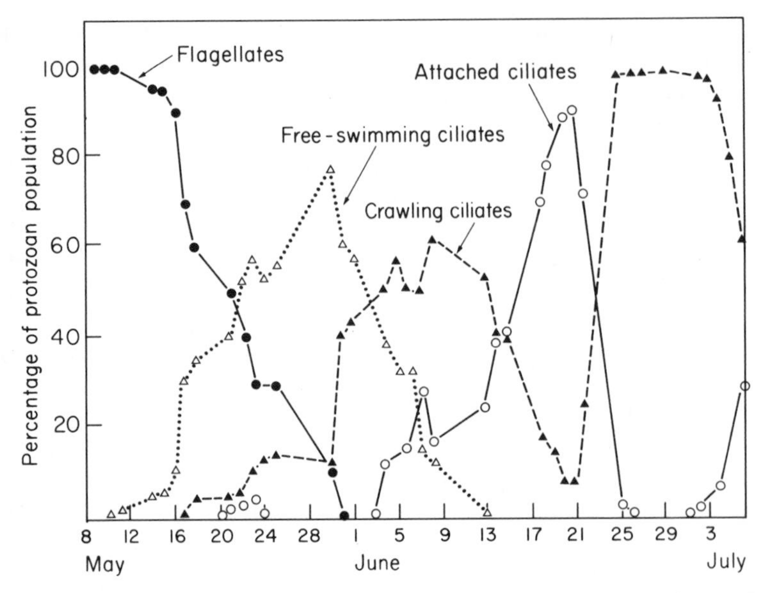

FIG. 9. Successions of protozoa in an activated-sludge plant (after Curds, 1966).

successions, but more recently Curds (1971c, 1973b) has been able to simulate, on a digital computer, successions similar to those found in practice simply from theoretical considerations of the growth kinetics, nutrition and settling properties of the organisms and upon the flow rates of an activated-sludge plant. These model successions will be described more fully later in this chapter.

During the development and succession of the various microbial populations of an activated sludge, the quality of the effluent improves, and it is this link between effluent quality and the species of protozoa present that has been responsible for the suggestion by various authors to use protozoa as indicators of activated-sludge effluent quality. Ardern and Lockett (1928, 1936) and Agersborg and Hatfield (1929) were the first to describe protozoan communities which they stated to be typical of four grades of activated-sludge efficiency. A sludge in a bad condition was noted to contain a preponderance of flagellates and rhizopods with few ciliates; a similar fauna was found in sludges described as being in an unsatisfactory condition, except that rather more ciliates were present. In a satisfactory sludge such ciliates as *Chilodonella* sp., *Colpoda* sp., *Colpidium* sp., *Aspidisca* sp. and some *Carchesium* sp. and *Vorticella* spp. were found; whereas in a good sludge where nitrification was well established there were very few flagellates and amoebae, and ciliates such as *Carchesium* sp., *Vorticella* spp., *Aspidisca* sp., *Loxophyllum* sp. and *Chaenea* sp. were dominant.

Dixon (1937) mentioned that the protozoa present could indicate the stage of purification likely to be achieved by the sludge, and Szabo (1960) suggested that the degree of purification could be predicted by an analysis of the relative abundance of protozoa belonging to the different grades of the saprobic system. Phelps (1944) observed that the changes in effluent quality were associated with the species of organisms present, and that large numbers of *Colpoda* sp., *Colpidium* sp., *Tetrahymena pyriformis* and *Metopus* sp. were associated with poor quality effluents. Reynoldson (1942) and Baines *et al.* (1953) all considered that the presence of peritrichous ciliates to the exclusion of other groups of protozoa indicated that the sludge was in prime condition. Furthermore Reynoldson (1942) and Pillai (1941) both stated that a negative correlation could be observed between the BOD of the effluent and the size of the vorticellid population in the sludge. However, Baines *et al.* (1953) found that whereas an inverse correlation could be detected for short periods, it did not apply over the entire year, and they concluded that the nature of the protozoan community was more important than the numbers of individual organisms. Hawkes (1963) pointed out the dangers of using biological indicators in the activated-sludge process, and he stated that the protozoan population as a whole should be taken into account rather than the numbers of individual species or types. It will be shown later in the mathematical approach to ecology (Section V) that provided there are no inhibitors present, then the numbers of organisms are to some extent irrelevant, whereas the habit of the organisms present is certainly of theoretical importance at least.

The survey of Curds and Cockburn (1970a) suggested that there could be a correlation between the species structure of an activated sludge and the quality of the effluent (Curds and Cockburn, 1970b). In order to test this hypothesis the effluents from the plants investigated were divided into the four following categories according to their BOD: very high quality (BOD range 0–10 mg/l), high quality (11–20 mg/l), inferior quality (21–30 mg/l) and low quality (above 30 mg/l). Initially the frequencies of each species occurring in plants delivering effluents within each of the above four categories were calculated on a percentage basis. Many species were associated with all categories of effluent, but there was generally a tendency for a given species to occur more frequently in plants delivering effluents within a particular quality category (Table IV): this indicated that the protozoan species found within the mixed liquor were in some way associated with the quality of the effluent. It is evident from Table IV that *Carchesium polypinum*, for example, was found principally in plants

TABLE IV. Percentage frequency of occurrence of some protozoa in plants producing effluents within four ranges of BOD (Curds and Cockburn, 1970b)

BOD range (mg/l)	Frequency of occurrence (%) and points awarded (in brackets)			
	0–10	11–20	21–30	>30
Vorticella convallaria	63 (3)	73 (4)	37 (2)	22 (1)
Vorticella fromenteli	38 (5)	33 (4)	12 (1)	0 (0)
Carchesium polypinum	19 (3)	47 (5)	12 (2)	0 (0)
Aspidisca costata	75 (3)	80 (3)	50 (2)	56 (2)
Euplotes patella	38 (4)	25 (3)	24 (3)	0 (0)
Flagellated protozoa	0 (0)	0 (0)	37 (4)	45 (6)

which produced good quality effluent, whereas flagellated protozoa were restricted to plants producing inferior effluents. An arbitrary total of 10 points was awarded to each species, and these were distributed between the four effluent categories so that the greatest number of points was given to the effluent category with which that species was most frequently found associated; for example, a ciliate species occurring with frequencies of 60, 80, 40 and 20%, in plants delivering effluents in the four categories, would be awarded the ten points in the ratio of 6:8:4:2, that is, 3, 4, 2 and 1 in the respective categories. Table IV lists the frequencies of some species in plants delivering effluent in the four categories of quality and illustrates the distribution of points for those species.

In order to predict the effluent quality of a particular plant the species of protozoa in the mixed liquor were identified and listed against the four effluent-quality categories (Table V). Using the comprehensive version of Table IV given by Curds and Cockburn (1970b), the appropriate number of points was awarded to each of the effluent categories for each species. The total number of points for each effluent category was then calculated and that category which gained the highest total number of points was the predicted effluent-quality category. An example is given in Table V, and the data was originally obtained from a full-scale activated-sludge plant. The method of Curds and Cockburn (1970b) was shown to be 85% correct on the original data and 83% correct when tested on a further 34 sites not included in the original survey.

Morishita (1969a, b, 1970) states that the protozoan populations of activated-sludge plants can be used to indicate (i) the type of sewerage system in use (e.g. combined or separate systems etc.), (ii) the BOD of the sewage, (iii) the BOD and COD of the effluent, (iv) the percentage removal of BOD and (v) the BOD loading of the sludge. (The reader is

TABLE V. Example of the use of association ratings to predict the quality of effluent from an activated-sludge plant (Curds and Cockburn, 1970b)

Protozoa in sludge	Effluent BOD ranges 0–10	11–20	21–30	>30
Trachelophyllum pusillum	3	3	3	1
Hemiophrys fusidens	3	4	3	0
Chilodonella cucullulus	4	4	1	1
Paramecium trichium	4	3	2	1
Vorticella communis	10	0	0	0
Vorticella convallaria	3	4	2	1
Vorticella fromenteli	5	4	1	0
Vorticella microstoma	2	4	2	2
Opercularia coarctata	2	2	3	2
Carchesium polypinum	3	5	2	0
Zoothamnium mucedo	10	0	0	0
Aspidisca costata	3	3	2	2
Euplotes moebiusi	3	3	3	1
Euplotes affinis	6	4	0	0
Euplotes patella	4	3	3	0
Total points	65	46	28	11

Highest total points awarded to effluent BOD range 0–10 mg/l. Measured BOD of effluent 8 mg/l.

recommended to consult the works of Morishita (1969a, b, 1970), since both papers are difficult to understand fully because of obvious translation difficulties.)

Many authors have pointed out the dangers of biological indicator systems, and Curds and Cockburn (1970b) state that all should be regarded with caution, since they oversimplify extremely complex ecological situations. However, perhaps in the future there will be sufficient fundamental information available to make it possible for us to interpret adequately the significance of biological communities.

C. Seasonal Incidence

The only paper which deals with the seasonal incidence of protozoa in used-water treatment processes, to the knowledge of the author, is that of Barker (1942). This paper deals exclusively with the protozoa found in a percolating filter for a period of one year, and in the present author's opinion such a short period is not sufficient to warrant conclusions concerning seasonal incidence. However, bearing this major criticism in mind, Barker (1942) made the following observations. In general, the protozoan population in percolating filters was more constant and abundant in spring and summer months, whereas it was erratic and there were fewer numbers in the autumn and winter. Certain species including *Amoeba limax*, *Vahlkampfia* sp., *Cochlipodium* sp., *Monas* sp., *Oicomonas* sp., *Pleuromonas* sp. and *Carchesium* sp. were represented throughout the year, but others like *Bodo* sp., *Cercobodo* sp., *Anisonema* sp., *Chilodonella* sp. and *Urotricha* sp., were found mainly during the winter months. Yet other species such as *Amoeba proteus*, *Arcella* sp., *Trepomonas* sp., *Cyclidium* sp., *Cinetochilum* sp. and *Podophrya* sp. were found during the spring months and some of these also exhibited secondary maxima during the autumn. Species of the genera *Cyathomonas*, *Litonotus* and *Uronema* were generally limited to the summer whilst the genera *Trichoda*, *Oxytricha* and *Euglypha* were in evidence in autumn.

D. Geographical Distribution

Although a specific study has not yet been made on the geographical distribution of protozoa in used-water treatment processes, the evidence which is available in the form of lists of protozoan species found in a variety of countries in different geographical locations, e.g. America, Europe, India and Japan, seems to indicate that the same species occur all over the world. This is not surprising since many species are able to encyst and are easily transmitted in the atmosphere or by birds etc. It is widely accepted that most free-living protozoan species have a world-wide distribution.

IV. The Role of Protozoa in Used-Water Treatment Processes

Whereas there is a large amount of literature pertaining to the role and importance of protozoa in the activated-sludge process, little work has been reported on the role of these organisms in percolating filters and, to the author's knowledge, none on anaerobic processes. There is, however, great similarity between the organisms found in the two aerobic habitats, and it is reasonable to suppose that they play a similar qualitative and quantitative part in both aerobic processes.

Protozoa were originally thought to be harmful to the activated-sludge process (Fairbrother and Renshaw, 1922). Most other authors, however, have since proposed that protozoa are of some benefit to the processes for a variety of reasons and to a variable extent. Pillai and Subrahmanyan (1942, 1944), for example, expressed the view that bacteria were of secondary importance in the purification processes since they claimed that the colonial peritrich *Epistylis* sp., even in the absence of bacteria, could account for a 70% reduction in the permanganate value, the albuminoid and ammoniacal nitrogen concentrations of sewage, and that without protozoa sludge would not grow nor was there any clarification. Furthermore, they claimed that *Epistylis* sp. was also responsible for 80% of the nitrification in activated sludge. Few of these claims have been substantiated by subsequent research workers, and now most authors agree that protozoa play a secondary but nevertheless an important role in aerobic used-water treatment processes.

Ardern and Lockett (1928) attempted to assess the effects upon effluent quality of selectively killing the protozoan populations of small-scale activated-sludge plants. Although these authors found that toluene treatment killed the protozoa and that there was a simultaneous deleterious effect upon the effluent quality, such methods are open to criticism since toluene is not a specific toxin to protozoa and the changes in effluent quality noted could be attributable to unintended changes in the other micro-organisms present. Curds *et al.* (1968) therefore repeated this type of experiment in reverse; they constructed six replicate bench-scale activated-sludge plants which were completely enclosed. The plants were sterilized chemically and inoculated with protozoa-free bacterial cultures isolated from activated sludge. The sludges which grew in the six replicate units were kept free from protozoa by dosing them with cool heat-treated sewage. These methods made it possible for the first time to study the effects of the subsequent addition of protozoa to the plants and to assess the role of protozoa and to quantify the magnitude of their effect upon effluent quality.

Under protozoa-free conditions all six plants produced highly turbid effluents of inferior quality and the turbidity was found to be significantly related to the presence of very large numbers of viable bacteria suspended in the effluent (hundreds of millions/ml). Without protozoa the BODs of the effluents were high as were their contents of organic carbon, suspended solids and other parameters (Fig. 10). Cultures of ciliated protozoa (*Pseudoglaucoma muscorum, Histriculus vorax, Opercularia coarctata* and *Opercularia microdiscum*) were then added to three of the plants while the other three continued to operate as control plants without protozoa. After a few days, during which time the protozoan population became properly established, there was a dramatic improvement in the effluent quality of the plants containing protozoa. Their clarity was greatly improved and this was associated with a significant decrease in the concentrations of viable bacteria in the effluents. Furthermore, the effluent BOD and concentrations of suspended solids and other parameters decreased significantly. The three units still operating without protozoa continued to deliver turbid, low-quality effluents. The histogram in Fig. 10 illustrates the ranges of effluent qualities obtained from these

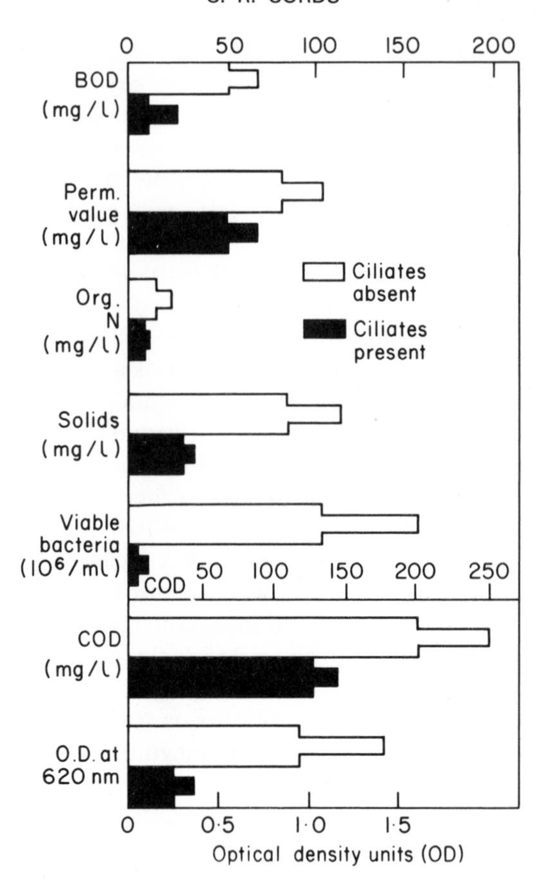

FIG. 10. Histogram of effluent qualities from experimental activated-sludge units operating in absence and presence of ciliated protozoa. Shoulders on blocks indicate ranges of means of effluent quality.

bench-scale pilot plants operating in the absence and presence of ciliated protozoa, and the photograph in Fig. 11 illustrates the appearance of the effluents about three days after protozoa had been added to three of the plants.

These results are in agreement with those of McKinney and Gram (1956), Ardern and Lockett (1928) and some of the results of Pillai and Subrahmanyan (1942, 1944), and it seems clear that the major role of the protozoa in the activated-sludge process is to clarify the effluents. Furthermore there is evidence from full-scale plants that when protozoa are not present, the effluent obtained is turbid and of inferior quality (Curds and Cockburn, 1970a). There are at least two ways in which protozoa might cause the improvement in effluent quality; either by flocculation or by predation.

There is a considerable amount of evidence in the literature to show that protozoa in pure culture are able to flocculate suspended particulate matter and bacteria; this can aid both clarification of the effluent and the formation of the sludge (Ardern and Lockett, 1928; Barker, 1943, 1946; Barritt, 1940; Curds, 1963a, b; Hardin, 1943; Hawkes, 1960, 1963; Jager, 1938; Jenkins, 1942; Pillai and Subrahmanyan, 1942; Sugden and Lloyd, 1950; Viehl, 1937; Watson, 1943, 1945). In certain species flocculation is thought to be brought about by the secretion of a mucous-like substance from the peristome region (Sugden and Lloyd, 1950; Watson, 1945). Hardin (1943) and Watson (1945) found that the flagellate *Oicomonas termo* and the ciliate *Balantiophorus minutus* were able to flocculate bacteria. Other workers (Curds, 1963a, b; Sugden and Lloyd, 1950) found that colloidal particles of DDT and Indian ink were flocculated by *Paramecium caudatum* and *Carchesium polypinum*. In the case of *P. caudatum*, Curds (1963a, b) found that at least two mechanisms were responsible; first the ciliate was able to secrete a soluble polysaccharide (a polymer of glucose and arabinose) into the medium, which changed the surface charge of the suspended colloidal particles present, and secondly, particles ingested during cyclosis were glued together by a mucin. It is well known, however, that many bacteria are able to flocculate or grow in flocculant forms (zoogloeal organisms) without the aid of protozoa (Jenkins, 1942; McKinney and Gram, 1956).

The reduction of the numbers of bacteria from sewage to effluent following aerobic treatment has been noted by many authors including Curds *et al.* (1968), Curds and Fey (1969), Horosawa (1950), Kisskalt (1915), Liebmann (1936), Richards and Sawyer (1922) and Viehl (1937). It is evident from the literature that the dominant types of protozoa found in these processes feed mainly on bacteria (Sandon, 1932), and so it has often been suggested that the predatory activities of the protozoa might

FIG. 11. Photograph of effluents from six replicate activated-sludge units. Units 2, 5 and 6 contained ciliated protozoa; units 1, 3 and 4 did not.

be responsible for bacteria removal. Curds and Vandyke (1966) showed that the ciliates from activated sludge could feed upon a number of strains of bacteria likely to be found in aerobic sewage-treatment processes. However, Barritt (1940) stated that since bacteria were only present as flocculated masses in activated sludge they were unavailable as a food source for the protozoa, and he suggested that under these circumstances these organisms utilize the soluble organic substances present. However, the work of Curds *et al.* (1968) has shown that non-flocculated bacteria do occur in large numbers in activated sludge but only when protozoa are not present. The quantitative feeding studies of Curds and Cockburn (1968, 1971) using batch and continuous-culture methods have led these authors to suggest that if the protozoa in activated sludge feed at rates similar to those of *Tetrahymena pyriformis* in pure culture, then the protozoan populations normally found in activated sludge could, by predation alone, easily remove sufficient quantities of bacteria to account for the reductions observed by Curds *et al.* (1968).

Protozoa are known to feed upon pathogenic bacteria, including those which cause diseases such as diptheria: cholera, typhus and streptococcal infections (Gemüd, 1916*a*, *b*; Heukelekian and Rudolfs, 1929; Kyriasides, 1931; Muller, 1919; Reploh *et al.*, 1963; Schepelewsky, 1910; Spiegel, 1913), as well as faecal bacteria such as *Escherichia coli* (Curds and Fey, 1969). In the case of *E. coli*, Curds and Fey (1969) found that 50% of these bacteria entering in the sewage were removed by

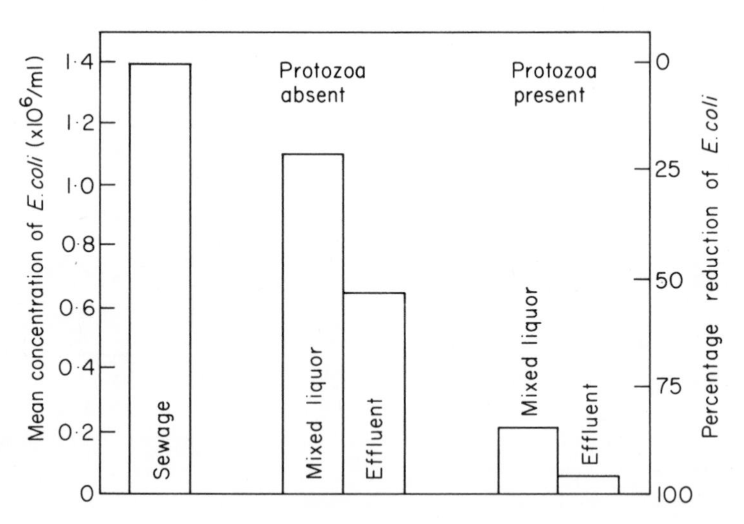

FIG. 12. Histogram illustrating the fate of *Escherichia coli* in experimental activated-sludge plants operating in absence and presence of ciliated protozoa.

unidentified processes when protozoa were absent, but after ciliates had been added a mean of 95% of the *E. coli* were removed from sewage. The histogram in Fig. 12 summarizes the populations of *E. coli* in sewage, mixed liquors and effluents during these experiments when protozoa were absent and present.

It seems likely from the above considerations that the major role of the ciliated protozoa in aerobic used-water treatment processes is the removal of dispersed growths of bacteria by predation, and it is unlikely that protozoa-induced flocculation is of any real importance. Flagellated protozoa and amoebae also feed upon bacteria and they no doubt will play a similar role; furthermore, the amoebae may have the ability to ingest flocculated bacteria, which would have the effect of reducing sludge production. It is well known that protozoa are also able to utilize soluble substrates, although to the author's knowledge no work has been carried out on the quantitative aspects of soluble substrate utilization when the organisms also have a plentiful supply of bacteria present as food.

V. A Mathematical Approach to the Ecology of Protozoa in Activated Sludge

Many of the observations made on protozoa in activated sludge can now be explained on the basis of growth kinetics, the settling properties of the organisms and the operational characteristics of the plant. A number of workers have derived mathematical models for the activated-sludge process (Grieves *et al.*, 1964; Herbert, 1961; McKinney, 1962; Reynolds and Yang, 1966; Schulze, 1964, 1965); all of these have considered sludge to consist of a single flocculating micro-organism, and a few (Downing and Knowles, 1966; Downing and Wheatland, 1962) have considered the dynamics of different organisms and the effects they might have upon plant performance.

The use of computers as tools in the testing of general ecological principles is still in its infancy, but the work of Garfinkel (1967*a*, *b*), Garfinkel and Sack (1964) and others indicate that this approach shows promise for the future. Bungay (1968) applied analogue computer-simulation techniques to studies of the dynamics of continuous mixed-culture systems, and he included an example of a predator and its prey growing together in a chemostat. He found that the populations of organisms oscillated in a regular, complementary manner. More recently Canale (1969) applied singular point analysis and linear approximation methods, and Curds (1971*b*) used a digital computer to simulate the dynamics of a similar culture system. Both of these latter authors found that there were three possible solutions to the equations dependent upon the kinetic constants of the two organisms, although generally a stable limit-cycle variation (as found by Bungay, 1968) would be expected.

While developing continuous-culture techniques to study the consumption of bacteria by ciliated protozoa, Curds (1970) found that when bacteria and ciliates were grown together in a reactor a steady state was not achieved, and instead a series of predator–prey oscillations was observed; later it was shown (Curds and Cockburn, 1971) that steady-state populations could be achieved when the two organisms were grown in separate vessels.

The mathematical modelling of protozoa and bacteria growing together has now been extended (Curds, 1971c) to consider the dynamics and fates of a variety of micro-organisms growing together in a completely-mixed activated-sludge reactor fitted with sludge feed-back. Curds (1971c) considered the plant would contain and/or receive a number of micro-organisms, which he defined by the following criteria:

(a) *dispersed sewage bacteria*, which are those borne in sewage in considerable quantities but on entry into the reactor do not flocculate; they remain in suspension in the sedimentation tank and are evenly dispersed throughout the effluent and recycle flows;

(b) *sludge bacteria*, which form the bulk of the sludge mass and are assumed not to be present in the sewage flow and always flocculate completely;

(c) *soluble substrate*, which for the purposes of the model is considered to be the limiting substrate contained in the sewage flow, and assuming that all other necessary elements are present in excess. This substrate is utilized by both types of bacteria but not by ciliated protozoa;

(d) *ciliated protozoa* are assumed to feed entirely on dispersed bacteria, since it is argued that the sludge bacteria are present only as flocculated masses. It is assumed that these organisms are not present in sewage and that there are two types. Firstly, free-swimming types which, because of their habit, never flocculate and remain evenly dispersed in the sedimentation tank, and secondly, sedentary protozoa which always flocculate since they are assumed to be attached directly to the flocculant sludge bacterial masses.

Of the above populations only the dispersed bacteria, free-swimming ciliates and soluble substrate are able to leave the plant in the effluent; sludge bacteria and sedentary protozoa never leave in the effluent, but some are continually pumped away at a fixed rate after sedimentation. One further important assumption is also made, that all organisms obey Michaelis–Menten growth kinetics, and that the specific growth rates of the two bacteria are related to the concentration of soluble substrate, and the specific growth rate of the ciliate population is related to the concentration of dispersed bacteria present in the reactor. All simulations were carried out on an IBM S/360 digital computer using the highly sophisticated simulation language "Continuous System Modelling Program" (CSMP S/360). The values of the various kinetic constants, the substrate and bacterial concentrations of the sewage used for these studies were thought to be feasible and in some cases were based on the results of experimental work. Arbitrary starting values were assigned to the populations and the computer was programmed to integrate these populations against time using the appropriate mass balance and kinetic equations.

Initially, two separate simulations were carried out to ascertain whether or not the activated-sludge population was dynamically stable; both simulations included dispersed-sewage and sludge bacteria, but in one the ciliate population consisted only of free-swimming forms and in the other the only ciliates were attached sedentary forms. The two simulations were therefore carried out as a direct comparison of the effect of ciliate habit on the dynamics and concentrations of the various microbial populations. In each case the initial starting values for the population sizes were assumed to be low and equal in each simulation. In both simulations as time proceeded the concentrations of sludge bacteria increased, whereas the concentrations of dispersed sewage bacteria and substrate concentration decreased. The attached ciliate population increased with time but the free-swimming population decreased with time. All populations, however, asymptotically approached steady-state conditions without signs of oscillation, so it was concluded that the model was dynamically stable. Under these circumstances it is valid to solve manually the mass balance equations for simple steady-state solutions at various rates. This has been done over a wide range of sludge specific wastage rates and sewage dilution rates (keeping the wastage rate arbitrarily at one tenth the dilution rate) under three situations; (a) when the protozoa are attached forms, (b) when the protozoa are free-swimming ciliates, and (c) when ciliated protozoa are not present; the results are illustrated in Fig. 13.

It can be shown by simple algebra that under steady-state conditions the specific growth rate of a settling organism (sludge bacteria and attached protozoa) is equal to the specific wastage rate of the sludge. This explains why in all three situations the concentration of substrate is precisely the same at any given wastage rate. The microbial populations, however, were different for each situation. From the mathematics it seems that substrate concentration is independent of ciliates; it is clear, however, that the turbidity of the effluent, due to dispersed sewage bacteria, is completely dependent upon protozoa. Large numbers of

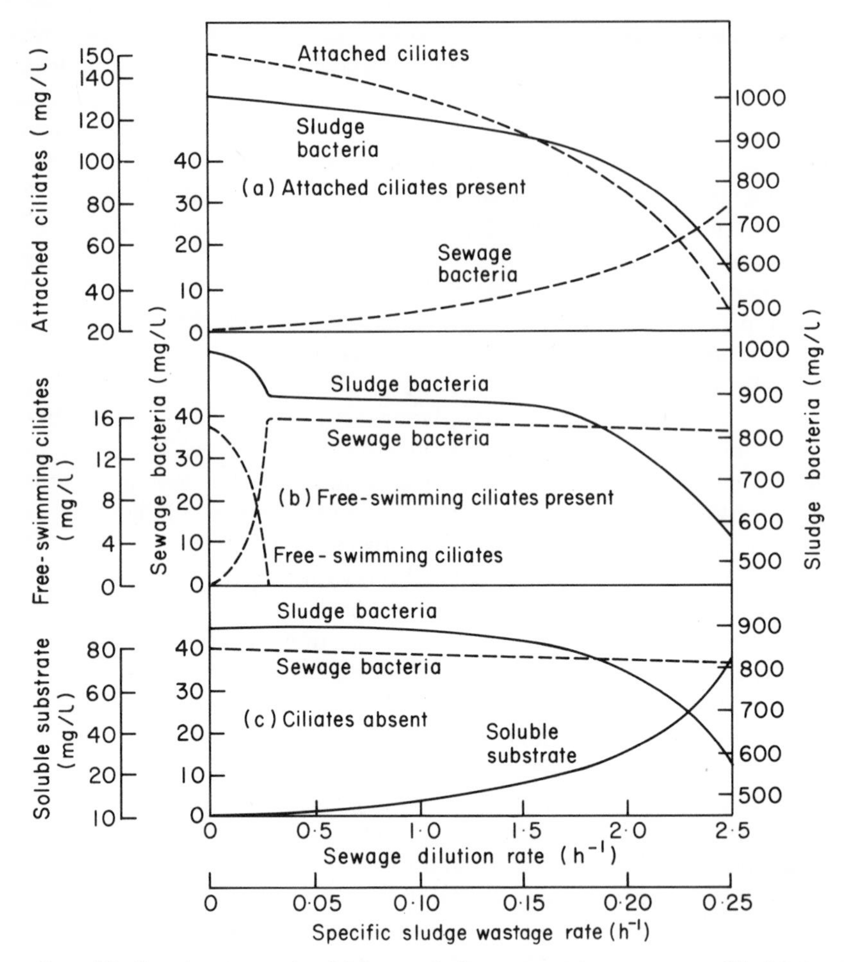

FIG. 13. Steady-state microbial populations at various sewage dilution and sludge wastage rates predicted by mathematical model of the activated-sludge process, (a) in presence of attached ciliated protozoa, (b) in presence of free-swimming ciliated protozoa and (c) in absence of ciliated protozoa (after Curds, 1971c).

dispersed bacteria were present when protozoa were absent (no predation), but were fewer when free-swimming ciliates were present and least when the ciliate population was composed of attached forms. The latter two cases are easily explained; the growth rate of the attached forms is very low and equals the specific wastage rate of the sludge, whereas the specific growth rate of the free-swimming forms is always much higher and equals the sewage dilution rate. It follows from Michaelis–Menten kinetics that higher growth rates require higher concentrations of substrate, which in this case is dispersed bacteria, to be present. Furthermore, free-swimming ciliates were washed out from the system at much lower dilution rates than were attached forms. From these theoretical observations an overall high-quality effluent would be expected to be obtained when attached ciliates are dominant, a slightly worse effluent quality when free-swimming ciliates are present, and a low-quality effluent when no ciliated protozoa are present. These ideas are in fact frequently expressed in the literature, being based on practical experience, and now mathematical modelling is supplying quantitative explanations for these phenomena. For example, free-swimming ciliates and flagellates were considered by McKinney and Gram (1956) to be indicative of a low-efficiency plant and the peritrichous ciliates (all attached forms) indicative of a high one. The indicator species studies of Curds and Cockburn (1970b) list the free-swimming ciliates as being those more frequently found in plants producing inferior effluents (BOD > 21 mg/l), whereas the attached species of the genera *Vorticella*, *Epistylis* and *Carchesium* were more commonly observed in plants producing good quality effluents (BOD < 20 mg/l).

The presence of protozoa and the habit of the dominant species therefore directly influence the concentration of dispersed bacteria present, and hence indirectly the concentration of sludge bacteria, since these two "types" of bacteria are in competition with each other for the soluble substrate. Thus the greater the concentration of dispersed bacteria, the smaller the population of sludge bacteria present.

It was stated earlier in this chapter that successions of protozoa are well documented in a variety of cultures (Barritt, 1940; Bick, 1958, 1960; Woodruff, 1912). Curds (1966) showed that when an activated-sludge plant is started without the addition of a sludge inoculum, a succession of protozoa appears with time (see Fig. 9). The usual explanation given for these successions may be summarized as being due to changes in the environment which successively favour the growth of new types. Evidence and experience of computer simulations suggests that successions of this nature might equally well be explained simply on a basis of growth kinetics and on the settling properties of the organisms. A computer programme was written (Curds, 1971c) to describe the situation in a hypothetical activated-sludge reactor which contained sludge bacteria, dispersed sewage bacteria, flagellated protozoa, free-swimming ciliates and attached ciliates. Flagellates have not been mentioned above, but for the purposes of this simulation they were assumed to utilize the soluble substrate, would not settle and would not be present in the inflowing sewage.

Figure 14 illustrates the results of this simulation and demonstrates that the use of the above criteria gave a succession of organisms which was qualitatively similar to that found in practice. Flagellated protozoa and dispersed bacteria were the first dominant organisms, but as time progressed both declined. The flagellates declined slowly owing to competition for substrate with the two bacteria, while the dispersed bacteria declined rapidly owing to competition for substrate and to the predatory activities of the ciliates present. The free-swimming ciliates reached a small peak after the dispersed bacteria but were soon displaced by the attached ciliate population owing to competition for bacteria. It is clear that any organism which settles and is hence returned back to the aeration tank has a distinct advantage over an organism which does not settle and is continually washed-out in the effluent. If crawling ciliates were assumed to settle, but less readily than attached forms, a peak of

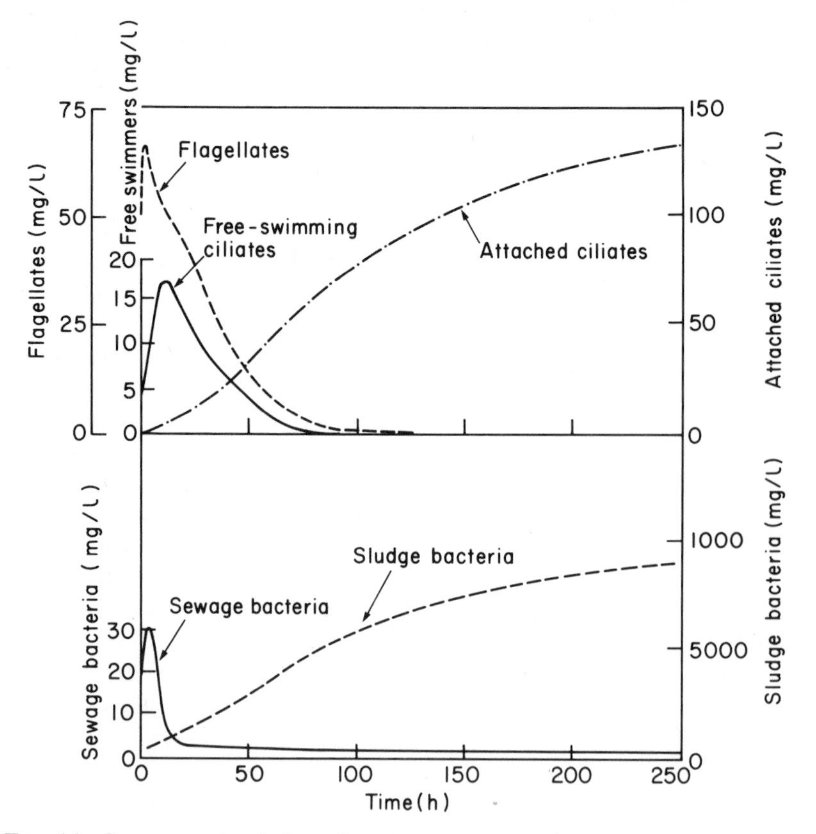

FIG. 14. Computer simulation of protozoan successions in an activated-sludge plant (after Curds 1971c).

crawling ciliates might be expected to occur between the other two ciliate peaks as is observed in practice.

It would appear from the above mathematical considerations that many of the observations made in the past concerning protozoa can be explained by the model, but this does not necessarily mean that it is correct. The activated-sludge community is highly complex and it is possible that the results of such simple models are chance. In the opinion of the author, however, the mathematical approach using simulation techniqes will in the future aid the ecologist to test his theories and to obtain a far greater depth of perception of the ecosystem than previously has been possible. Simple, yet highly sophisticated, computer simulation languages will enable the ecologist to manipulate highly complex dynamic ecological situations without the need for an intimate knowledge of mathematics and computer programming. In activated sludge, for example, Curds (1973b) has already been able to investigate the effects of diurnal variations in the flow and strength of sewage upon microbial population dynamics; furthermore the effects of predatory carnivorous ciliates have been examined. Such investigations are giving possible explanations as to why the protozoan populations of activated-sludge plants vary to the extent they are observed to do. As the models more closely approximate to the activated-sludge process, perhaps more and more of the observations on all forms of microbial life will be explained.

List of Taxonomic Works

BOURRELLY, P. (1970). "Les Algues d'Eau Douce. Initiation à la Systématique. III. Algues Bleues et Rouges. Les Eugléniens, Peridinens et Cryptomadines." Bourbée et Cie, Paris.

CASH, J. (1905–1909). "The British Freshwater Rhizopoda and Heliozoa." Vol. I (1905), Vol. II (1909). Ray Society, London.

CASH, J. and WAILES, G. H. (1915–1918). "The British Freshwater Rhizopoda and Heliozoa" Vol. III (1915), Vol. IV (1918). Ray Society, London.

CLAPAREDE, E. and LACHMANN, J. (1858–1861). *Mém. Inst. nat. Genèv.* **5,** 1–260; **6,** 261–482; **7,** 1–291.

KAHL, A. (1930–1935). *In* "Die Tierwelt Deutschlands" (Ed. F. Dahl), Teil 18 (1930), 21 (1931), 25 (1932), 30 (1935). Gustav Fischer, Jena.

KAHL, A. (1934). *In* "Die Tierwelt der Nord–und Ostsee" (Eds G. Grimpe and E. Wagler), Lief 26 (Teil II, C5). Leipzig.

KENT, W. S. (1880–1882). "A Manual of the Infusoria" Vols I–III. David Bogue, London.

NOLAND, L. E. and FINLEY, H. E. (1931). *Trans. Am. microsc. Soc.* **50,** 81–123.

PASCHER, A. (1914–1927). "Die Süsswasserflora Deutschlands, Österreich und der Schweiz." Vol. I (1914), Vol. II (1923), Vol. III (1927). Gustav Fischer, Jena.

PENARD, E. (1896). *Mém. soc. phy. et l'hist. nat.* Vol. **31.** Geneva.

PENARD, E. (1922). "Études sur les Infusories d'Eau Douce." Georg et Cie, Geneva.

READ, R. (1967). *Trans. Am. microsc. Soc.* **86,** 486–493.

WEST, G. S. and FRITSCH, F. E. (1927). "A Treatise on British Freshwater Algae." Cambridge University Press, Cambridge, England.

References

AGERSBORG, H. P. K. (1929). *Trans. Am. microsc. Soc.* **48,** 158–180.

AGERSBORG, H. P. K. and HATFIELD, W. D. (1929). *Sewage Wks J.* **1,** 411–424.

ALEXEIEFF, A. (1917). *Zh. Mikrobiol.* **4,** 97–102.

ALEXEIEFF, A. (1926). *Russk. Arkh. Protist.* **4,** 127–132.

ARDERN, E. and LOCKETT, W. T. (1928). *Manchester Rivers Dept, Ann. Rept.* App. **1,** 41–46.

ARDERN, E. and LOCKETT, W. T. (1936). *Wat. Pollut. Control* **35,** 212–215.

BAINES, S., HAWKES, H. A., HEWITT, C. H. and JENKINS, S. H. (1953). *Sewage ind. Wastes* **25,** 1023–33.

BARK, T. (1971). *Annls Stn biol. Besse* **6–7,** 241–260.

BARKER, A. N. (1942). *Ann. appl. Biol.* **29,** 23–33.

BARKER, A. N. (1943). *Naturalist, Hull* 65–69.

BARKER, A. N. (1946). *Ann. appl. Biol.* **33,** 314–325.

BARKER, A. N. (1949). *Wat. Pollut. Control* **48,** 7–27.

BARRITT, N. W. (1940). *Ann. appl. Biol.* **27,** 151–156.

BICK, H. (1957). *Vom Wass.* **24,** 224–227.

BICK, H. (1958). *Arch. Hydrobiol.* **54,** 506–542.

BICK, H. (1960). *Arch. Hydrobiol.* **56,** 378–394.

BICK, H. (1972). "Ciliated Protozoa." World Health Organization, Geneva.

BICK, H. and SCHOLTYSECK, E. (1960). *Arch. Hydrobiol.* **57,** 196–216.

BODINE, J. H. (1923). *J. exp. Biol.* **37,** 115–125.

BRINK, N. (1967). *Int. Revue ges. Hydrobiol. Hydrogr.* **52,** 51–122.

BROWN, T. J. (1964). "Ecological and taxonomic studies on certain ciliates in activated sludge." Ph.D. Thesis, London University.

BROWN, T. J. (1965). *Wat. Pollut. Control* **64,** 375–378.

BUNGAY, H. R. (1968). *Chem. Engng Prog. Symp.* Ser. **64,** 19–22.

BUNGAY, H. R. and BUNGAY, M. L. (1968). *Adv. appl. Microbiol.* **10,** 269–290.

BURBANCK, W. D. (1942). *Physiol. Zool.* **15,** 342–362.

BUSWELL, A. M. and LONG, H. L. (1923). *J. Am. Wat. Wks Ass.* **10,** 309–321.

Butterfield, C. T., Purdy, W. C. and Theriault, E. J. (1931). *Publ. Hlth Rep., Wash.* **46,** 393–426.

Cairns, J. and Dickson, K. L. (1970). *Trans. Kans. Acad. Sci.* **73,** 1–10.

Cairns, J., Beamer, T., Churchill, S. and Ruthven, J. (1971). *Hydrobiologia* **38,** 193–205.

Calaway, W. T. and Lackey, J. B. (1962). "Waste Treatment Protozoa. Flagellata." *Fla Engng Ser.* No. 3.

Canale, R. P. (1969). *Biotechnol. Bioengng.* **11,** 887–907.

Clay, E. (1964). "The fauna and flora of sewage processes. 2. Species list." I.C.I. Ltd, Paints Division. Research Dept Memorandum PVM 45/A/732.

Coleman, G. S. (1964). *J. gen. Microbiol.* **37,** 209–223.

Contois, D. E. (1959). *J. gen. Microbiol.* **21,** 40–50.

Cooke, W. B. (1959). *Ecology* **40,** 273–291.

Cooke, W. B. and Hirsch, A. (1958). *Sewage ind. Wastes* **30,** 138–156.

Cramer, R. and Wilson, J. A. (1928). *Ind. Engng Chem. ind. Edn* **20,** 4–9.

Crozier, W. J. (1923). *Science, N.Y.* **58,** 424–425.

Crozier, W. J. and Harris, E. S. (1923a). *Anat. Rec.* **24,** 403.

Crozier, W. J. and Harris, E. S. (1923b). *Bull. New. Jers. agric. Exp. Stn* **43,** 503–516.

Curds, C. R. (1963a). "Studies on the ecology of ciliated protozoa in activated sludge and their role in the process of sewage purification." Ph.D. Thesis, London University.

Curds, C. R. (1963b). *J. gen. Microbiol.* **33,** 357–363.

Curds, C. R. (1966). *Oikos* **15,** 282–289.

Curds, C. R. (1969). "An illustrated key to the British freshwater ciliated protozoa commonly found in activated sludge." Water Pollution Technical Paper No. 12. H.M. Stationery Office, London.

Curds, C. R. (1970). *In* "Proc. Symp. on Methods of Study of Soil Ecology" (Ed. J. Philipson), pp. 127–129. UNESCO, Paris.

Curds, C. R. (1971a). *In* "Journées de l'Eau." Centre de Perfectionnement Technique, Rueil–Malmaison, Paris.

Curds, C. R. (1971b). *Wat. Res.* **5,** 793–812.

Curds, C. R. (1971c). *Wat. Res.* **5,** 1049–1066.

Curds, C. R. (1973a). *Amer. Zool.* **13,** 161–169.

Curds, C. R. (1973b). *Wat. Res.* **7,** 1269–1284.

Curds, C. R. and Cockburn, A. (1968). *J. gen. Microbiol.* **54,** 343–358.

Curds, C. R. and Cockburn, A. (1970a). *Wat. Res.* **4,** 225–236.

Curds, C. R. and Cockburn, A. (1970b). *Wat. Res.* **4,** 237–249.

Curds, C. R. and Cockburn, A. (1971). *J. gen. Microbiol.* **66,** 95–108.

Curds, C. R. and Fey, G. J. (1969). *Wat. Res.* **3,** 853–867.

Curds, C. R. and Vandyke, J. M. (1966). *J. appl. Ecol.* **3,** 127–137.

Curds, C. R., Cockburn, A. and Vandyke, J. M. (1968). *Wat. Pollut. Control* **67,** 312–329.

Cutler, D. W. and Bal, D. V. (1926). *Ann. appl. Biol.* **13,** 516–534.

CUTLER, D. W. and CRUMP, L. M. (1924). *Biochem. J.* **18,** 905–912.
CUTLER, D. W., CRUMP, L. M. and DIXON, A. (1932). *J. Anim. Ecol.* **1,** 141–151.
DARBY, H. H. (1929). *Arch. Protistenk.* **65,** 1–37.
DAWSON, J. A. (1929). *Proc. Soc. exp. Biol. Med.* **26,** 335–340.
DAWSON, J. A. and HEWITT, D. C. (1931). *Am. Nat.* **65,** 181–186.
DAWSON, J. A. and MITCHELL, W. H. (1929). *Am. Nat.* **63,** 476–478.
DIXON, A. (1937). *Rep. Proc. 2nd. Int. Congr. Microbiol.* 1936, p. 250 London.
DOBELL, C. and O'CONNER, F. W. (1921). "The Intestinal Protozoa of Man." John Bale, Sons and Danielson, London.
DOWNING, A. L. and KNOWLES, G. (1966). *Proc. 3rd Int. Conf. Wat. Pollut. Res.* **2,** 117–136.
DOWNING, A. L. and WHEATLAND, A. B. (1962). *Trans. Instn Chem. Engrs* **40,** 91–103.
FAIRBROTHER, T. H. and RENSHAW, A. (1922). *J. Soc. chem. Ind., Lond.* **41,** 134–144.
FRYE, W. W. and BECKER, E. R. (1929). *Sewage Wks J.* **1,** 286–308.
GANCZARCZYK, J. and DOMANSKI, J. (1953). *Gaz Woda Tech. sanit.* **27,** 326–329.
GARFINKEL, D. (1967a). *J. theor. Biol.* **14,** 46–58.
GARFINKEL, D. (1967b). *J. theor. Biol.* **14,** 325–327.
GARFINKEL, D. and SACK, R. (1964). *Ecology* **45,** 502–507.
GELEI, J. V. (1925). *Arch. Protistenk.* **52,** 404–417.
GEMÜD, W. (1916a). *Hyg. Rdsch.* **26,** 489–496.
GEMÜD, W. (1916b). *Hyg. Rdsch.* **26,** 521–529.
GIESE, A. C. (1938). *Trans. Am. microsc. Soc.* **57,** 245–255.
GIESE, A. C. and ALDEN, R. H. (1938). *J. exp. Zool.* **78,** 117–134.
GRAY, E. (1952). *J. gen. Microbiol.* **6,** 108–122.
GRIEVES, R. B., MILBURY, W. F. and PIPES, W. O. (1964). *J. Wat. Pollut. Control Fed.* **36,** 619–635.
GROSCOP, J. A. and BRENT, M. M. (1964). *Can. J. Microbiol.* **10,** 579–584.
HADJIPETROU, L. P., GERRITS, J. P., TEULINGS, F. A. G. and STOUTHAMMER, A. H. (1964). *J. gen. Microbiol.* **36,** 139–150.
HALL, R. P. (1953). "Protozoology." Prentice-Hall, New York.
HAMILTON, R. D. and PRESLAN, J. E. (1969). *J. exp. mar. Biol. Ecol.* **4,** 90–99.
HAMILTON, R. D. and PRESLAN, J. E. (1970). *J. exp. mar. Biol. Ecol.* **5,** 94–104.
HARDIN, G. (1943). *Nature, Lond.* **151,** 642.
HARDIN, G. (1944). *Ecology* **25,** 192–201.
HARDING, J. P. (1937). *J. exp. Biol.* **14,** 422–430.
HARKNESS, N. (1966). *Wat. Pollut. Control* **65,** 542–557.
HAUSMAN, L. A. (1923). *Bull. New Jers. agric. Exp. Stn* **390,** 28–39.
HAWKES, H. A. (1960). *In* "Waste Treatment" (Ed. P. C. G. Isaac), pp. 52–98. Pergamon Press, Oxford.

Hawkes, H. A. (1963). "The Ecology of Waste Water Treatment." Pergamon Press, Oxford.

Heal, O. W. (1967a). *In* "Progress in Soil Biology" (Eds O. Graff and J. E. Satchell), pp. 120–126. North-Holland, Amsterdam.

Heal, O. W. (1967b). *In* "Proc. Symp. on Methods of Study of Soil Ecology" (Ed. J. Philipson), pp. 120–126. UNESCO, Paris.

Herbert, D. (1961). *In* "Continuous Culture of Micro-organisms." Monograph No. 12, pp. 21–53. Society of Chemical Industry, London.

Hervey, R. Y. and Greaves, Y. E. (1941). *Soil Sci.* **51,** 85–100.

Hetherington, A. (1933). *Arch. Protistenk.* **80,** 255–280.

Heukelekian, H. (1945). *Sewage Wks J.* **17,** 23–38.

Heukelekian, H. and Rudolfs, W. (1929). *Sewage Wks J.* **1,** 561–567.

Hoare, C. A. (1927). *Parisatology* **19,** 154–222.

Holtje, R. H. (1943). *Sewage Wks J.* **15,** 14–29.

Honigberg, B. M., Balamuth, W., Bovee, E. C., Corliss, J. O., Gojdics, M., Hall, R. P., Kudo, R. R., Levine, N. D., Loeblich, A. R., Weiser, J. and Wenrich, D. H. (1964). *J. Protozool.* **11,** 7–20.

Horosawa, I. (1950). *J. Jap. Sewage Wks Ass.* **148,** 62–67.

Hutner, S. H., Provasoli, L., Stokstad, E. L. R., Hoffman, C. E., Belt, M., Franklin, A. L. and Jukes, T. H. (1949). *Proc. Soc. exp. Biol. Med.* **70,** 118–120.

Ingols, R. S. and Heukelekian, H. (1940). *Sewage Wks J.* **12,** 849–861.

Ingram, W. T. and Edwards, G. P. (1960). *In* "3rd Conference on Biological Waste Treatment." Manhattan College, New York.

Issel, R. (1906). *Atti Soc. ligust. Sci. nat. geogr.* **17,** 3–72.

Jager, G. (1938). *Z. Hyg. InfektKrankh.* **120,** 620–625.

Javornicky, P. and Prokesova, V. (1963). *Int. Revue ges. Hydrobiol. Hydrogr.* **48,** 335–350.

Jenkins, S. H. (1942). *Nature, Lond.* **150,** 607.

Johannes, R. E. (1965). *Limnol. Oceanogr.* **10,** 434–442.

Johnson, J. W. H. (1914). *J. econ. Biol.* **9,** 105–124.

Kidder, G. W. and Dewey, V. C. (1951). *In* "Biochemistry and Physiology of Protozoa" (Ed. A. Lwoff), Vol 1, pp. 323–400. Academic Press, New York and London.

Kidder, G. W. and Stuart, C. A. (1939). *Biol. Bull. mar. biol. Lab., Woods Hole* **75,** 336–343.

Kisskalt, K. (1915). *Z. Hyg. InfektKrankh.* **80,** 57–65.

Kitching, J. A. (1939). *Biol. Bull. mar. biol. Lab., Woods Hole* **77,** 339–353.

Koffmann, M. (1924). *Arch. mikrosk. Anat. EntwMech.* **103,** 168–181.

Kolkwitz, R. (1926). *Kleine Mitt. Mitgl. Ver. WassVersorg.* **3,** 70–74.

Kolkwitz, R. (1950). *Kleine Mitt. Mitgl. Ver. Wass.- Boden- u. Lufthyg.* **4,** 1–64.

Kolkwitz, R. and Marsson, M. (1908). *Ber. dt. bot. Ges.* **26a,** 505–519.

Kolkwitz, R. and Marsson, M. (1909). *Int. Revue ges. Hydrobiol. Hydrogr.* **2,** 126–152.

KSHIRSAGAR, S. R. (1962). *Envir. Hlth, India.* **4,** 156–160.
KUDO, R. P. (1932). "Protozoology." Charles C. Thomas, Springfield, Illinois, U.S.A.
KYRIASIDES, K. (1931). *Z. Hyg. InfektKrankh.* **112,** 350–364.
LACKEY, J. B. (1924). *Bull. New Jers. agric. Exp. Stn* **403,** 40–60.
LACKEY, J. B. (1925). *Bull. New Jers. agric. Exp. Stn* **417,** 1–39.
LACKEY, J. B. (1926). *Bull. New Jers. agric. Exp. Stn* **427,** 30–41.
LACKEY, J. B. (1932a). *Biol. Bull mar. biol. Lab., Woods Hole* **463,** 287–295.
LACKEY, J. B. (1932b). *Bull. New Jers. agric. Exp. Stn* **463,** 5–29.
LACKEY, J. B. (1938a). *Publ. Hlth Rep., Wash.* **53,** 2037–58.
LACKEY, J. B. (1938b). *Ecol. Monogr.* **8,** 501–527.
LACKEY, J. B. (1941). *In* "A Symposium on Hydrobiology." Madison, Wisconsin, U.S.A.
LACKEY, J. B. and DIXON, R. M. (1943). *Sewage Wks J.* **15,** 1139–52.
LESLIE, L. D. (1940). *Physiol. Zool.* **13,** 430–438.
LIEBMANN, H. (1936). *Z. Hyg. InfektKrankh.* **118,** 555–561.
LIEBMANN, H. (1949). *Vom Wass.* **17,** 62–82.
LIEBMANN, H. (1951). "Handbuch der Frischwasser- und Abwasserbiologie." Gustav Fischer, Jena.
LILLY, D. M. (1953). *Ann. N.Y. Acad. Sci.* **56,** 910–920.
LILLY, D. M., HERBST, B. A. and STILLWELL, R. H. (1960). *Am. Zool.* **1,** 459.
LLOYD, L. (1945). *Wat. Pollut. Control* **44,** 119–139.
MAGUIRE, B. (1959). *Ecology* **40,** 312.
MAGUIRE, B. (1963). *Ecol. Monogr.* **33,** 161–185.
MARTIN, D. (1968). "Microfauna of Biological Filters." Univ. of Newcastle upon Tyne, Dept of Civil Engng Bull. No. 39. Oriel Press, Newcastle upon Tyne.
McKINNEY, R. E. (1962). *J. sanit. Engng Div., Am, Soc. civ. Engrs* **88,** SA3, 87–113.
McKINNEY, R. E. and GRAM, A. (1956). *Sew ind. Wastes* **28,** 1219–37.
MEIKLEJOHN, J. (1932). *Ann. appl. Biol.* **19,** 584–608.
MINISTRY OF TECHNOLOGY (1968). Notes on Water Pollution No. 43, H. M. Stationery Office, London.

MOHR, J. L. (1952). *Science, N.Y.* **74,** 7–9.
MORISHITA, I. (1968). *Jap. J. Hydrobiol.* **4,** 12–20.
MORISHITA, I. (1969a). *Jap. J. Protozool.* **2,** 17–18.
MORISHITA, I. (1969b). "Protozoa in Polluted Water (Polluted Water Protozoology)." Japan (in Japanese).
MORISHITA, I. (1970). *Jap. J. Protozool.* **3,** 1–13.
MOSER, H. (1958). *Carnegie Inst. Washington* Publ. No. 614, 4. Washington.
MULLER, P. T. (1919). *Arch. Hyg. Bakt.* **75,** 320–352.

Nasir, S. M. (1923). *Ann. appl. Biol.* **10,** 122–133.
Nikitinsky, J. and Mudrezowa-Wyss, F. K. (1930). *Zentbl. Bakt. ParasitKde* (Abt. 2) **81,** 167–198.
Nikoljuk, V. F. (1969). *Acta Protozoologica* **7,** 99–109.
Noc, F. (1914). *C. r. Séanc. Soc. Biol.* **76,** 166–168.
Noland, L. E. (1925). *Ecology* **6,** 437–452.
Noland, L. E. and Gojdics, M. (1967). *In* "Research in Protozoology" (Ed. T-T. Chen), pp. 216–266. Pergamon Press, Oxford.
Painter, H. A. and Viney, M. (1959). *Biotechnol. Bioengng* **1,** 143–162.
Pantin, C. F. A. (1930). *Proc. R. Soc. Ser. B* **105,** 538–550.
Pattamapirat, W. (1963). University of Durham. Report by Publ. Hlth Engng Section 1962, Bull **27,** 3–6.
Phelps, A. (1946). *J. exp. Zool.* **102,** 277–292.
Phelps, A. (1961). *Am. Zool.* **1,** 467.
Phelps, E. B. (1944). "Stream Sanitation." Wiley, London.
Picken, L. E. R. (1937). *J. Ecol.* **25,** 368–384.
Pillai, S. C. (1941). *Curr. Sci., Bangalore* **10,** 84–85.
Pillai, S. C. and Rajagopalan, R. (1948). *Curr. Sci., Bangalore* **17,** 399.
Pillai, S. C. and Subrahmanyan, V. (1942). *Nature, Lond.* **150,** 525.
Pillai, S. C. and Subrahmanyan, V. (1944). *Nature, Lond.* **154,** 179.
Powell, E. O. (1958). *J. gen. Microbiol.* **18,** 259–268.
Proper, G. and Garver, J. C. (1966). *Biotechnol. Bioengng* **8,** 287–296.
Read, R. (1966). "The vorticellid protozoa of activated sludge and their significance, an experimental and ecological analysis. Comparative studies on one-year survey on the protozoa populations at two sewage works operating activated-sludge plants." Ph.D. Thesis, London University.
Reploh, H., Gamel, G. and Nehrkora, A. (1963). *Forsch Ber. Londes N. Rhein-Westf.* **11,** 856–862.
Reynolds, I. D. and Yang, J. T. (1966). *Proc. 21st ind. Waste Conf., Purdue Univ. Engng Extn Ser.* **121,** 696–713.
Reynoldson, T. B. (1939). *Wat. Pollut. Control* **38,** 158–172.
Reynoldson, T. B. (1942). *Nature, Lond.* **149,** 608–609.
Richards, H. and Sawyer, G. C. (1922). *J. Soc. Chem. Ind., Lond. Trans.* **41,** 62–72.
Rudolfs, W., Campbell, F. L., Hotchkiss, M. and Lackey, J. B. (1924). *Bull. New Jers. agric. Exp. Stn* **403,** 60–81.
Rudolfs, W., Hotchkiss, M., Fischer, A. J. and Lackey, J. B. (1926). *Bull. New Jers. agric. Exp. Stn* **427,** 50–74.
Salt, G. W. (1967). *Ecol. Monogr.* **37,** 113–144.
Salt, G. W. (1968). *J. Protozool.* **15,** 275–280.
Salt, G. W. (1969). *Ecology* **50,** 135–137.
Sandon, H. (1932). "The Food of Protozoa." *Egyptian Univ. Cairo Publ. Fac. Sci.* No. 1.
Schaeffer, A. A. (1910). *J. exp. Zool.* **8,** 75–132.

SCHEPELEWSKY, E. (1910). *Arch. Hyg. Bakt.* **72,** 73–90.
SCHOENBORN, H. W. (1947). *J. exp. Zool.* **105,** 269–277.
SCHOENBORN, H. W. (1949). *J. exp. Zool.* **111,** 437–444.
SCHOFIELD, T. (1971). *Wat. Pollut. Control* **70,** 32–47.
SCHULZE, K. L. (1964). *Wat. Sewage Wks* **111,** 526–538.
SCHULZE, K. L. (1965). *Wat. Sewage Wks* **112,** 11–17.
SINGH, B. N. (1941). *Ann. appl. Biol.* **28,** 65–78.
SINGH, B. N. (1945). *Br. J. exp. Path.* **26,** 316–325.
SLADECEK, V. (1969). *Verh. Int. verein Limnol.* **17,** 546–559.
SMITH, J. A. (1940). *Biol. Bull. mar. biol. Lab., Woods Hole* **79,** 379–384.
SPIEGEL, H. (1913). *Arch. Hyg. Bakt.* **80,** 183–201.
ŠRÁMEK-HUŠEK, R. (1954). *Arch. Protistenk.* **100,** 246–267.
ŠRÁMEK-HUŠEK, R. (1956). *Arch. Hydrobiol.* **51,** 376–390.
STOUT, J. D. (1955). *Trans. R. Soc. N.Z.* **82,** 1165.
STOUT, J. D. (1956). *Ecology* **37,** 178–191.
STRONG, F. M. (1944). *Science, N.Y.* **100,** 287.
SUGDEN, B. and LLOYD, L. R. (1950). *Wat. Pollut. Control* **49,** 16–23.
SYDENHAM, D. H. J. (1968). "The ecology of protozoa and other organisms in activated sludge." Ph.D. Thesis, University of London.
SZABO, Z. (1960). *Öst. Wasserw.* **12,** 224–226.
TARTAR, V. (1961). "The Biology of *Stentor*." Pergamon Press, New York.
TAYLOR, H. (1930). *Proc. Ass. Mgrs Sewage Disp. Wks* **3,** 108–119.
TOMLINSON, T. G. (1941). *Wat. Pollut. Control* **40,** 39–76.
TOMLINSON, T. G. (1946). *Wat. Pollut. Control* **45,** 168–178.
TUROBOYSKI, L. (1953). *Gaz Woda Tech. sanit.* **27,** 326–328.
VIEHL, K. (1932). *Zentbl. Bakt. ParasitKde* **86,** 34–43.
VIEHL, K. (1937). *Z. Hyg. InfektKrankh.* **119,** 383–412.
WAGTENDONK, W. J. VAN (1955). *In* "Biochemistry and Physiology of Protozoa" (Ed. A. Lwoff), Vol. 2, pp. 85–90. Academic Press, New York and London.
WATSON, J. M. (1943). *Nature, Lond.* **152,** 693–694.
WATSON, J. M. (1945). *Nature, Lond.* **155,** 171–172.
WOODRUFF, L. L. (1912). *J. exp. Zool.* **12,** 205–264.
WOODRUFF, L. L. (1913). *J. exp. Zool.* **14,** 575–582.
YASUKO, I. (1960). *Jap. J. Ecol.* **10,** 207–214.
ZIMMERMAN, P. (1961). *Schweiz. Z. Hydrol.* **23,** 1–81.

6

Nematoda

F. Schiemer

Limnologisches Institut
Österreichische Akademie der Wissenschaften
Berggasse 18/19
Vienna A-1090
Austria

I. Introduction

Nematodes are widely distributed and are frequently dominant in a variety of aquatic and terrestrial habitats. Although mass occurrences of nematodes have been recorded in some used-water purification processes, research on the taxonomy, ecology and functional importance of these organisms in such habitats is limited. The explanations for this are that microbial activities are considered to be of much greater importance for decomposition processes and that taxonomic and methodological problems have been encountered. The first point is true in the case of the activated-sludge process, where the role of metazoa is likely to be insignificant. In percolating filters, however, large populations of metazoa, particularly nematodes, can develop, and their importance in decomposition processes should be considered. During the last 20 years nematode taxonomy has received a good deal of attention and since the fauna in used-water processes proves to be highly specific and of low diversity, taxonomic problems should not hamper future work.

The following discussion on the ecology and functional role of used-water nematodes is partly hypothetical in character, since it was necessary to use information and data obtained on species other than those from waste-treatment habitats. Nevertheless it is hoped that the following approach is valid to determine the mode of action and the effect of nematodes from a used-water management point of view.

For a further introduction to the work on nematodes the following selected bibliography is recommended: Chitwood and Chitwood (1950), Crofton (1966), De Coninck (1965), Florkin and Scheer (1969), Lee (1965), Sasser and Jenkins (1960) and Tarjan (1961).

II. Taxonomy

In general, free-living nematodes are very similar in appearance. Their body is cylindrical, tapering at both ends, and the length of most species lies within the range of 0·5–3 mm. Despite this overall similarity, large numbers of free-living species have been recognized. A general treatment of the systematics of this group is given in "Traité de Zoologie" (see De Coninck, 1965). The authors there followed the partition in Secernentea (syn. Phasmidia) and Adenophorea (syn. Aphasmidia) as amended by Chitwood (1958), and based on the form of the excretory organ and the presence or absence of phasmids (glandulo-sensory structures) in the tail region. Higher taxa—from families up to orders—are defined mainly by differences in stoma, pharynx and amphids (the latter are sense organs situated in the head region, functioning possibly as chemoreceptors). Taxonomically characteristic structures indicated in the text are presented in Fig. 1. For the definition of taxa from genera upwards, and the species list of each genus, the reader is referred to Goodey's "Soil and Freshwater Nematodes" (1963). Of the ten orders recognized by Goodey, five are represented in used-water treatment plants. These are Rhabditida (for details see below), Enoplida (*Tobrilus, Tripyla, Monochus*), Dorylaimida (*Dorylaimus* s. l., Araeolaimida (*Plectus*) and Monhysterida (*Monhystera*). While the species of the four last mentioned orders are only rarely encountered and are mainly chance invaders from neighbouring biotopes, several species of the order Rhabditida are common and characteristic

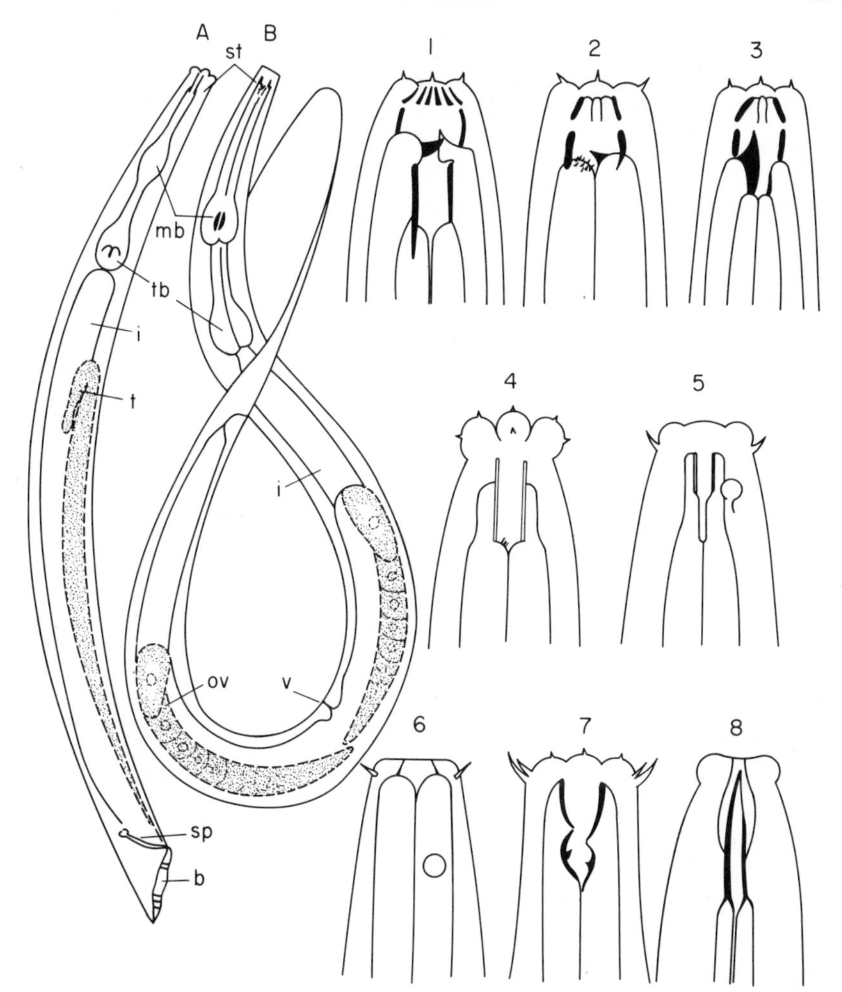

FIG. 1. General structure of nematodes and main types occurring in used waters. (A) Rhabditinae; (B) Diplogasterinae. Abbreviations: st = stoma, mb = median pharyngeal bulb, tb = terminal pharyngeal bulb, i = intestine, t = testis, ov = ovary, v = vulva, sp = spiculae, b = bursa. STOMA TYPES: Diplogasterinae: 1 = *Mononchoides*, 2 = *Paroigolaimella*, 3 = *Diplogasteritus*. Rhabditinae: 4 = *Rhabditoides*. Araeolaimida: 5 = *Plectus*. Monhysterida: 6 = *Monhystera*. Enoplida: 7 = *Tobrilus*. Dorylaimida: 8 = *Dorylaimus*.

representatives of sewage-treatment plants. They will be considered here in greater detail. Rhabditids are characterized by the possession of a terminal pharyngeal bulb, pore-like amphids and the absence of caudal glands.

Two sub-families (Rhabditinae and Diplogasterinae), both belonging to the family Rhabditidae, contain the main representatives. Differential characteristics of these two subfamilies are as follows: Rhabditinae can be recognized by a median pharyngeal bulb without valves, a terminal bulb with valves, a tubular stoma, and a "bursa" (cuticula faults at the male's tail) (Fig. 1). The main genera belonging here are *Rhabditis*, *Pelodera*, *Rhabditoides*, *Caenorhabditis*, *Mesorhabditis*, and *Rhabditonema*, all of which have been found or can be expected in the habitat type considered here. In Diplogasterinae the arrangement of pharyngeal valves is opposite to that in Rhabditinae, i.e. the median bulb has valves and the terminal bulb has not. The stoma is variable, but in many genera they are very big and armed with teeth. The structure of the stoma provides the best differential characters for genera (Fig. 1), of which *Mononchoides*, *Diplogasteritus*, *Paroigolaimella*, *Demaniella*, *Fictor*, *Butlerius* and *Rhabditolaimus* include characteristic used-water species. Representatives of other families of the order Rhabditida include Cylindrocorporidae (*Cylindrocorpus*), Cephalobidae (*Cephalobus*) and Panagrolaimidae (*Panagrolaimus*), but their frequency of occurrence is negligible compared with those of the two sub-families mentioned above.

Although the taxonomy of free-living nematodes has improved considerably during recent years, a modern key for species determination is still lacking. For a first allocation, Goodey's treatise can be recommended (Goodey, 1963). Meyl (1960) described and illustrated the characteristics of about 900 soil and freshwater nematodes in the series "Die Tierwelt Mitteleuropas". A more modern account of European soil nematodes including the main sewage fauna is in preparation (Andrássy, in prep.). While the determination of genera is relatively simple and should be possible with the literature cited, that of species is often problematic and only possible with specialized literature and original descriptions at hand. Species determinations should preferably be checked by specialists. For the determination of Diplogasterinae the careful, but taxonomically outdated, treatment by Weingärtner (1955) can be recommended. More recently, there have been accounts of some of the relevant genera: *Mononchoides* (Calaway and Tarjan, 1973), *Demaniella* (Pillai and Taylor, 1968b) and *Butlerius* (Pillai and Taylor, 1968c).

A systematic treatment of Rhabditinae has been carried out by Osche (1952) and a nomenclatorial analysis by Dougherty (1955). Since representatives of the other systematical groups have not been found to be of quantitative significance, species determinations seem to be of less importance. For the common genera and groups the following literature can be recommended: *Plectus*, Maggenti (1961); *Tobrilus*, Andrássy (1964) and Schiemer (1971); and Dorylaimida, Ferris (1971) (taxonomic review of the order), Andrássy (1959, 1960, 1969) and Thorne (1939).

III. Occurrence of Nematodes in Different Processes

Nematodes have been recorded from practically all aerobic phases of the various used-water treatment processes. Their numbers, however, are highly variable. In general they are of great significance in the microbial film of percolating filters and are often the most abundant metazoan group here (Peters, 1930; Lloyd, 1945; Calaway, 1963). Nematodes are usually abundant in oxidation ponds (Chang and Kabler, 1962) and in older purification processes such as Imhoff and digestion tanks (Lackey, 1924; Agersborg, 1929; Agersborg and Hatfield, 1929). They are regularly present in activated-sludge tanks, but their numbers on the organic flocs are low. Compared with percolating filters their number in the activated-sludge process is much smaller. Although nematodes are frequently mentioned as a group, only a few species determinations have been made (Table I), which indicate that a limited number of highly specific used-water species may exist. The species diversity is low, i.e. a limited number of species occur of which one or a few show a marked predominance.

Information on species composition is more or less restricted to percolating filters. Here, representatives of the sub-family Diplogasterinae are generally the most important, with *Mononchoides* spp., *Diplogasteritus nudicapitatus* and *Paroigolaimella bernensis* as the typical members. These are followed by the Rhabditinae, of which *Rhabditoides longispina* is a characteristic form. However, several other species of Rhabditinae have also been encountered. Species belonging to other systematic categories are usually of less significance and are mainly chance invaders from surrounding biotopes. This is particularly true in the case of the Secernentea, with the exception of *Tobrilus gracilis*, which according to its ecology may be a more regular member of the sewage-treatment fauna. However, in a quantitative sense this species is of no importance. The fauna in percolating filters, as discussed above, is identical or similar to that of effluents and influents of the whole filter system. Weninger (1964), for example, mentioned a predominance of *Paroigolaimella bernensis* and *Diplogasteritus nudicapitatus* in these habitats, and Pillai and Taylor (1968a) found *P. bernensis* to be common in the effluents of percolating filters. Faunistic information on sewage-treatment habitats other than percolating filters is limited. Scherb (1968) mentioned "*Diplogaster* sp." (i.e. Diplogasterinae) from bench-scale experiments with activated sludge. Weninger (1964) listed "*Diplogaster* sp.", *Paroigolaimella* sp., and *Rhabditis* sp. from similar experiments. Murad (1965, 1970) listed nematodes occurring in drying beds of the Texas disposal plant. It is interesting to note that rhabditids (*Rhabditis axei*, *Pelodera chitwoodi*) were always

TABLE I. Occurrence of nematode species in percolating filters,[a] according to different authors. Weninger (a, b, c): filters processing mainly domestic sewage in three small towns in Austria, Peters; Pillai and Taylor: filter in Urbana, Illinois, U.S.A.

Species	Synonyms	Weninger (1964, 1971a,b) a	b	c	Peters (1930)	Pillai and Taylor (1968a)
DIPLOGASTERINAE						
Butlerius sp.						*
Demaniella spp.						*
Diplogaster rivalis (Leydig, 1854)	(*D. micans*)					*
Diplogasteritus nudicapitatus (Steiner, 1914)	(*Acrostichus, Diplogaster*)	**	**	***		***
Fictor anchicoprophaga (Paramonov, 1952)						**
Mononchoides striatus (Bütschli, 1876)	(*Diplogaster, Eudiplog.*)	*		***	***	
Mononchoides spp.						**
Paroigolaimella bernensis (Steiner, 1914)			***	***		***
Rhabdontolaimus sp.						***
RHABDITINAE						
Pelodera teres (Schneider, 1866)				*		*
Rhabditis (*Chloriorhabditis*) *producta* (Schneider, 1866)				*	*	
Rhabditis sensu lato			**	*		***
Rhabditoides longispina (Reiter, 1928)		***	**			
OTHER RHABDITIDA						
Cephalobus spp.						*
Cylindrocorpus curzii (T. Goodey, 1935)						*
Diploscapter coronata (Cobb, 1893)					*	**
Panagrolaimus sp.						**
Myolaimus sp.					*	
Turbatrix aceti (Müller, 1783)						*
NON RHABDITIDA						
Dorylaimidae spp.		*		*	*	*
Monhystera spp.						*
Mononchus sp.						*
Plectus sp.				*		*
Tobrilus sp.				*		*
Tripyla sp.				*		

[a] Murad (1965) listed the following nematodes from drying beds: *Rhabditis axei* (Cobbold, 1884), *Pelodera chitwoodi* (Bassen, 1940), *Mesorhabditis* sp., *Plectus* sp., *Diplogaster* sp., *Dorylaimus* sp. and *Rhabditis* sp.

found to predominate during a seven-month sampling period and that diplogasterids were of much less importance. Chang and Kabler (1962) and Chaudhuri *et al.* (1965) mention "*Diplogasteroides*", "*Diplogaster*" and *Rhabditis* as predominant in the final effluent of used-water treatment plants.

While the fauna in the different sewage-treatment processes is characterized by the marked predominance of Diplogasterinae and Rhabditinae, nematode associations in sand-filter beds, as used for drinking water purification, rather resemble a natural psammic "cleanwater" fauna. Rhabditids are of minor importance and representatives of different systematic categories, mainly belonging to the Secernentea, are present. In general, species diversity is higher and no typical association can be named. This may be explained by the wider range of ecological factors in sand filters governing the composition and population dynamics of the fauna (Cobb, 1918; Husmann, 1958). An occasional occurrence of nematodes in potable water supply, therefore, should not be taken to be indicative of pollution of sanitary significance (Calaway, 1963).

IV. Ecology of Used-Water Nematodes and Factors Determining their Presence and Abundance

Most of the species reported from sewage-treatment plants are well known from a variety of natural saprobic habitats (compost, dung etc.) and polluted freshwater biotopes. (For the terminology of the saprobity system see Sládeček, 1973.) Meiofauna, i.e. microscopically small metazoa, are usually abundant in zones of medium pollution (α,β mesosaprobic) but tend to avoid strongly polluted (polysaprobic) zones. There are, however, a few records of nematodes inhabiting polysaprobic environments, though extensive surveys of nematode distribution are lacking. Steinmann and Surbeck (1918) found *Diplogasteritus nudicapitatus* and *Paroigolaimella bernensis* together with *Tobrilus gracilis* in highly polluted rivers, and Hirschmann (1952), who studied nematodes in freshwater habitats of different degrees of pollution, listed mainly Rhabditinae and Diplogasterinae as polysaprobic. Among those she included were *Mononchoides striatus* and the two rhabditid species (*D. nudicapitatus* and *P. bernensis*) mentioned above. Species associations proved to be distinctly different in zones of varying degrees of pollution, and the author concluded that nematodes may be well suited as indicator organisms of saprobic valency. From the limited amount of information available, it seems that the fauna in used-water treatment plants is more or less identical, at least in its main components, to other polysaprobic freshwater habitats.

Some of the used-water treatment processes, especially percolating filters, provide ideal conditions for saprobic species, i.e. high food levels combined with sufficient oxygen supply and fine structured substrata. Deoxygenation processes (particularly in filters that are ponding), changes in load and composition of sewage, and low temperatures, have adverse effects on maximal population development.

Food conditions appear to be the major factor governing the colonization and growth of nematodes in used-water treatment plants. This, together with oxygen, chemistry and substrate conditions, will be discussed in more detail below. Temperature effects on metabolism, growth and population dynamics are treated in Section VI.

A. Food

The extreme nature of the environment in used-water treatment plants, high concentration of bacteria and organic material, produce specific faunal associations. Two feeding types are usually present; bacteria feeders form the major proportion, while predators feeding on other meiofauna, such as nematodes and rotifera, are usually less abundant. All members of the sub-family Rhabditinae and several of the common Diplogasterinae (e.g. *Paroigolaimella bernensis* and *Diplogasteritus nudicapitatus*) belong to the first category. Other representatives of this sub-family, e.g. *Mononchoides*, *Fictor* and *Butlerius*, are facultative predators. They may be successfully cultured on bacteria–amoeba media as well as on nematodes (Pillai and Taylor, 1968a). Other predators, like *Mononchus* sp. and *Labronema ferox* (dorylaimid), can be cultured only with nematodes as food (Pillai and Taylor, 1968a). In both types of feeding food uptake is the same; bacteria, dissolved organic matter or meiofauna are sucked in by sudden dilatation of the pharynx. For a detailed description of the feeding mechanisms within the Rhabditida, the papers by Doncaster (1962) and Mapes (1965a,b) are recommended.

Several species of nematodes have been reared in axenic cultures (i.e. in a medium without other organisms present), but under these conditions it was always necessary to add some growth factor, usually tissue extract, to obtain growth and reproduction. The required growth factor is a heat-sensitive substance, either a protein or a protein-bound material (for a review of axenic cultures of nematodes see Rothstein and Nicholas, 1969). Of the species of direct interest here, *Diplogasteritus nudicapitatus* and *Diplogasteroides* sp. have been reared in a medium consisting of casein, gelatin, thiamine and buffer, supplemented by rabbit serum (Chaudhuri, 1964). However, it was found that growth and reproduction in all axenic cultures were considerably lower than in cultures fed with bacteria (Cryan *et al.*, 1963). This implies that under ecological conditions the importance of nematodes in mineralizing dissolved organic material is low, and bacteria form their main food supply. This is in accordance with the ecological studies of Sachs (1950) and earlier experiments of Dotterweich (1938) and Hirschmann (1952) on different saprobic species. Little information can be obtained from the literature on the food specificity of saprobic nematodes. The population growth of several rhabditids, including *D. nudicapitatus*, feeding on different nutrient agars and a broad spectrum of monoxenic bacterial cultures has been examined by Sohlenius (1968a,b). Good growth was obtained with several bacterial species (*Escherichia coli, Proteus vulgaris, Bacillus megatherium*), while others produced poor (*Pseudomonas fluorescens*) or no growth (*Bacillus* sp., unidentified, blue-green alga and *Vitreoscilla* sp.). Comparing these results with those from similar experiments on the suitability of different micro-organisms as food for the free-living saprobic stages of gastro-intestinal nematodes (McCoy, 1929; Wang, 1970), it appears that (a) saprobic species can use a wide, but not unlimited range of foods, but that (b) different nematode species may react differently to a microbial species, although several bacteria, such as *E. coli* and *Aerobacter aerogenes*, appear to be suitable for a variety of nematodes.

So far, little emphasis has been put on quantitative relationships between the bionomics of a species and the food concentration. The experiments reported above were simply qualitative, and no information was given on the influence of varying food concentrations on nematode growth and survival rates. Quantitative relationships between bacterial concentration and growth and reproduction of *Plectus palustris*, a freshwater benthic nematode feeding on bacteria, have been assessed by Schiemer *et al.* (in press). Since the ecology of the Plectidae is similar to that of the rhabditids, the results may be applicable to or at least indicative of used-water species. Growth and reproduction of *P. palustris* only occurred above a bacterial concentration of *ca.* 10^8 cells/ml (*Acinetobacter*) at 20°C. Above this value production rates increased with increasing bacterial numbers but asymptotic production was reached at 10^{10} cells/ml. Bacterial concentrations in percolating-filter slimes have been estimated to be in the order of 10^{10}–10^{11} cells/ml (Pike and Curds, 1971), which indicates that the food supply in percolating filters is at the optimal level for sewage nematodes. This assumes, however, that the bacterial species present are all digestible and that they are of high nutritive value. So far, the discussion has been based on results from laboratory cultures and experiments. To define the influence of single factors from ecological observations on used-water treatment plants is admittedly much more complicated and the results are far less easy to interpret. The few observations that have been made indicate that nematode densities increase with higher sewage strengths, and that nematodes occur at higher strengths than rotifers (mean permanganate values 84 mg/l against 58 mg/l) (Weninger, 1971a,b). More detailed data on optimal sewage strengths for nematode development have not yet been established.

B. Oxygen

An object of aerobic biological used-water treatment processes is to increase rates of decomposition by aeration. Despite this, anaerobic conditions may develop due to excessive bacterial oxidation. For example, O_2 gradients can occur in percolating filters, oxidation ponds and even in activated-sludge tanks (Liebmann, 1968; Sládeček, 1973). This is particularly true in older types of percolating filters, which are characterized by low sewage loads and low dosing rates, and O_2 depletion occurs in the inner parts. This is often correlated with the formation of ooze.

Within the metazoan fauna associated with used-water treatment plants, nematodes have been described as being the most resistant to O_2 depletion (Weninger, 1971a,b). On the other hand, only a few freshwater species seem to be well adapted to anaerobic conditions (review in Ott and Schiemer, 1973) and the majority of free-living nematodes is rather sensitive (Rogers, 1962). This is particularly true in rhabditids, judging from some well studied examples. In *Caenorhabditis briggsae*, for example, the oxygen demand is high and growth and reproduction rates decline with decreasing oxygen concentration. Under anaerobic or microaerobic ($<5\%$ O_2) conditions, individuals can survive for up to a maximum of 80 h at 20°C (Nicholas and Jantunen, 1966; Cooper and Van Gundy, 1970). Similar results were obtained for other rhabditids (Bair, 1955).

Sensitivity towards low O_2 concentration may explain the lower importance of nematodes in older used-water treatment processes, e.g. Imhoff and digestion tanks. The decline of population densities within percolating filters is also associated with low oxygen concentrations in connection with decreased food supply. Strong O_2 depletion in such zones may be indicated by the occurrence of *Tobrilus gracilis*, a species with high tolerance to anaerobic conditions (Schiemer *et al.*, 1969).

C. Chemistry

The chemical ecology of sewage nematodes has not yet been studied in detail, and a discussion of this topic has therefore to be based on a wider level of experience. Behavioural aspects, namely the reactions of species to chemical stimuli, have been reviewed by Croll (1970) and will not be discussed here. Of greater relevance is the question, how do chemical conditions other than oxygen influence the growth and population development of nematodes? The chemical environment in which a sewage nematode has to live, is characterized by high and fluctuating concentrations of inorganic salts, a variety of dissolved organic materials and bacterial metabolites. High concentrations of ammonia, nitrates, phosphates, chlorides and other ions are the result of decomposition processes or are due to industrial and agricultural wastes. Hence the question arises, how do the concentrations, composition and fluctuations of inorganic ions influence nematode populations?

Bench-scale experiments by Weninger (1971a) indicated that various inorganic salts may have different effects on used-water treatment-plant biocenosis. He found that an addition of nitrate to domestic sewage caused an increase in the nematode population, while addition of phosphates resulted in higher rotifer densities. The same author was able to show that nematode peaks, occurring during the winter months, could be correlated with high concentrations of nitrate. Tolerance of change in osmotic pressure is generally high in nematodes (Bird and Wallace, 1969), and sapropelic species living for example in faeces have to be adapted to osmotic pressures up to 10 atm and occasionally even higher (Osche, 1952).

Both inhibition and stimulation of hatching, moulting and growth by microbial and plant exudates are well known from plant-parasite forms (Bird and Wallace, 1969), but are less well studied for non-parasitic nematodes. A possible control of sewage nematodes by microbes was discussed by Murad (1966), who described an antibiotic action of *Aspergillus* sp. against a sewage nematode (*Pelodera chitwoodi*) under experimental conditions. It may be expected that tolerance limits for such substances will be higher in saprobic forms than in species living in microbially less intensive environments.

D. Substratum

Nematodes occur in a wide range of substratum types but never live exclusively in a free-swimming mode. *Sphaerotilus* growths, filter film, sand and sedimented ooze all provide a fine-structured interstitial substratum adequate for good population growth. Only in the activated-sludge process is a convenient mechanical environment missing. Organic flocs appear to be too small and unstable, and locomotory movements may be too costly, in terms of energy, to allow maximum population development. This, together with high organic load and an unfavourable relationship between the population doubling time of a species and sewage retention time and sludge age (see Chapter 7 on Rotifera) may be the main reason for the small importance of nematodes in this used-water treatment type (Chaudhuri *et al.*, 1965).

V. Seasonal and Spatial Distribution Patterns

It appears that population dynamics are linked mainly with changes in the quantity and quality of the applied sewage and with operational characteristics of a specific filter. Strong population fluctuations have been recorded from trickling filters (Hausmann, 1923; Weninger, 1964, 1971b; Murad and Bazer, 1970), drying beds (Murad, 1970) and sewage plant effluents (Chaudhuri *et al.*, 1965). Murad and Bazer (1970) and Weninger (1971b) observed the population densities to increase with decreasing temperatures. Population maxima occurred at what might be considered to be a sub-optimal range (7–10°C), considering the results of laboratory experiments on population growth (see Section VI). Chaudhuri *et al.* (1965), however, found maximal population development in aerobic used-water treatment plants at temperatures between 17–18°C, i.e. a better agreement with experimental results. Winter peaks (Weninger, 1971b) coincided with increasing dosing rates, higher concentrations of nitrates and stronger growths of *Sphaerotilus*, and seem to be the result of multifactorial effects rather than a simple temperature relationship. Population decrease during spring was linked with sloughing processes. Strong population variations could be observed after re-operating filters which had been closed for some time. Nematodes proved to be part of the pioneering community and reached peaks two weeks after commencement (Weninger, 1964). *Paroigolaimella bernensis*, in particular, tends to develop short-term maxima on such occasions, and also after marked changes in the ecological conditions of the filter.

Spatial distribution patterns can be expected in treatment processes where strong ecological gradients can become established. This is the case in some types of percolating filters and oxidation ponds, though to a lesser extent in activated-sludge tanks. In percolating filters, vertical zonation patterns form due to the development of a film and deoxygenation processes in the inner parts of the filter, which result in gradients of pO_2, pH and microbial activity. Such gradients are more apparent in low-rate filters than in high-rate filters. In the latter, the formation of slime and the deoxygenation processes may be reduced by recirculation of processed used water.

Highest numbers of nematodes appear at the surface of a filter, particularly in the *Sphaerotilus* growth (Weninger, 1964). Population densities decrease towards the inner parts of the filter but the species composition remains similar.

VI. The Role of Nematodes in Purification Processes

The effect of nematodes on decomposition processes is primarily dependent on their population densities and the specific rates of feeding, respiration and growth. In the case of population density, only sundry data are available, and more information is both desirable and necessary for an understanding of the importance of nematodes in used-water treatment. Weninger (1964) reported maximal values of 180 individuals per ml from *Sphaerotilus* growth on a percolating-filter surface and 32 individuals/ml from inside the filter, but maximal values occurred during the winter at low temperatures. Scherb (1968) recorded up to 240 individuals/ml during bench-scale experiments with activated sludge. On agar-bacteria plates, however, population densities of up to 1000 sewage species per ml have been found (Pillai and Taylor; 1968a), and we can make the hypothetical assumption that population densities in percolating filters may approach similar values.

Information on specific rates of feeding, respiration and growth is more extensive, if it is valid to apply data of species not commonly found in used water but with an otherwise similar ecology; for example, studies on *Plectus palustris*, a freshwater benthic nematode, provide information on the ingestion rates of a bacteria-feeding species (Duncan *et al.*, 1974). As the feeding ecology is similar to rhabditids, and pharyngeal pumping rates are in the same order of magnitude (*P. palustris* 200 pumps per min, *Rhabditis axei* 300, *Panagrellus silusiae* 120–240 (the last two rates according to Mapes, 1965a,b)), the quantitative data obtained may be used as a first approximation for used-water species. In a feeding medium of $5–10\times10^9$ bacteria cells per ml, and at experimental temperatures of 20°C, adult females of *Plectus* ingested approximately 5000 cells per min, which is equivalent to $1{\cdot}94\times10^{-6}$ g dry weight per day and equal to 650% of the animal's body weight. However, only 12% of the ingested material was assimilated (Duncan *et al.*, 1974). Respiration rates have been determined for a number of free-living nematodes, using mainly Cartesian diver techniques (Holter and Zeuthen, 1966). A respiration (in units of O_2 $\mu l\times10^{-3}$ per individual per hour)–body weight (fresh weight in μg) regression calculated by Klekowski *et al.* (1972) for free-living nematodes from a wide range of habitats and using data from different authors, gave $R = 1{\cdot}40\ W^{0{\cdot}72}$ for a temperature of 20°C. The intercept of the regression line on the y-axis (the constant 1·4) indicates the oxygen consumption of an individual per hour when W equals 1 μg fresh weight.

Respiration rates of representatives of the order Rhabditida range from 1·1–3·5 with a highest value of 4·3 for *Panagrellus redivivus* (in units of O_2 $\mu l\times10^{-3}/\mu g$ fresh weight per hour). A value of two might therefore be more appropriate as a first approximation of respiration rates of sewage species from well aerated habitats. This value conforms with the respiration–body weight regression for the saprobic rhabditid *Panagrolaimus rigidus:* $R = 2{\cdot}01\,W^{0{\cdot}67}$ for 20°C and including all life stages (Klekowski *et al.*, 1974). Species adapted to oxygen-deficient strata, such as *Tobrilus gracilis,* tend to have lower metabolic rates (Schiemer and Duncan, 1974). The value of the power factor, giving the weight dependency of respiration, ranges in regression on life-cycles of individual species from 0·65–0·85 (Schiemer and Duncan, 1974), and has to be applied when respiration rates for different weight classes of nematodes are calculated.

Although there is little published data on the influence of temperature on nematode metabolism, there appears to be no adaptive response (Overgaard-Nielsen, 1949, 1961). Therefore one is justified in assuming that respiration rates are related to temperature by Kroghs's "normal curve" (for a tabulation see Winberg, 1971, p. 54). Food concentration seem to have no significant influence on respiration, at least not within "ecological ranges" (Klekowski *et al.*, in press).

Rhabditid nematodes are easy to culture and doubling times have been established for several species under controlled temperature conditions (*Diplogasteritus nudicapitatus, Pelodera teres*—Sohlenius, 1968a; *Paroigolaimella bernensis, Fictor anchicoprophaga*—Pillai and Taylor, 1968a; *Demaniella basili*—Pillai and Taylor, 1968b) (Table IIA). In the majority of species the doubling time is as short as 3–4 days at 20°C. The relationship of temperature to development time has been found to vary in different species. For example, in *P. bernensis* strongest dependence is between 20 and 25°C, in *F. anchicoprophaga* between 15 and 20°C, and in *D. basili* between 10 and 15°C. Upper and lower temperature limits of development differ correspondingly. Furthermore, one has to take into account that different development processes may have different temperature limits. In *Rhabditis terricola,* for example, egg production stops at 25°C, growth at 25–28·5°C and hatching at 28·5–30°C (Sohlenius, 1968). These results are interesting from a management point of view, since they allow the prediction of optimal temperature ranges for nematode growth. The occurrence of population maxima at low temperatures (8·5–10°C) in percolating filters (Weninger, 1971a,b), however, shows, that the application of laboratory data to field situations has to be carried out with caution.

The fecundity of rhabditids is high when compared with other free-living nematodes, with the exception of certain species of *Plectus.* For example, daily egg production per female of *Rhabditis terricola* was 18 at 15°C and 40 at 22°C, but the reproductive period in this species is limited

TABLE II. (A). Population doubling time (hours) of sewage nematodes in relation to temperature. (* denotes temperatures at which no growth occurred)

Species	Author	Temperature °C 10	15	20	25	30	35
Paroigolaimella bernensis	Pillai and Taylor (1968a)	*	90–100	76–80	54–57	46–48	*
Fictor anchicoprophaga	Pillai and Taylor (1968a)	*	160–178	80–90	54–56	46–48	40–44
Demaniella basili	Pillai and Taylor (1968b)	432–480	216–240	120–156	120–144	108–120	*

(B). Time of population maxima (days) in bacteria–agar cultures after inoculation with gravid females. (* denotes low population maxima)

Species	Author	Temperature °C 15	20	22	25	30	35
P. bernensis	Pillai and Taylor (1968a)		11		8	5	
F. anchicoprophaga	Pillai and Taylor (1968a)		10		10	5	5*
D. basili	Pillai and Taylor (1968b)	35*	35		35	14	
Pelodera teres	Sohlenius (1968a)			8–10			
Diplogasteritus nudicapitatus	Sohlenius (1968a)			10–12			

to 1–2 weeks (Sohlenius, 1968a). A much more extended reproductive period was observed in *Plectus palustris*, where egg production continued for about 8 weeks at a very regular daily production rate. This rate was dependent on food concentration and was (at maximum) 38 eggs/day at 10^{10}–10^{11} bacteria/ml at 20°C (Schiemer *et al.*, in press). Because of the short doubling times and the high fecundity, mass populations can develop within 5–10 days under laboratory conditions (Table IIB) and within 2–3 weeks in the field (Weninger, 1964, 1971a,b), exploiting periods of good conditions.

Data on nematode feeding and metabolism (as discussed above), combined with results on bacterial growth and population densities, can be used to estimate the possible influence of nematodes on the efficiency of used-water treatment plants. One positive influence resulting from the mechanical action of loosening the filter slime may be to make better aeration possible (Peters, 1930). However, the major influence of nematodes is probably on decomposition processes by their metabolism, and on bacteria by their grazing activities. This may be discussed in a simplified energy-flow diagram of a percolating filter (Fig. 2). Three main

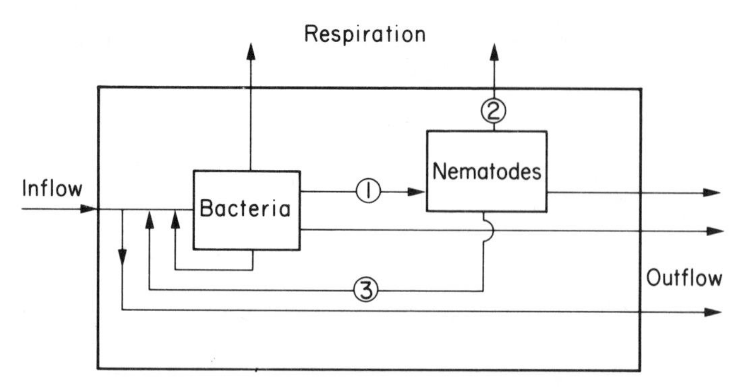

FIG. 2. Simplified energy-flow diagram of a percolating filter containing bacteria and nematodes. Numerals refer to links mentioned in text.

links are evident: (1) nematode grazing on bacteria and the effect on bacterial density and growth; (2) decomposition of organic material by nematode metabolism; (3) recycling of energy-rich substances (excretion products, faeces, dead body tissues) inside the filter and loss due to outflow.

Let us first consider the possible role of nematode grazing on`bacterial growth. The rate of change of a bacterial population is governed by the balance between gains and losses. This may be conveniently expressed as

$$\frac{dx}{dt} = \mu x - (Gx + Dx + Mx)$$

where μx represents the gain by bacterial growth, and the bracketed terms the compound losses due to grazing (G), outflow (D) and mortality (M). All rates (μ, G, D, M) are given as log units per hour (Uhlmann, 1971).

Assuming steady state conditions $\left(\frac{dx}{dt} = 0\right)$ at a level of 10^{10} cells/ml, the rate of bacterial growth and hence their activity will be determined by the population losses. The specific grazing rate of nematodes per unit volume (ml) sewage, was calculated using the data obtained for *Plectus* (see above) and assuming an exponential relationship between grazing and bacterial density. These rates were 0·0305 $\log_e$ units/h for a nematode population density of 1000 individuals/ml, and 0·003 loge units/day for a population density of 100 individuals/ml. To cover these losses the bacterial growth rates have to equal those values. Assuming logarithmic population growth, the necessary doubling time (t_d) can be calculated by

$$t_d = \frac{\log_e 2}{\mu}$$

Thus, the doubling time would have to be 22·8 h in the first case and 228 h in the second, to keep the bacterial concentration at a level of 10^{10} cells/ml. Maximal values of bacteria densities under laboratory conditions are similar to those found in sewage plants (10^{10}–10^{11} cells ml^{-1}). As the bacterial population approaches these levels, growth rates decrease and are mainly dependent on population losses. Therefore grazing, which tends to reduce these maximal values, clearly has a stimulating effect on bacterial growth rates.

Finally we must discuss the question, to what extent are nematodes directly responsible for decomposition of organic material via respiration. Let us assume a filter has a passage time of 1 hour and a community respiration of 50 mg O_2/l during this period. Assuming that there is a mean nematode fresh weight of 1 μg and using the respiration rate for 20°C as discussed above (2·8 $\mu g \times 10^{-3}/\mu g$/h), the oxygen consumption due to nematodes would be only 2·8 μg O_2/h at a population density of 1000 nematodes/ml and 0·28 μg O_2/h with 100 nematodes/ml. Since nematode densities tend to be be lower within the filter, the second value appears to be the more realistic one. This means that the oxygen consumption due to nematodes will be in the order of 0·001% of the community respiration in percolating filters. From calculations of this type it is possible to demonstrate that meiofauna in general, and nematodes in particular, may have important effects on bacterial activity, but that the direct role of nematodes in used-water decomposition is likely to be negligible.

References

Agersborg, H. P. K. (1929). *Trans. Am. microsc. Soc.* **48,** 158–180.

Agersborg, H. P. K. and Hatfield, W. D. (1929). *Sew. Wks J.* **1** (4), 411–424.

Andrássy, I. (1959). *Acta. zool. Acad. Sci. Hung.* **5,** 191–240.

Andrássy, I. (1960). *Acta. zool. Acad. Sci. Hung.* **6,** 1–28.

Andrássy, I. (1964). *Ann. Univ. Sci. Budapest Sect. Biol.* **7,** 3–18.

Andrássy, I. (1969). *Opusc. zool. Budapest* **9,** 187–233.

Andrássy, I. (in prep.). *In* "Bestimmungsbücher zur Bodenfauna Europas" (Ed. Franz).

Bair, T. D. (1955). *J. Parasit.* **41,** 613–623.

Bird, A. F. and Wallace, H. R. (1969). *In* "Chemical Zoology" (Eds M. Florkin and B. T. Scheer), Vol. 3, pp. 561–592. Academic Press, New York and London.

Calaway, W. T. (1963). *J. Wat. Pollut. Control Fed.* **35,** 1006–1016.

Calaway, W. T. and Tarjan, A. C. (1973). *J. Nematol.* **2,** 107–116.
Chang, S. L. and Kabler, P. W. (1962). *J. Wat. Pollut. Control Fed.* **34,** 1256–1261.
Chaudhuri, C. (1964). Thesis, University of Illinois, Urbana, Illinois, U.S.A.
Chaudhuri, N., Engelbrecht, R. S. and Austin, J. H. (1965). *J. Am. Wat. Wks Ass.* **57,** 1561.
Chitwood, B. G. (1958). *Bull. Zool. Nomencl.* **15,** 860–895.
Chitwood, B. G. and Chitwood, M. B. (1950). "An Introduction to Nematology." Monumental Printing Co., Baltimore, Maryland, U.S.A.
Cobb, N. A. (1918). *Contr. Sci. Nematol.* **7,** 189–212.
Cooper, A. F. and Van Gundy, S. D. (1970). *J. Nematol.* **2,** 305–315.
Crofton, H. D. (1966). "Nematodes." Hutchinson University Library, London.
Croll, N. A. (1970). "The Behaviour of Nematodes." Edward Arnold, London.
Cryan, W. S., Hansen, E. L., Martin, M., Bayre, F. W. and Yarwood, E. A. (1963). *Nematologica,* **9,** 213–319.
De Coninck, L. (1965). "Traité de Zoologie, Némathelminthes (Nématodes)" (Ed. P. Grassé), Tome IV, Fascicule II, 3–217. Masson et Cie, Paris.
Doncaster, C. C. (1962). *Nematologica* **8,** 313–320.
Dotterweich, H. (1938). *Zool. Anz.* **122,** 266–268.
Dougherty, E. C. (1955). *J. Helminth.* **29,** 105–152.
Duncan, A., Schiemer, F. and Klekowski, R. Z. (in press). *Pol. Arch. Hydrobiol.*
Ferris, V. R. (1971). *In* "Plant Parasitic Nematodes" (Eds B. M. Zuckermann, W. F. Mai and R. A. Rohde), Vol. 1, pp. 163–189. Academic Press, New York and London.
Florkin, M. and Scheer, B. T. (1969). "Chemical Zoology" Vol. III. Academic Press, New York and London.
Goodey, T. (1963). "Soil and Freshwater Nematodes." Methuen, London.
Hausmann, L. A. (1923). *New Jersey Agric. Exp. Stn Bull.* **390,** 28–39.
Hirschmann, H. (1952). *Zool. Jb.* (*Syst.*) **81,** 313–436.
Holter, H. and Zeuthen, E. (1966). *In* "Physical Techniques in Biological Research" (Eds G. Oster and A. W. Pollister), Vol. 3, pp. 251–317. Academic Press, New York and London.
Husmann, S. (1958). *Abh. Braunschw. Wiss. Ges.* **10,** 93–116.
Klekowski, R. Z., Wasilewska, L. and Paplinska, E. (1972). *Nematologica* **18,** 391–403.
Klekowski, R. Z., Schiemer, F. and Duncan, A. (in press). *Pol. Arch. Hydrobiol.*
Lackey, J. B. (1924). *New Jersey Agric. Exp. Stn Bull.* **403,** 40–60.
Lee, D. L. (1965). "The Physiology of Nematodes". Oliver and Boyd, Edinburgh and London.

LIEBMANN, H. (Ed.) (1968). "Tropfkörper und Belebungsbecken." R. Oldenbourg, Munich and Vienna.
LLOYD, L. (1945). *J. Inst. Sew. Purif.* (1), 3–22.
MAGGENTI, A. R. (1961). *Proc. helminth. Soc. Wash.* **28,** 138–166.
MAPES, C. J. (1965a). *Parasitology* **55,** 269–284.
MAPES, C. J. (1965b). *Parasitology* **55,** 583–594.
MCCOY, D. R. (1929). *Am. J. Hyg.* **10,** 140–156.
MEYL, A. H. (1960). *In* "Die Tierwelt Mitteleuropas" (Eds P. Brohmer, P. Ehrmann and G. Ulmer). Quelle and Mayer, Leipzig.
MURAD, J. L. (1965). Ph.D. Dissertation. Dept. of Biology, Texas A and M Univ., College Station, Texas, U.S.A.
MURAD, J. L. (1966). *Tex. J. Sci.* **18,** 90–91.
MURAD, J. L. (1970). *Proc. helminth. Soc. Wash.* **37** (1), 10–13.
MURAD, J. L. and BAZER, G. T. (1970). *J. Wat. Pollut. Control Fed.* **42** (1), 105–114.
NICHOLAS, W. L. and JANTUNEN, R. (1966). *Nematologica* **12,** 328–336.
OSCHE, G. (1952). *Zool. Jb.* (*Syst.*) **81,** 175–312.
OTT, J. and SCHIEMER, F. (1973). *Neth. J. Sea Res.* **7,** 233–243.
OVERGAARD-NIELSEN, C. (1949). The soil inhabiting nematodes. *Nat. Jutl.* **2,** 1–131.
OVERGAARD-NIELSEN, C. (1961). *Oikos* **12** (1), 18–35.
PETERS, B. G. (1930). *J. helminth.* **8,** 165–184.
PIKE, E. B. and CURDS, C. R. (1971). *In* "Microbial Aspects of Pollution" (Eds G. Sykes and F. A. Skinner), pp. 123–147. Academic Press, London and New York.
PILLAI, J. K. and TAYLOR, D. P. (1968a). *Nematologica* **14,** 89–93.
PILLAI, J. K. and TAYLOR, D. P. (1968b). *Nematologica* **14,** 159–170.
PILLAI, J. K. and TAYLOR, D. P. (1968c). *Nematologica* **14,** 285–294.
ROGERS, W. P. (1962). "The Nature of Parasitism." Academic Press, New York and London.
ROTHSTEIN, M. and NICHOLAS, W. L. (1969). *In* "Chemical Zoology" (Eds M. Florkin and B. T. Scheer), Vol. 3, pp. 289–328. Academic Press, New York and London.
SACHS, H. (1950). *Zool. Jb.* (*Syst.*) **79,** 209–320.
SASSER, J. N. and JENKINS, W. R. (1960). "Nematology, Fundamentals and Recent Advances with Emphasis on Plant Parasitic and Soil Forms." University of North Carolina Press, Chapel Hill, North Carolina, U.S.A.
SCHERB, K. (1968). *In* "Tropfkörper und Belebungsbecken" (Ed. H. Liebmann), pp. 158–206. R. Oldenbourg, Munich and Vienna.
SCHIEMER, F. (1971). *Carinthia* **II,** 147–157.
SCHIEMER, F. and DUNCAN, A. (1974). *Oecologia* **15,** 121–126.
SCHIEMER, F., LÖFFLER, H. and DOLLFUSS, H. (1969). *Verh. int. Ver. Limnol.* **17,** 201–208.
SCHIEMER, F., DUNCAN, A. and KLEKOWSKI, R. Z. (in press). *Pol. Arch. Hydrobiol.*

SLÁDEČEK, V. (1973). *A. Hydrobiol. Beih. Ergebu. Limnol.* **7,** 1–218.
SOHLENIUS, B. (1968a). *Pedobiologia* **8,** 137–145.
SOHLENIUS, B. (1968b). *Pedobiologia* **8,** 340–344.
STEINMANN, P. and SURBECK, G. (1918). "Preisschr. der Schweiz. Zool. Ges.". Bern.
TARJAN, A. C. (1961). "Check-list of Plant and Soil Nematodes: A Nomenclatorial Compilation." Univ. of Florida Press; Gainesville, Florida, U.S.A.
THORNE, G. (1939). *Capit. Zool.* **8,** 1–261.
UHLMANN, D. (1971). *Mitt. int. Ver. Limnol.* **19,** 100–124.
WANG, G.-T. (1970). *J. Parasit.* **56,** 753–758.
WEINGÄRTNER, I. (1955). *Zool. Jb.* (*Syst.*) **83,** 248–317.
WENINGER, G. (1964). *Wasser und Abwasser* **4,** 96–167.
WENINGER, G. (1971a). *Sitz.-Ber. Öst. Akad. d. Wiss.* **179,** 1–32.
WENINGER, G. (1971b). *Sitz.-Ber. Öst. Akad. d. Wiss.* **179,** 129–158.
WINBERG, G. G. (1971). "Methods for the Estimation of Production of Aquatic Animals." Academic Press, London and New York.

Note added in proof

Recently some papers on feeding and energetics of saprobic nematodes have been published:

MARCHANT, R. and NICHOLAS, W. L. (1974). *Oecologia* **16,** 237–252.
MERCER, E. K. and CAIRNS, E. J. (1973). *J. Nematol.* **5,** 201–208.
NICHOLAS, W. L., GRASSIA, A. and VISWANATHAN, S. (1973). *Nematologica* **19,** 411–420.

7

Rotifera

Margaret Doohan*

Department of Zoology
Royal Holloway College
Englefield Green
Surrey
England

I. Introduction

The role of Rotifera in used-water treatment processes has received very little attention, probably because of their low numbers relative to the protozoa. Yet their biology, their opportunistic exploitation of a suitable environment and their relative inability to respond even to small environmental changes, make them ideal indicator species (Arora, 1966; Ruttner-Kolisko, 1971).

Sydenham (1971) found on the basis of volume that rotifers, ciliates and rhizopods occurred in the proportion 1:4:5 in activated sludge and suggested that further work on rotifers and rhizopods was called for. Prior to this, McKinney (1967) suggested that Rotifera indicated 95–100%

* Present name and address: Margaret Frayne, *Avery Hill College of Education, Bexley Road, Eltham, London SE9 2PQ.*

purification of sewage in activated sludge and that they appear after the protozoa have died off. Calaway (1968) considers that rotifers can appear earlier in the purification sequence and he gives a good summary of rotifer distibution and their role in used-water treatment. The classification system he used is out-dated but this does not detract from the useful, if speculative, discussion.

Klimowicz (1968, 1970a,b,c,d, 1972) has produced several species lists of rotifers occurring in oxidation ponds and activated sludge and has reviewed some of the literature (Klimowicz, 1973). His 1970 papers also describe some experimental work with activated sludge using different sewage loadings and aeration times but the results tend to be inconclusive due to the number of variables involved. Arora (1966) provides a species list for oxidation ponds, and Godeanu (1966) attempts to group the rotifers into those species characteristic of treatment plants and those appearing and surviving in low numbers and for brief periods. None of the published work gives reliable quantitative information on rotifera and most studies on the fauna of used-water treatment either omit them altogether or mention them only briefly.

II. Taxonomy

According to Voigt (1957) the class Rotifera can be subdivided into three orders of which one, the Seisonidae, is entirely marine. The other two, the Bdelloidea and Monogononta, are widely represented in used-water treatment processes but in differing proportions (see Tables I and II). The Bdelloidea consists of four families, three of which have been recorded in activated sludge and percolating filters. The Monogononta consists of eleven families of which the majority are planktonic, though there is a clear developmental series within the group, from attached, creeping forms resembling Bdelloidea, to completely free-floating species. This group is represented in all three treatment processes but particularly in oxidation ponds.

At the species level rotifer taxonomy tends to be confused and different authorities may not agree. In Monogononta sexual reproduction is rare and it is completely lacking in the Bdelloidea. Consequently the most reliable criterion for species definition cannot be applied to the group and classification even at the species level is largely based on the morphology and anatomy of the female.

The most important and comprehensive taxonomic works on Rotifera have been published in German, the most recent being that of Voigt (1957). Ruttner-Kolisko (1972a) presents an excellent simplified key, and the survey of Romanian Rotifera by Rudescu (1960) also contains a useful review of the taxonomy. Monographs on particular genera and

TABLE I. A species list of Rotifera reported to occur in used-water treatment processes

Order BDELLOIDEA
Family Habrotrochidae
Habrotrocha tripus (Murray)
H. flava Bryce
H. bidens (Gosse)
H. rosa Donner
H. thienemanni Hauer
H. constricta (Dujardin)
Family Philodinidae
Philodina citrina Ehrb.
P. eurythrophthalma Ehrb.
P. acuticornis Murray
P. flaviceps Bryce
P. megalotrocha Ehrb.
P. roseola Ehrb.
Dissotrocha macrostyla (Ehrb.)
Macrotrachela concinna (Bryce)
Rotaria citrina Ehrb.
R. rotatoria (Pallas)
R. trisecata (Weber)
R. tardigrada (Ehrb.)
Family Adinetidae
Adineta vaga (Davis)
Order MONOGONONTA
Sub-order PLOIMA
Family Brachionidae
Sub-family Brachioninae
Epiphanes senta (Müll.)
Brachionus angularis Gosse
B. calyciflorus Pallas
B. quadridentatus Hermann
B. urceolaris Müll.
B. rubens Ehrb.
Keratella cochlearis (Gosse)
Euchlanis dilatata Ehrb.
Sub-family Colurinae
Colurella colurus (Ehrb.)
C. adriatica Ehrb.
C. bicuspidata (Ehrb.)
C. geophila Donner
Lepadella acuminata (Ehrb.)
L. patella (Müll.)

TABLE I (*Continued*)

Family Lecanidae
Lecane stichaea Harring
L. tenuiseta Harring
L. clara (Bryce)
L. inermis (Bryce)
L. arcuata (Bryce)
L. galeata (Bryce)
L. pyriformis (Daday)
L. lunaris (Ehrb.)
L. bulla (Gosse)
L. closterocerca (Schmarda)
L. hamata (Stokes)
L. decipiens (Murray)
**L. pideis* (Harring & Myers)
Proales decipiens (Ehrb.)
Family Notommatidae
Notommata cyrtopus Goose
N. glyphura Wulfert
Cephalodella catellina (Müll)
C. gibba (Ehrb.)
C. gracilis (Ehrb.)
C. gracilis var. *lenticulata* Wulfert
Pleurotrocha petromyzon Ehrb.
Eosphora sp.
Family Dicranophoridae
Dicranophorus forcipatus (Müll.)
D. grandis (Ehrb.)
Encentrum lupus Wulfert
Family Asplanchnidae
Asplanchna priodonta Goss
Sub-order FLOSCULARIACEA
Family Testudinellidae
Filinia longiseta (Ehrb.)

* First record in Europe if identification is correct

families are available but these mainly deal with planktonic Monogononta and are of greatest use in determining Rotifera in oxidation ponds. Unfortunately the monographs tend to emphasize morphological detail at the expense of ecological significance (Ahlstrom 1940, 1942; Berzinš 1951, 1954; Petjler 1957a,b, 1962; Parise, 1961). The most useful simple keys in English are those of Donner (1966) and Edmondson

(1959). Bartos (1966) contains an English summary of the literature relating to Bdelloidea which are notoriously difficult to identify, and Donner (1965) deals particularly with soil Bdelloidea. A list of species reported to be found in used-water treatment processes is given in Table I.

III. Factors Governing Rotifer Distribution in Used-Water Treatment Processes

Table II is a summary of information from various authors arranged in families. Not all the available data were used, since some covered several years' samples and so gave undue importance to incidental occurrences of some species. Godeanu (1966) presents the results of several years' work, but Table II only records the species which occur with some regularity in his samples.

In activated sludge and percolating filters both Bdelloidea and Monogononta are represented, with only minor differences in the number of species from each order. In oxidation ponds, however, Monogononta

TABLE II. Numbers of Rotifer species found in different treatment processes

	Activated sludge		Percolating filters	Oxidation ponds	
	Calaway (1968)	Godeanu (1966)	Dommer (1966)	Arora (1966)	Klimowicz (1968)
BDELLOIDEA					
Adinetidae	1		1		
Philodinidae	4	6	4	1	3
Habrotrochidae	1	4	4		3
Totals:	6	10	9	1	3
MONOGONONTA					
Sub-order Ploima					
Brachionidae	2	1	2	2	7
Lecanidae	2	7	2		
Notommatidae	1	1	2	2	1
Dicranophoridae			1		
Asplanchnidae				1	1
Sub-order Floscularіaceae					
Testudinellidae				3	1
Totals:	5	9	7	7	10

predominate with only few representatives of the Bdelloidea. This difference can be accounted for by a variety of factors, among which the physical nature of the habitat and the food substances available are of considerable importance.

A. Physical nature of the habitat

Rotifera may be grouped into benthic forms, partially planktonic, or euplanktonic according to the presence and relative importance of the foot. All Bdelloidea (e.g. *Philodina* sp.; Fig. 1c) have a well-developed foot with 0–4 toes or an attachment disc. Although they are capable of swimming, propelled by the ciliary currents produced by the corona, their usual method of locomotion is a leech-like crawling. They therefore tend to dominate localities where the ratio of free-water mass to substratum is

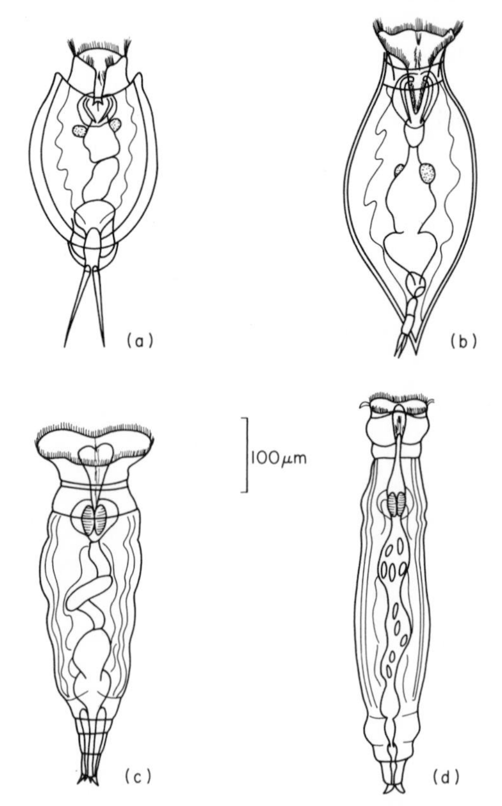

FIG. 1. Rotifera common in percolating filters and activated sludge. (a) *Lecane* sp., (b) *Notommata* sp., (c) *Philodina* sp., (d) *Habrotrocha* sp.

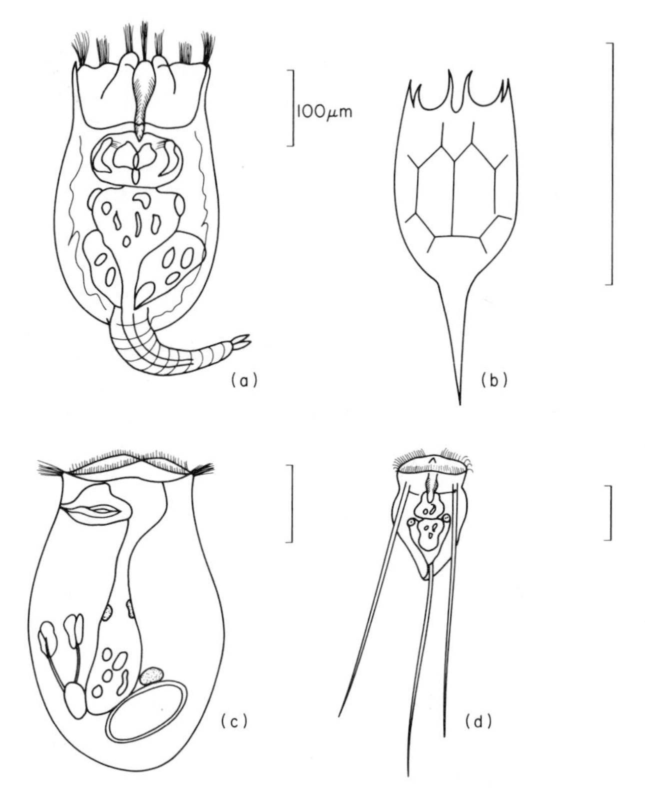

FIG. 2. Rotifera recorded from oxidation ponds. (a) *Brachionus calyciflorus*, (b) *Keratella cochlearis*, (c) *Asplanchna* sp., (d) *Filinia longiseta*.

relatively low. They are able to withstand long periods of desiccation; some Bdelloids secrete a protective cyst when environmental conditions become rigorous. Some remain in an encysted state for as long as three or four years and resume their active life in periods varying from some minutes to several hours after the return of suitable conditions (Hyman, 1951). This group is therefore particularly well suited to survive in transient environments.

The Monogononta contains genera showing all the gradations in development of the foot. It may be completely absent (e.g. *Keratella cochlearis*, *Asplanchna* sp.; Fig. 2b,c); reduced and used mainly for attachment rather than locomotion (e.g. *Brachionus* sp.; Fig. 2a); or it may be mainly locomotory (e.g. *Lecane* sp., *Notommata* sp.; Fig. 1a,b). Genera without a foot are euplanktonic, occurring in the pelagic zone of lakes, whilst those with a reduced foot are more common in the littoral zone and often dominate the zooplankton of small ponds. The predominant family of Monogononta in oxidation ponds is the Brachionidae, particularly the genus *Brachionus*, which contributes five of the seven species recorded from this family. *Brachionus* (Fig. 2a) is a loricate genus with a reduced foot such as one would expect in an environment where the free-water body is of some importance but not sufficient to exclude the influence of the substratum.

In activated sludge, where there is a substantial amount of suspended material, and in percolating filters with a large substratum area, the predominant family of Monogononta is the Lecanidae. Though morphologically this family closely resembles the Brachionidae, the most important genus in used-water treatment processes, *Lecane*, make extensive use of the foot in locomotion. This rotifer is most often found crawling over the surfaces of stones or algal filaments. It is therefore predictable, even from superficial observations, that it would assume greater importance in activated sludge then in oxidation ponds.

B. Food availability

Table III represents a regrouping of the information already presented in Table II, but this time the basis of the arrangement is feeding habit as demonstrated by the anatomy of the mastax (jaws). Voigt (1957) has made the mastax the basis of classification at the level of family. This classification therefore gives a good assessment of the ecological role of a particular rotifer species or genus and is often as valuable as the species definition itself. Quantitative work at the family level would have given more information regarding the role of Rotifera in used-water treatment in far less time than the detailed non-quantitative species lists which have so far been available.

TABLE III. Feeding groups of Rotifera

	Activated sludge		Percolating filters	Oxidation ponds	
	Calaway (1968)	Godeanu (1966)	Donner (1966)	Arora (1966)	Klimowicz (1968)
1. Groups with biting or grinding mastax	8	17	12	1	3
2. Brachionidae (chewing mastax)	2	1	2	5	8

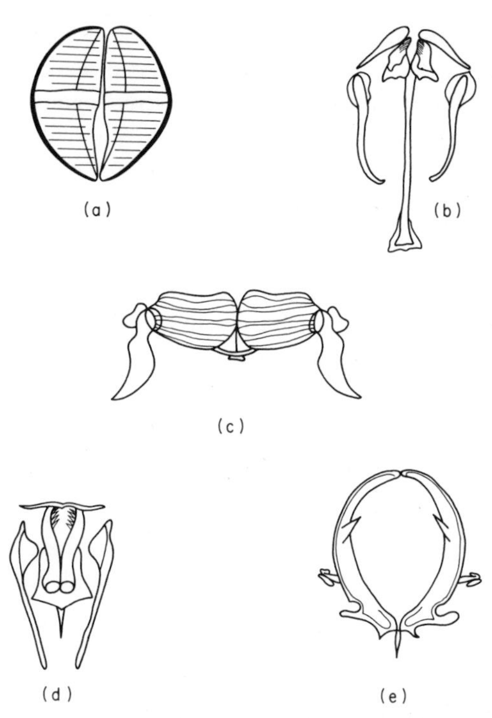

FIG. 3. Rotifer mastax types. (a) Ramate (grinding), (b) virgate (sucking), (c) malleate (chewing). Predatory forms: (d) forcipate, (e) incudate.

The various mastax types found in rotifers from used-water treatment are given in Fig. 3. Bdelloidea feed on fine suspended particulate matter brought to the jaws by ciliary currents produced by the corona. Alternatively, the corona is laid against the substratum so that the current releases particles which are then brought to the jaws by centrifugal action. The jaws are of the ramate type which grind larger particles before digestion. Meadow and Barrows (1971) have kept *Philodina acuticornis* var. *odiosa* in bacterial suspension only, and generally assumed that Bdelloidea feed on very fine particles. The stomach in this group is a syncytium with no lumen at all in the Habrotrochidae and only a narrow one in the Philodinidae. The digestive process is therefore in a sense intracellular and hence resembles the feeding process of the ciliate protozoa which have a similar ecological role to the Bdelloidea.

The Lecanidae have a malleate mastax but a slightly modified version of the more primitive Brachionid type. Besides chewing particles of food, the Lecanidae can also produce a certain amount of suction (Donner, 1966), augmenting the centrifugal action of the corona. Unlike the

Brachionid mastax, that of the Lecanidae is not at the base of a funnel-shaped buccal area. It can therefore be applied directly to large pieces of edible material which are then held by suction against the jaws to be chewed before entering the stomach. *Lecane* sp. is seldom seen upright and feeding by ciliary currents alone. More often the corona is applied to the substrate and the jaws clip off pieces of organic material.

The sucking mastax reaches the culmination of its development in the Notommatidae, where the development of a piston-like muscle attached to the broad rami, and the reduced area of other parts of the jaws, produce suction supplementing ciliary activity. Since the teeth of the unci are reduced, the Notommatidae and some other pelagic groups (e.g. Synchaetidae) are able to grasp large but soft particles which are sucked into the pharynx (Pourriot, 1965). Other groups feeding on large particles are predatory: the Dicranophoridae with a forcipate mastax, and the Asplanchnidae with an incudate mastax. These two groups are clearly less restricted in the matter of food particle size than the Bdelloidea, and the Dicranophoridae have been recorded from filters where they were preying on other rotifers. *Asplanchna*, on the other hand, is a highly adapted planktonic predator with a thin saccate body and no foot, totally unsuited to an environment rich in suspended material or with substrate predominating over the free-water mass.

The Brachionidae have a malleate, chewing mastax, but unlike that of the Lecanidae, it is not extrusible and can apply no suction. This group is therefore dependent on particulate matter brought to the jaws by the centrifugal action of the corona. The relative strength or weakness of this action limits the size of particle reaching the jaws to less than about 12 μm (Pourriot, 1965). This is larger than the usual particle size ingested by Bdelloidea.

On the basis of feeding mechanisms, as well as locomotor adaptations, the Lecanidae therefore show greater similarity to the Bdelloidea than to the Brachionidae which are nevertheless taxonomically their closest relatives. The nature of the available food material may also play some part in restricting the distribution of rotifers. Bdelloidea, as mentioned earlier, can survive on bacteria alone, but these and the Lecanidae are largely detritivores, feeding on suspended and flocculated organic matter. Brachionidae are generally algivorous and would therefore thrive in oxidation ponds, where surface illumination over a wide, shallow body of water provides ideal conditions for high levels of primary productivity. These high levels may not become apparent as a high standing crop of algae, due to the efficiency of the grazing population. Pennington (1941) showed the dramatic effect of grazing by *Brachionus calyciflorus* on a population of *Diogenes rotundus* in a closed system. But where nutrient levels are maintained by an external supply, the rotifer population will increase only to such proportions as can be sustained by the rapid turnover of a very active phytoplankton population. This is essentially the situation in oxidation ponds.

C. Other Factors

Though food and substratum play a major role in determining which groups of rotifers colonize which types of treatment plant, particular species or genera may be limited by specific biochemical demands. So far this aspect of rotifer ecology has not been investigated, but several studies have been made in other habitats and in cultures. Most rotifers have very fine tolerance limits for the variety of minerals which affect the pH of their habitat. Myers (1931) gives a list of apparently acidophilous benthic species, though Pejler (1957a) considers that planktonic forms may be influenced by pH only indirectly through variations in primary producers on which they feed. Lansing (1942) has correlated calcium deposition in cell membranes with aging in the rotifers *Euchlanis triquetra* and *Proales* spp. The fertility and longevity of the former increases in alkaline solutions over cultures maintained at pH 6 and 7, whilst the longevity of *Proales* was increased in low calcium concentrations. Clearly no generalizations can be made.

Calaway (1968) mentions temperature as a possible limiting factor for rotifers in used-water treatment at different latitudes. Green (1972) has found latitudinal associations of planktonic rotifers, finding some species confined to the tropics, some to arctic and temperate zones, and others cosmopolitan. No similar work has been attempted with bdelloids, but many of them and a number of Monogononta are noted for their resistance to high temperatures even in the active state, and resistance to extreme cold has also been recorded in Bdelloidea though only in an inactive, resistant condition (Hyman, 1951).

In addition, the metabolism of planktonic rotifers at least seems to be relatively insensitive to temperature above 15°C where the Q_{10} of respiration is less than 2 (Doohan, 1973). Pourriot and Deluzarches (1971) found greater temperature sensitivity in egg development rates, but Halbach (1970) found that mean population densities in the field were independent of temperature between 12 and 28°C. It seems that in the field situation, food availability exerts far greater influence than temperature on the birth rate of rotifers (Doohan, 1973). It is therefore unlikely that temperature limits rotifer colonization of used-water treatment plants.

IV. Population Growth Rates and Sludge Wastage Rates

The establishment of an animal community in an environment with a flowing medium is dependent on the relationship between the rate of flow of the medium and the specific growth rate of the population. The latter may be defined as the variable, *r*, in the logistic equation $N_t = N_o e^{rt}$, in which N_o is the population at time *o*, N_t the population after time *t*, *e* is the base of natural logarithms, and *r* is the resultant of the birth rate and death or elimination rates within the specified time *t*. In order that the population may survive under the conditions described, *r* must be at least equal to the rate at which the population is depleted by substrate extraction. If *r* is less than this, the population will decline. If it is greater, the population will build up until it is limited by other factors in the environment, such as food depletion or physico-chemical changes.

Most of the work on population dynamics of rotifers has dealt with planktonic forms, so that no further information on field populations of sessile rotifers can be added to that reviewed by Edmondson (1945). However, Edmondson's paper suggests that the population dynamics of sessile forms may be essentially similar to those of planktonic species. If this can be assumed, then certain generalizations can be made about the dynamics of sessile rotifer populations.

The birth rates of several field populations of planktonic rotifers have been calculated using the method devised by Edmondson (1960). This makes use of the fact that some rotifers carry their eggs, and that the development of the eggs can be followed. The birth rate is then equivalent, at any one temperature, to $\frac{E}{D}$ where E is the number of eggs per female in the population or sample, and $\frac{1}{D}$ the development rate. The latter is temperature-sensitive whilst egg number is more closely related to food availability. Using the work reviewed by Hutchinson (1967) and other data both in the literature and newly calculated, Doohan (1973) has suggested that the relationship between $\frac{1}{D}$ and temperature for most planktonic rotifers can be adequately described by the equation $y = 0{\cdot}044x - 0{\cdot}11$, where x is ambient temperature and y is $\frac{1}{D}$. This means that at 20°C the birth rate of a population in which each female is carrying one egg would be 0·77.

To give a more accurate picture of the growth of a rotifer population, one must also take into account the fact that rotifers have a short post-embryonic development period. Pourriot and Deluzarches (1971) give several measurements of this post-embryonic development time and from their data it appears that it is, on average, 1·59 times the egg

TABLE IV.

Sewage retention time:	High rate 2 h	Conventional rate 8 h	Low rate 36 h
Minimum specific growth rate for survival of free-swimming forms (h^{-1})	0·5	0·125	0·028
Maximum doubling time (h)	1·38	5·5	28·75
Minimum specific growth rate for the survival of sedentary forms ($days^{-1}$)	0·04	0·02	0·012
Maximum doubling time (days)	17	35	58

development time. One can therefore deduce that the average doubling time of a rotifer population at 20°C, assuming no deaths have occurred, would be 1·22 days. Substitution for t in the logistic equation when $N_t = 2N_o$ then gives a birth rate of 0·57 day^{-1} or 0·023 h^{-1}.

Table IV has been produced by Curds and Vandyke (1966) using equations proposed by Downing *et al.* (1964) and assuming that three plants with different sewage retention times are treating sewage of biochemical oxygen demand (BOD) 250 mg l^{-1} with a sludge concentration of 3000 mg l^{-1} and return sludge ratio of 1:1. An additional reason for the predominance of sessile or benthic rotifera in sludge may easily be deduced from this Table. If the birth rate of rotifers is only 0·023 h^{-1} then free-swimming forms can only survive in plants where the aeration period is unusually extensive. Extended aeration may therefore account for occasional records of Brachionidae and Notommatidae in sludge. With sessile or benthic forms the picture is quite different. With a minimum sewage retention time of 2 h the birth rate is well above the minimum specific growth rate for population survival and would still be so even if the entire reproducing population were removed or died after the eggs hatched. At temperatures higher than 20°C the birth rate would be accelerated according to the equation mentioned earlier, as it would also be if the food situation allowed the production of more than one egg per day by each female in the population.

V. Role of Rotifera in Used-Water Treatment Processes

The best available treatment of the role of rotifers is that of Calaway (1968) which deals mainly with activated sludge. Most of this Section is modified from his paper.

The two major feeding methods described earlier perform two distinct functions in the process of used-water treatment. They break up floc particles, providing nuclei for further floc formation, and they clear the supernatant by removing non-flocculated bacteria still in suspension. They also contribute to floc formation through the production of discrete faecal pellets consisting of undigested material bound together by a mucus secretion. In the Habrotrochidae, these particles are trapped in the mucus which forms the tube in which these sedentary rotifers live. In this way particles are effectively removed from suspension.

Calaway suggests that bdelloids are in a better competitive position than free-swimming protozoa regarding their energy requirements because they are usually stationary. This is not necessarily true. The energy available from digested food depends on the assimilation efficiency and the maintenance cost of the animal. The assimilation efficiency of rotifers varies inversely with the concentration of food in the medium (Galkovskaya, 1963). The greatest energy requirement of adult rotifers, even of free-swimming forms, is egg production, since rotifers grow only slightly after hatching (Ruttner-Kolisko 1972b; Pilarska, 1973; Doohan, in prep.). Clearly this is not so in the case of protozoa where reproductive costs must be low and closely related to growth. It seems far more likely that the appearance of rotifers late in the development of activated sludge is due to reduced grazing pressure from protozoa as progressive flocculation produces particles too large to be available to protozoa. Secondly, the stronger ciliary currents of the rotifers would be more effective in sedimenting the diminished numbers of suspended bacteria than the relatively weak currents produced by the vorticellids which precede them in sludge development. It seems that this gives the Rotifera a better competitive position rather than reduced energy requirements.

Similar principles would also govern the role of rotifera in percolating filters. In oxidation ponds rotifers have quite a different role (Klimowicz, 1968). Primary producers release oxygen which promotes bacterial decomposition of organic matter. The dominant rotifers (Table II) feed mainly on the primary producers and play only a secondary role in the removal of organic wastes. Nevertheless their contribution to the overall balance of the ecosystem cannot be overlooked.

VI. Concluding Remarks

Many generalizations have necessarily had to made in the compilation of this Chapter, due to the paucity of data. An essential requirement in future work is for good quantitative data on adult rotifers and egg numbers, preferably on a seasonal basis, allied with a careful monitoring of environmental conditions. From these basic studies, population dynamics of benthic rotifers could be related to ambient temperature and to food availability, as well as to the physico-chemical conditions, and some understanding of their role in used-water treatment processes might be obtained.

References

AHLSTROM, E. H. (1940). *Bull. Am. Mus. Nat. Hist.* **77,** 143–212.

AHLSTROM, E. H. (1942). *Bull. Am. Mus. Nat. Hist.* **80,** 411–457.

ARORA, H. C. (1966). *Hydrobiologia* **27,** 146–159.

BARTOS, E. (1966). *Vestnik Ceskoslovenské Zool. Spolecnosti—Acta Soc. Zool. Bohemoslovenicae, Svazak* **15,** 241–500.

BERZINŠ, B. 1951). *Ark. Zool.* **2,** 565–592.

BERZINŠ, B. (1954). *Ark. Zool.* **8,** 549–559.

CALAWAY, W. T. (1968). *J. Wat. Pollut. Control Fed.* **40,** 412–422.

CURDS, C. R. and VANDYKE, J. M. (1966). *J. appl. Ecol.* **3,** 127–137.

DONNER, J. (1965). Ordnung Bdelloidea (Rotatoria, Rädertiere). *In* "Bestimmungsbücher zur Bodenfauna Europas" (Lief 6). Akademie Verlag, Berlin.

DONNER, J. (1966). "Rotifera." Frederick Warne, London.

DOOHAN, M. (1973). "Energetics of planktonic rotifera applied to populations in reservoirs." Ph.D. Thesis, University of London.

DOWNING, A. L., PAINTER, H. A. and KNOWLES, G. (1964). *Proc. Inst. Sew. Purif.* **63**(2), 3–31.

EDMONDSON, W. T. (1945). *Ecol. Monogr.* **15,** 141–172.

EDMONDSON, W. T. (1959). *In* "Freshwater Biology" (Eds H. B. Ward and G. C. Whipple). John Wiley, New York.

EDMONDSON, W. T. (1960). *Mem. Ist. ital. Idrobiol.* **12,** 21–77.

GALKOVSKAYA, G. A. (1963). *Zool. Zh.* **42,** 506–516 (in Russian). Translation: *J. Fish. Res. Bd. Can.* 997.

GODEANU, S. (1966). *Studii de Protectia si Epurarea Apelor* **7,** 569–599 (in Romanian).

GREEN, J. (1972). *J. Zool. Lond.* **167,** 31–39.

HALBACH, U. (1970). *Oecologia (Berl.)* **4,** 176–207 (in German).

HUTCHINSON, G. E. (1967). "A Treatise on Limnology." John Wiley, New York.

HYMAN, L. H. (1951). "The Invertebrates. Vol. III: Acanthocephala, Aschelminthes and Entoprocta." McGraw-Hill, New York.

KLIMOWICZ, H. (1968). *Pol. Arch. Hydrobiol.* **28,** 225–235.

KLIMOWICZ, H. (1970a). *Gaz, Woda i Techn. Sanit.* **1,** 30–33 (in Polish).

KLIMOWICZ, H. (1970b). *Przegl. Inform. Wod. i Kanal.* **2,** 74–76 (in Polish).

KLIMOWICZ, H. (1970c). *Gaz, Woda i Techn. Sanit.* **6,** 345–349 (in Polish).

KLIMOWICZ, H. (1970d). *Acta Hydrobiol.* **12,** 357–376.

KLIMOWICZ, H. (1972). *Acta Hydrobiol.* **14,** 19–36.

KLIMOWICZ, H. (1973). *Acta Hydrobiol.* **15,** 167–188.
KUTIKOVA, L. A. (1970). *Akad. Nauk CCCR Izd. "Nauka".* **104,** 1–744.
LANSING, A. I. (1942). *Biol. Bull. mar. biol. Lab., Woods Hole* **82,** 392–400.
MCKINNEY, R. E. (1967). *Appl. Microbiol.* **5,** 167–173.
MEADOW, P. and BARROWS, N. D. (1971). *J. Gerontol.* **26,** 302–309.
MYERS, F. J. (1931). *Am. Mus. Novit.* **494,** 1–12.
PARISE, A. (1961). *Hydrobiol.* **18,** 121–135 (in French).
PEJLER, B. (1957a). *K. svenska VetenskAkad. Handl.* **6**(5), 1–68.
PEJLER, B. (1957b). *K. svenska VetenskAkad. Handl.* **6**(7), 1–52.
PEJLER, B. (1962). *Zool. Bidr. Upps.* **35,** 307–319.
PENNINGTON, W. (1941). *J. Ecol.* **29,** 204–211.
PILARSKA, J. (1972). *Polskie Archwm Hydrobiol.* **19,** 265–277.
POURRIOT, R. (1965). Recherches sur l'écologie des Rotifères. (Thesis). *Vie Milieu* No. 21.
POURRIOT, R. and DELUZARCHES, M. (1971). *Ann Limnol.* **7,** 25–52.
RUDESCU, L. (1960). Fauna R.P.R. Trochelminthes II, Part II: Rotatoria. Editura Academiei R.P.R., Bucharest. (In Romanian.)
RUTTNER-KOLISKO, A. (1971). *Sber. Akad. Wiss. Wien.* **179,** 283–298.
RUTTNER-KOLISKO, A. (1972a). Die Binnengewasser **26.** Das Zooplankton der Binnengewasser 1. Tiel Schweizerbart'sch. Verlagsbuchhandlung, Stuttgart.
RUTTNER-KOLISKO, A. (1972b). *Verh. dt. zool. Ges.* **65,** 89–95.
SYDENHAM, D. H. J. (1971). *Hydrobiologia* **38,** 553–563.
VOIGT, M. (1957). "Die Rädertiere Mitteleuropas." Vol. I: Text, Vol. II: Figures. Gebrüder Borntraeger, Berlin.

8

Annelida

J. F. de L. G. Solbé

Water Research Centre
Stevenage Laboratory
Elder Way
Stevenage
Herts
England

I. Introduction

Only two families of the phylum Annelida, the Enchytraeidae and the Lumbricidae, have received more than brief attention in the literature of used-water treatment. The phylum is almost unknown in the activated-sludge process, although some oligochaetes, for example certain naids and lumbriculids which are active swimmers, may have been overlooked. Nevertheless, it is in the percolating filter or bacteria bed where annelids will most frequently be found and where they have been studied. In the United Kingdom one of the first centres of research into the biology of filters was at Leeds University, followed by Minworth Sewage Works, Birmingham and by the Water Pollution Research Laboratory (WPRL). Because of the paucity of data from elsewhere this chapter relies heavily on the work of these three but, where relevant, information on annelids from studies of streams and the soil has been included.

II. Taxonomic Survey

Brinkhurst (1971) enables identification to family level in the Oligochaeta and Mann (1954) in the Hirudinea. References to works of identification to species are given for each family. For each species listed its percentage occurrence in percolating filters, as found in a recent survey by Learner (in prep), is noted and references given to relevant papers.

Phylum ANNELIDA
Class OLIGOCHAETA
Order Plesiopora

1. Aelosomatidae (see Brinkhurst, 1971)
One species, *Aelosoma hemprichi* Ehrenberg, has been recorded from filters (21%) (Rudolfs, 1924; Hawkes, 1963; Learner, in prep.) and from aeration tanks in the activated-sludge process, where numbers may be so great that a reddish colour is found on the surface.

2. Enchytraeidae (see Nielsen and Christensen, 1959)
Eight species are known from filters:
Enchytraeus albidus Henle (Reynoldson, 1947a,b, 1948; Learner, in prep.) was once thought to be frequently present in filters but is now considered less common (9%)
Enchytraeus buchholzi Vejdovsky (Learner, in prep.) (57%)
Enchytraeus coronatus Nielsen & Christensen (22%) found at W.P.R.L. with
Enchytraeus minutus Nielsen & Christensen (1%) (Solbé *et al.*, 1967; Williams *et al.*, 1969; Learner, in prep.)
Lumbricillus rivalis Levinsen, very common (91%) (Solbé *et al.*, 1967; Williams and Taylor, 1968; Williams *et al.*, 1969; Solbé *et al.*, 1974)
Lumbricillus lineatus Müller, now considered very rare in filters (0%) but previously the subject of several studies (Reynoldson, 1939a,b, 1948; Lloyd *et al.*, 1940; Lloyd, 1943a; Hawkes, 1955). (It may be that these *L. lineatus* were in fact *L. rivalis*, as some doubt had been expressed by Reynoldson in the discussion of Solbé *et al.* (1967), and Dr Christensen had confirmed that worms provisionally identified at W.P.R.L. as *L. lineatus* were *L. rivalis*.)
Fredericia sp. (4%) (Learner, in prep.)
Buchholzia appendiculata (Buchholz) (1%) (Learner, in prep.), found only at WPRL. but in very high numbers.

3. Naididae (see Brinkhurst, 1971)
No species of this family were found in filters in this country until the

survey by Learner (in prep.), although Reynoldson (1939a) had observed naids in the effluent channels at Knostrop (Leeds); they now seem to be of widespread occurrence. Rudolfs (1924) recorded *Pristina* in experimental filters in the U.S.A. The species found by Learner are given below in order of ascending frequency:
Nais simplex Piguet (8%)
Pristina foreli Piguet (15%)
Chaetogaster diaphanus (Gruithuisen) (22%)
Pristina aequiseta Bourne (42%)
Chaetogaster langi Bretscher (43%)
Pristina idrensis Sperber (48%)
Nais elinguis Müller (52%)
Nais variabilis Piguet (54%)

4. Tubificidae (see Brinkhurst, 1971)
As in the case of the Naididae, this family and the next were rarely mentioned in studies on filters until 1967. Learner (in prep.) recorded *Tubifex tubifex* Müller (16%) and *Limnodrilus udekemianus* Claparede (8%).

5. Lumbriculidae (see Brinkhurst, 1971)
Lumbriculus variegatus (Müller) was found in only 3% of percolating filters (Learner, in prep.) but in one filter it occurred in very high numbers, almost to the exclusion of any other macro-invertebrates.

Order Opisthopora

6. Lumbricidae (see Gerard, 1964)
Eisenia foetida (Savigny) (31%) represents the sub-family Eiseninae (Terry, 1951; Solbé, 1971; Learner, in prep.). Four species of the Lumbricinae have been noted to date:
Dendrobaena rubida (Savigny) f. *subrubicunda* (Eisen) (43%) (Tomlinson, 1946; Hawkes, 1963; Solbé *et al.*, 1967; Solbé, 1971; Learner, in prep.);
Eiseniella tetraedra (Savigny) f. *typica* (52%) (Solbé *et al.*, 1967; Solbé, 1971; Learner, in prep.);
Lumbricus castaneus (Savigny) (1%), rare in filters (Learner, in prep.);
Lumbricus rubellus Hoffmeister (18%) (Reynoldson, 1939a; Terry, 1951; Hawkes, 1963; Learner, in prep.).

Class HIRUDINEA
Order Gnathobdellida

7. Hirudidae (see Mann, 1954)
Haemopsis sanguisuga (Linnaeus) (3%) (Terry, 1951; Learner, in prep.).

Order Pharyngobdellidae
8. Erpobdellidae (see Mann, 1954)
Trocheta subviridis Dutrochet (10%) (Harding, 1910; Learner, in prep.).

III. Factors Affecting the Abundance and Distribution of Annelids in Percolating Filters

A. Abiotic

1. Flow

The rate of flow of sewage on to the filter surface and the means of distributing it over the medium may affect the annelids in several ways. Indirectly, the flow pattern largely determines the lateral and vertical distribution of the film in the filter. The film is both a food supply and a source of protection for the macrofauna. Flow may also act directly, washing out numbers of worms or cocoons differentially depending on their behaviour within the medium and their powers of adhesion to the wetted surfaces. An example of a difference in rate of discharge resulting from a difference in the vertical distribution of species was given by Williams *et al.* (1969). *Enchytraeus coronatus* was found to deposit cocoons towards the base of the filter and they were washed out three times more readily than the cocoons of *Lumbricillus lineatus* deposited in the upper half of the filter.

In one of the earliest papers on the biology of high-rate filtration, Reynoldson (1941) showed that enchytraeids were very rare in a primary filter treating the industrial Huddersfield sewage at a rate of $1{\cdot}01-1{\cdot}12\ m^3/m^2$ per day (9–10 million gal/acre per day), although there were quite large populations of *Lumbricillus lineatus* and *Enchytraeus albidus* in the secondary filters being dosed at $0{\cdot}22-0{\cdot}34\ m^3/m^2$ per day (2–3 million gal/acre per day). Comparison with nearby filters receiving a similar sewage at $0{\cdot}08\ m^3/m^2$ per day ($\frac{3}{4}$ million gal/acre per day) suggested that the lack of worms in the high-rate filters was due to the flow, rather than the toxicity of the feed. In a later paper, however, (Reynoldson, 1942) it was found that the eggs of *L. lineatus* were unable to develop in the sewage or filter effluent and that there was a similar (slightly lower) toxicity to *E. albidus*.

Reynoldson (1948), studying the tenuous hold of *Enchytraeus albidus* in filters treating a somewhat toxic sewage, considered that the abrupt loss of the worms from the filter in spring was caused by their being flushed out of the filter on flocs of humus during sloughing. High concentrations of solids in filter effluent at any time of year were correlated with high numbers of worms. On the other hand, Solbé *et al.*

(1967) compared the number of enchytraeid and lumbricid worms in a filter per gram of dry film with the number in the effluent per gram of suspended solids, and found that the worms, especially the enchytraeids, were always preferentially retained within the filter. Nevertheless, the enchytraeid population was at risk during sloughing, and the lumbricids, having a long maturation period, even more so—often more than 10% of the filter population being discharged in a month.

Hawkes (1955) has shown that by controlled frequency of dosing, the fly population of an alternating double filtration plant could be replaced with one of *Lumbricillus.* In contrast with filters whose distributors rotated at 1–5 min/rev, in which high and greatly fluctuating film growth occurred, encouraging large populations of *Psychoda* and *Anisopus fenestralis* (one of the "nuisance" species), the *Lumbricillus* populations flourished in filters where the distributor rate was slowed to 42–55 min/rev and film growth was reduced in both mass and in fluctuation. The fly populations were virtually eliminated. Hawkes considered that *Lumbricillus* survived the high rate of flushing in the medium because the worms possess strong, curved, prehensile setae and also because the cocoons are firmly attached to the medium.

Bruce and Merkens (1970) reported that no enchytraeid worms were found in six high-rate filters, four of which had plastics and two mineral media, treating settled domestic sewage at $6\,m^3/m^3$ per day (a surface loading of $0{\cdot}52\,m^3/m^2$ per h). However, at another site a high-rate plant having a coarse rock medium and receiving $17{\cdot}9\,m^3/m^3$ per day ($1{\cdot}35\,m^3/m^2$ per h), one-third of which was settled-sewage, the remainder recirculated effluent, was found to contain considerable numbers of *Lumbricillus rivalis.* It is possible that the rate of flushing and the strength of the sewage were too great in the first case, but it may simply be that insufficient time had elapsed to allow colonization by enchytraeids to occur.

2. Temperature

The temperature of a percolating filter is largely determined by that of the sewage (Bayley and Downing, 1963) and, while temperature is seldom a severely limiting factor in temperate countries, probably the major fluctuations of film or fauna in a mature percolating filter are primarily temperature-dependent. Temperature affects the rate of growth of the film although an increase in temperature does not necessarily cause an increase in biomass. The fecundity, survival and growth of annelids are also closely linked with temperature, both directly, and indirectly through the food supply.

The enchytraeid *Lumbricillus lineatus* has been shown to migrate from temperatures at or below 5–6°C, both in filters outside (Reynoldson, 1939a) and in the laboratory (Reynoldson, 1939b). As soon as the temperature rose in winter, large numbers of worms were found in the surface growths. In the former paper it was shown that cocoon production, rate of development and viability were adversely affected by temperatures below 6°C; nevertheless breeding in the filters between November and May is enhanced by the reduced fluctuations of temperature compared with the conditions in more natural habitats. Reynoldson (1939b) showed that there was an increase in the percentage of cocoons per worm as temperature rose from 1·5–20·2°C; that sterility could be greater than 30% outside the range 7·4–15°C; that the percentage of worms hatching increased from 66% at 4·4°C to 100% at 9·9°C and then reduced only slightly towards 20·6°C; and finally that the rate of development increased with a rise in temperature from 4·4–20·6°C.

Reynoldson (1943) compared the life-cycles of *Enchytraeus albidus* and *Lumbricillus lineatus* and found the fecundity of the first species to be greater over the range 1–25°C. While the incubation period of *E. albidus* was longer, maturation was quicker, and in general *E. albidus* with its greater tolerance of temperature extremes was better suited to its natural terrestrial habit than was *L. lineatus.* In certain filter insects the average body size decreased as temperature rose (Golightly, 1940), but neither the size of adult worms nor the length of the setae were related to temperature.

Reynoldson (1947b, 1948) considered that high temperatures reduced the toxicity of a sewage by decreasing the period which the worms *Enchytraeus albidus* and *Lumbricillus lineatus* spent in the more susceptible early stages of development. Increases in temperature may also reduce the toxicity of poisons which the animal is capable of detoxifying, particularly if the rate of absorption of the poison is independent of temperature while detoxification is linked to a general, temperature-dependent rate of metabolism.

Temperature may control a series of events, involving in turn several populations in the community. Lloyd (1943a) found that a warm winter enabled *Lumbricillus lineatus* to breed more successfully and to accelerate the off-loading of the surface growth. This deprived the larvae of the flies *Metriocnemus longitarsus* and *Psychoda severini* of food. The former species especially is predaceous and in suitable conditions almost completely destroys the eggs of the summer-breeding flies *Metriocnemus hirticollis* and *Hydrobaenus minimus* (*Spaniotoma minima*).

Hawkes (1957) concluded that winter accumulations of film occurred because low temperatures had a greater retarding effect on grazers than on the bacterial film developed in filters receiving domestic sewage.

Williams and Taylor (1968), applying an artificial sewage of constant composition to experimental filters at 15°C, found that *Lumbricillus rivalis*, alone or in the presence of *Psychoda alternata*, could prevent the blocking of the interstices. The dominant organism in the film was the fungus *Mucor hiemalis* Whemer. Solbé *et al.* (1974) studied film accumulation and filter performance at constant temperatures of 7, 10 and 13°C, again using an artificial sewage. Three filters at each temperature were inoculated with *Psychoda* and enchytraeids and three were kept free of macro-invertebrates. It was shown that the rise in temperature from 10 to 13°C had a much more pronounced effect on performance than the rise from 7 to 10°C. Oxidation of ammoniacal nitrogen particularly was seriously impaired below 13°C. The balance between the enchytraeid species changed over the 280 days of the study; the inoculum was a more or less equal mixture of *L. rivalis* and *Enchytraeus coronatus* and this balance was maintained to the end of the study at 13°C. At 7 and 10°C, however, *E. coronatus* became very much the dominant species. This difference between the genera agrees with the findings of Reynoldson (1943). Enchytraeids established larger populations than the *Psychoda* larvae at all temperatures. At 7°C the enchytraeid density rose from the inoculum concentration of 30 worms/dm^3 to 9400/dm^3. At higher temperatures the numbers of worms were a little lower, but this may be explained by the greater success of the *Psychoda* larvae.

The direct effect of temperature on metabolism of *Lumbricillus rivalis* was demonstrated by Williams *et al.* (1969) and of the lumbricids *Dendrobaena subrubicunda* and *Eiseniella tetraedra* by Solbé (1971). The oxygen consumption of the enchytraeids rose by a factor of 1·6 with a rise in temperature from 10° to 20°C, while the Q_{10} of the lumbricids varied with the size of the animal (especially in *E. tetraedra*), so that worms of 0·3 mg dry weight had a Q_{10} of 3·6 while those of 30 mg had a Q_{10} of 1·8.

The success of hatching of lumbricid cocoons in tap-water was dependent on temperature but differed for the two species. The hatching of *Eiseniella tetraedra* became less successful as temperatures rose from 7 to 20°C, but a higher proportion of *Dendrobaena subrubicunda* cocoons hatched as temperatures rose. The optimum temperature for the production of cocoons of both species was 15°C, and the weight of cocoons and the incubation period decreased with increasing temperature.

3. *Medium*

Very little work has been done on the relationships between the type of filter medium and any member of the macro-invertebrate community in filters. Terry (1951) showed that *Eisenia foetida* and *Lumbricus rubellus* were more abundant in filters having small-grade media (19 mm, $\frac{3}{4}$ in.)

than in larger grades. It was considered that this preference might have been caused by the thigmotactic response of the earthworms. If this is so it is unlikely that earthworm populations will ever become successful in the simpler structures of plastics media, particularly since plastics are often used in high-rate plants and it is thought that the larger worms will not be able to resist the flushing action of the sewage.

4. *Light*

Light has little effect on the percolating filter except in enabling algae such as *Phormidium* sp. to flourish on the surface in some filters (Reynoldson, 1939a) and in controlling some of the local migrations of annelids and other invertebrates. Reynoldson reported that large numbers of *Lumbricillus lineatus* appeared on the surface of a filter during summer nights. Terry (1951) has described similar behaviour for the lumbricids *Lumbricus rubellus* and *Eisenia foetida.* In both cases it seems that the behaviour is the result of a photofugic reaction, but in natural habitats *E. foetida* mates on the surface of the ground at night.

Solbé (1971) found that the numbers of *Dendrobaena subrubicunda* and *Eiseniella tetraedra* discharged in the effluent followed a circadian rhythm, high numbers being lost from the filter at night. Darkening the underdrains had, in an earlier study (Department of Scientific and Industrial Research, 1963), prevented the formation of this rhythm, supporting the theory that photofugic behaviour was responsible. Also Solbé found that the extent of the period when higher numbers were found in the effluent was more restricted in the shorter summer nights. It was found that the average size of *E. tetraedra* appearing in the effluent was 85% greater than the average within the filter, pointing either to a gradual development of the photofugic habit, or to the superimposing of another behaviour pattern, possibly once connected with reproduction, although *E. tetraedra* is now an obligatory parthenogen. While maximum numbers of worms were discharged at around 2300 h, the largest worms appeared between 0100 h and 0500 h, but the significance of this difference is not yet understood.

5. *pH*

The degree of sensitivity to extremes of pH values has been related to the distribution of earthworms in soils (Laverack, 1963; El-Duweini and Ghabbour, 1964), and similar results have been obtained using species from filters (Solbé, unpublished). Those species of lumbricids found to be less sensitive (by delayed withdrawal of the prostomium) to immersion in solutions of low pH were also found to be present in both the primary and the secondary filters of a works treating a mixture of industrial and domestic sewage, while intolerant species were restricted to the secondaries. It is not suggested that this is of prime importance in filters, but rather that resistance to high or low pH may indicate general physiological robustness in a species.

Reynoldson (1947b) found that both *Lumbricillus lineatus* and *Enchytraeus albidus* were occasionally killed by low pH. Adults were killed in 50–75 min in sewage having a pH value of 3·5, in which young worms survived only 18 min. The effects of low pH were confined to the top of the filter, the pH of the effluent being 6·1 even during periods when the sewage had pH values around 2·8.

6. *Toxicity*

Rudolfs *et al.* (1950) have reviewed the literature on toxic materials affecting used-water treatment processes, streams and BOD determinations, but very little toxicological work appears to have been performed on the annelids of percolating filters.

The salinity preferences of estuarine tubificids and littoral enchytraeids have been examined by Palmer (1968) and Tynen (1969) respectively. The downstream limit of *Tubifex tubifex* in the River Thames was determined by salinity, the maximum tolerated being 10% of sea-water. *Lumbricillus lineatus* was shown by Tynen (1969) to withstand salinities of 0–150% of Menai Straits sea-water, although in filters it is not considered a particularly tolerant species. Reynoldson (1947a) found that *L. lineatus* was very scarce in the filters at Huddersfield but that *Enchytraeus albidus* was more abundant than usual. The enchytraeid population as a whole was reduced and the depth-distribution was abnormal, worms being found only rarely in the top 30 cm (12 in) of the filter. In a second paper (1947b) the sewage was shown to be toxic to both species but especially to *L. lineatus*, whose eggs would not develop even in the reduced toxicity of the effluent from the filter. Resistance to the toxic material increased as the worms aged. The rare occurrence of *L. lineatus* in the filter was attributed to worms being washed in during periods of high rainfall.

Whitley (1968) examined the resistance of *Tubifex tubifex* and *Limnodrilus hoffmeisteri* to zinc sulphate, lead nitrate and sodium pentachlorophenate (PCP), but the test conditions, in particular the Knop solution in which the poisons were dissolved, make interpretation of the data difficult. Twenty-four-hour median tolerance limits at pH 7·5 were >50 mg Pb/l, 46 mg Zn/l and 0·31 mg PCP/l. It is unlikely that such high concentrations of heavy metals would be retained in solution, in a potentially toxic form, in sewage or filter effluent, since precipitation as carbonates or complexing with organic substances would probably occur. The toxicity of the metals was increased considerably by changes in pH from neutral. The survival of worms in PCP at pH 10·5 was always over 85%, while in the uncontaminated Knop solution only 38% survived a pH of 10·0 and none survived pH 11·0. The reaction of the worms in producing mucus in response to the presence of heavy metals was said to cause a lethal reduction in the diffusion of oxygen across the body wall. PCP affected respiration and energy production at the cellular level. Reynoldson (1948) found that mucus was produced by *Enchytraeus albidus* and especially by *Lumbricillus lineatus*, and suggested that the mucus was protective and had a marked buffering effect against acids. Nevertheless, too much mucus would obviously have the detrimental effect described above.

The data in Table I have been drawn from the Information Service on Toxicity and Biodegradability (INSTAB) at the Water Pollution Research Laboratory; no information is available for enchytraeids or lumbricids. Apparently certain substances (e.g. cadmium, copper and mercury) are highly toxic to tubificids while others (e.g. manganese, nickel, sulphate and chloride) have a lower toxicity, at least in the short term. Many of these materials are found in filters treating industrial wastes and there is an obvious lack of data on their effects on filter organisms.

7. Dissolved Oxygen and the Organic Strength of the Sewage

The temperature of the sewage is largely responsible for the seasonal fluctuations in the accumulation of film in the filter, but Hawkes (1965) found that the seasonal fluctuation in sewage strength was more important in those parts of filters from which grazing animals had been excluded. Solbé *et al.* (1974) showed that although filters maintained with or without animals at 7°C accumulated much more film than others at 10°C receiving the same feed, filters at 13°C did not contain less film than those at 10°C.

Sewage strength is traditionally measured as its capacity for absorbing oxygen. The half-life of the biochemical oxygen demand (BOD) of the sewage entering a percolating filter is less than eleven min (Eden *et al.*, 1964), and despite a rate of ventilation greatly in excess of the theoretical requirements (Mitchell and Eden, 1963) it is possible that certain annelids are restricted in filters by low concentrations of dissolved oxygen. This restriction is not imposed on some of the other species to be found in filters, the larvae of *Psychoda*, *Eristalis* or *Tipula*, which may immerse themselves in the microbial film leaving only the respiratory siphon exposed at the surface (see Fig. 9 in Williams and Taylor, 1968). In compensation, the blood of many of the annelids found in this habitat contains haemoglobin, generally dissolved in the plasma. In addition, *Tubifex* for instance has a well-developed capillary system in the body wall (Palmer, 1966) which probably helps it to maintain its rate of oxygen consumption, independent of the oxygen concentration of its environment, down to very low levels. *Limnodrilus hoffmeisteri* has no such capillary system and this may contribute to its apparent lack of success in filters and to its inability to compete with *Tubifex* and *Chironomus* larvae in a grossly polluted stream (Brinkhurst and Kennedy, 1965). It may be that filter animals exhibit respiratory acclimatization and this may be a seasonal phenomenon, as Mann (1958) found for the leech *Erpobdella testacea*. The leech became acclimatized to low concentrations of oxygen in summer only so that the rate of oxygen consumption was constant down to ambient concentrations of one-third of the air saturation value.

Fox and Taylor (1955), while studying the tolerance of aquatic invertebrates to oxygen, found that more young *Tubifex* survived and were bigger in water exposed to gas mixtures containing 4% oxygen than in 21% oxygen. Adult *Tubifex* did not show this difference but worms of all ages were quickly killed in water exposed to 100% oxygen. Alsterberg (1922) found that *Tubifex* could survive 2–3 days in anaerobic conditions during which time it consumed glycogen (Dausend, 1931) and after which the uptake of oxygen was enhanced as though an oxygen debt were being repaid. The data of Ralph (1957) on *Lumbricus* also suggest that an oxygen debt can be incurred.

Holtje (1943) considered that the presence of the genera *Tubifex* and *Limnodrilus* in filters indicated gross overloading, while Lloyd (1945) thought that to feel at home in filters *Tubifex* required a high concentration of solids. Of the remaining common filter annelids, their natural habitats are generally both extremely humid and rich in organic matter. Thus *Eiseniella tetraedra* is described as amphibious and mostly to be found covered with water, and *Dendrobaena subrubicunda*, though terrestrial, is common in sites such as compost heaps (Gerard, 1964). Enchytraeids known in filters are to be found in mud flats (Lloyd, 1945) and on the sea-shore in rotting seaweed (Tynen, 1969), but genera such as *Hemihenlea*, *Cognettia* or *Achaeta* from forest soils (O'Connor, 1963) have not yet been recorded from percolating filters.

B. Biotic

A great deal of the literature on macro-invertebrates in percolating filters has been concerned with the interactions within and between species, ranging from predator-prey relationships (Lloyd, 1943a,b) through physiological differences (Reynoldson, 1948) to density-dependent factors

TABLE I. Summary of data on the toxicity of materials to annelids

Substance	Species	Concentration (mg/l) to cause: Injury	Death	Period of study	Temp. (°C)	Reference
Ammonia	*Tubifex tubifex*	>2·7 (free NH_3) 45 (total NH_4^+)			12–14	Stammer (1953)
Cadmium	*T. tubifex*	0·3	5	7 days		Schweiger (1957)
Calcium (as $CaCl_2$)	*T. tubifex*	>500, <1000			18–22	Zelinka (1957)
Chloride (as NaCl)	*T. tubifex*	>1576				Kanagina and Lebedeva (1957)
	Limnodrilus hoffmeisteri	>2000				
	Nais communis }	5000		2 days	20	Learner and Edwards (1963)
	N. variabilis }		10 000	36 min	20	
Cobalt (as $CoCl_2$)	*T. tubifex*	90	400			Schweiger (1957)
Copper	*T. tubifex*		4	10 days		Boch (1960)
			0·8 (LC_{30})	8 days		
	T. tubifex		0·25–0·5 (LC_{50})	7 days	10	Liepolt and Weber (1958)
	Nais elinguis, *N. communis*, *N. variabilis*		1 (in hard water, less in soft)	<6 h		Learner and Edwards (1963)
Cyanide	*Limnodrilus*	8	11			Gillar (1962)
DDT	*T. tubifex*, *Limnodrilus*	>100		96 h	20	Whitten and Goodnight (1966)
	"True worms"	>10		10 days		Richards and Cutkomp (1946)

Ethanol	*T. tubifex*	>500 <1000				Zelinka (1957)
Hydroxylamine	*T. tubifex*	2		2–3 wks	10	Seibold (1955)
	Stylaria lacustris	2		2–3 wks	8	Seibold (1955)
	Herpobdella octoculata	5		2–3 wks	13	Seibold (1955)
Iron	*H. octoculata*		3 (LC_{50})	3 days		Walter (1966)
Lead	*T. tubifex* *L. hoffmeisteri*		see text			Whitley (1968)
Malathion	*T. tubifex*		21 (LD_{50})	48 h	20	Whitten and Goodnight (1966)
	Limnodrilus		17 (LD_{50})	96 h	20	Whitten and Goodnight (1966)
Manganese (as $MnCl_2$)	*T. tubifex*	200	700			Schweiger (1957)
Mercury (as $HgCl_2$)	*T. tubifex*	0·1	0·3	7 days		Schweiger (1957)
Nickel (as $NiCl_2$)	*T. tubifex*	20	30	7 days		Schweiger (1957)
Sulphate (as Na_2SO_4)	*T. tubifex*	>500 <1000			18–22	Zelinka (1957)
Zinc	*T. tubifex* *Limnodrilus*		see text			Whitley (1968)
	H. octoculata	5*		25 days		Herbst (1967)
	Glossiphonia complanata		5* (LC_{45})	16–25 days		Herbst (1967)

* 3 mg/l did not affect either leech.

within a species (Solbé, 1971). All these phenomena may have an effect on the abundance and distribution of species in filters.

1. *Interspecific Effects*

Eisenia foetida would not develop from cocoon to maturity in sterilized soil which had been recolonized with soil fungi and bacteria until protozoa had also been added (Miles, 1963). Similarly, Backlund (1945) showed that *Enchytraeus albidus* died in less than a month when kept on sterile seaweed. Such extreme effects are not likely to be of importance in the microbiologically rich community of the filter film, but they show a dependence of metazoa on the abundance of micro-organisms. The effects of annelids on bacteria and fungi will be discussed in Section VI of this Chapter.

Reynoldson (1939b) reported the presence of a ciliate of the Anoplophryinae in the fore-gut of *Lumbricillus lineatus*, and found sporozoan parasites in the segments containing the testis. *Glaucoma* sp. (Ciliatea) sometimes attacked the cocoons and destroyed the eggs.

Lloyd *et al.* (1940) examined predation on the worms and cocoons of *Lumbricillus lineatus* and found that the latter were often eaten by the larvae of *Psychoda severini*, *Spaniotoma minima* and *Metriocnemus longitarsus*. Worms were also eaten, but in lower proportion. On the other hand, it is suggested that while the worms do not harm the flies directly, they do compete with them for food and also loosen the filter film, freeing many of the fly eggs deposited in it and causing them to be washed down the filter.

Some of the interrelationships in the filter community described by Lloyd (1943a) have already been discussed under the heading Temperature (Section III A2), but in a later paper Lloyd (1945) emphasizes the results of competition between both *Psychoda severini* and *Metriocnemus longitarsus*, and *Lumbricillus lineatus*. The fly populations were found to be above average after hard winters unfavourable to the worms, and below average after milder winters in which the worms flourished. The deprivation of food was more marked since worms and flies occupied the same depth-distribution, near the surface of the filter.

The enchytraeid worm population (*Lumbricillus lineatus*, *Enchytraeus coronatus* and *Enchytraeus minutus*) in a maturing 12 m^3 percolating filter, although present in April, did not increase until July–August after the *Psychoda severini* population had been eliminated by adverse temperatures (Solbé *et al.*, 1967). Similarly, the lumbricids in the filter did not increase until a decline in numbers of both enchytraeids and psychodids after the following spring. A general decline in the maximum densities of enchytraeids and psychodids after the first year was attributed to the influx of lumbricids and to the increasing populations of the common species within the filter community as it matured.

Williams and Taylor (1968) found that enchytraeid populations increased more rapidly in experimental filters at 15°C when with psychodids (*Psychoda alternata*) than when kept as single-species cultures. They considered that the difference was due to "conditioning" of the film by the fly larvae, whose life-cycle was shorter than that of the worms. Eventually, the density of the worm population on its own was found to be 30% greater than when in competition with *Psychoda*. The species balance between enchytraeids and psychodids at three relatively low temperatures (Solbé *et al.*, 1974) has already been discussed.

2. Intraspecific Effects

Apart from Lloyd (1943b) in his studies on *Psychoda alternata*, there have been few reports on competition within a species in percolating filters. Solbé (1971), however, has reported the occurrence of an apparently density-dependent factor in filter lumbricids. The calculated survival of *Eiseniella tetraedra* cocoons was related to the observed density of the worms in such a way that when the population exceeded 40 worms/dm^3 there was less than a 40% survival of cocoons, and only when populations were 20/dm^3 or less did survival reach 90–100%.

IV. Seasonal Incidence

The seasonal incidence of any species in a percolating filter is the resultant of those forces capable of maintaining the population (breeding and colonization) and of those causing reductions (incipient mortality, predation, parasitism, starvation and wash-out).

Reynoldson (1939a) described the seasonal distribution of the alga *Phormidium* sp. at the surface of a filter and explained annual changes in the density of *Lumbricillus lineatus* at various depths as resulting from changes in the abundance of the algae. The growth of *Phormidium* sp. began in late May, replacing *Ulothrix* sp. and *Chlorella* sp. which had been scantily covering the pebble medium. The *Phormidium* grew to a maximum by August and maintained its weight until the following spring, when the algae were washed down through the filter and out in the effluent. During the year, enchytraeids had spasmodically attacked the film, giving it a spongy appearance. At the time of sloughing the worms were said to follow their food down and eventually to be washed out of the filter. It is difficult to judge whether the movements of the worms were active migrations or the result of the inability of the worms to cling to the medium during the sloughing period. The sampling technique used to assess the population density at 30 cm and 76 cm (12 and 30 in) consisted of attracting worms into muslin bags containing 20 g of scalded *Phormidium*. It is possible that the increase in numbers at these depths during sloughing may have been caused by the greater attraction of the bait when the film elsewhere was being depleted by sloughing. The general impression of seasonal incidence of *L. lineatus* was of high numbers somewhere in the filters from April to October and very restricted populations for the rest of the year (Fig. 1a).

Hawkes (1955) found maximum numbers of *Lumbricillus* sp. in February in filters whose distributors rotated at 42–55 min/rev. Minimum densities were found in August (Fig. 1b), and these data agree with those of Solbé *et al.* (1967) for *Lumbricillus rivalis*, *Enchytraeus coronatus* and *Enchytraeus minutus* (Fig. 1c). Neither of these studies used baited bags, instead, perforated metal canisters filled with medium were inserted in

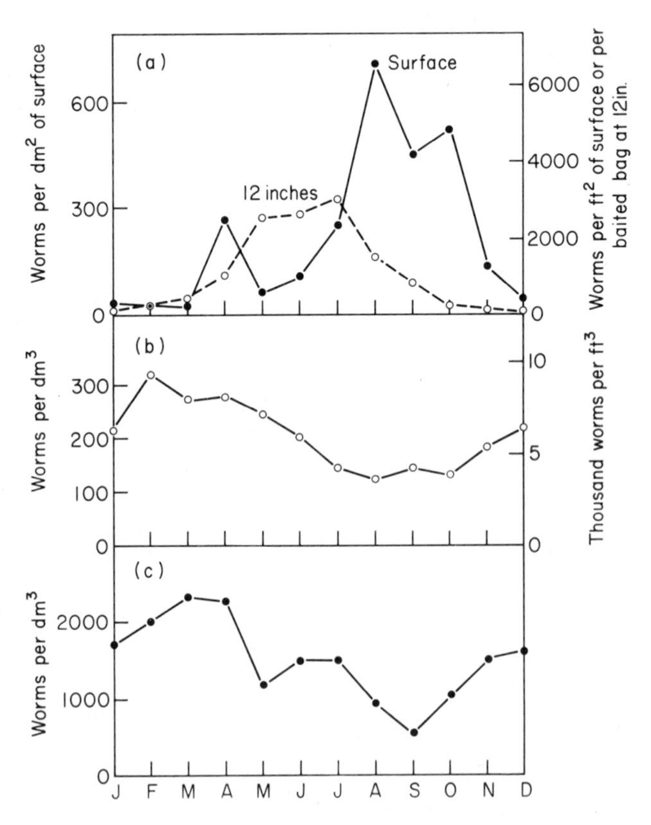

FIG. 1. Seasonal incidence of enchytraeid worms.

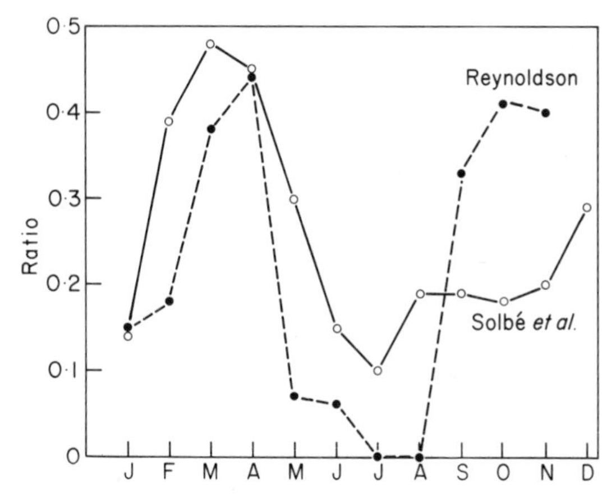

FIG. 2. Seasonal rhythm of the cocoon–worm ratio of enchytraeid worms.

the filter. Sampling involved the removal of canisters from 0–76 cm by Hawkes and 0–152 cm (5 ft) by Solbé *et al.* Hawkes' data are particularly interesting since there was very little change in the weight of film throughout the 33-month study, providing relatively constant conditions of food supply and physical environment, except for the effects of temperature and light. The phasing of the rhythm cannot be expected to be identical to that of soil populations of enchytraeids since the conditions in the filter are generally much more amenable. Williams *et al.* (1969) found that the biomass and oxygen consumption of enchytraeids in a filter were 100 times greater than in a coniferous forest soil studied by O'Connor (1963). O'Connor showed that two of the three species present in the highly-organic, acid soil declined in numbers abruptly at the onset of winter.

Reynoldson (1947a) demonstrated seasonal rhythms in cocoon production of *Enchytraeus albidus* by comparing the week to week variation in cocoon–worm ratio from his baited-bag data (Fig. 2). Cocoons were particularly abundant in February–April and September–November. Combined data from Solbé *et al.* (1967) for three species of enchytraeids have given a curve of similar shape (Fig. 2), suggesting periods of cocoon deposition in spring and autumn. Williams *et al.* (1969) separated the species referred to in the previous paper and calculated the production of cocoons of the two major species *Lumbricillus rivalis* and *Enchytraeus coronatus*), using the equation $P = (N \times T/t) + D$, where P = cocoon production, N = number of cocoons in filter at a given time, T = interval between samples, during which time the number of cocoons discharged in the effluent is D and the development time is t. The cocoon production in each species was compared with the number of mature (clitellate) worms.

This technique showed that there was a single peak in the number of mature worms each year, between March and June, and that on the whole cocoon production matched these peaks but was extended on some occasions until September.

Terry (1951) demonstrated that *Eisenia foetida* and *Lumbricus rubellus* underwent a seasonal variation in numbers in the medium near the surface of filters. There was a small difference in the phasing of the rhythms, the maximum and minimum being found in May and November repectively for *L. rubellus* but in June–July and November–February for *E. foetida*. Terry concluded that the rhythm was caused by the depletion of food in the upper medium in spring and summer. The earlier phasing of the minimum density in *L. rubellus* was said to occur at a works where the rest of the grazing fauna was more abundant. Solbé (1971) found that *Eiseniella tetraedra* and *Dendrobaena subrubicunda* also both exhibited seasonal rhythms in numbers near the surface, but the maxima were found to occur slightly earlier than described above.

Lumbricids were found to deposit cocoons at all times of year (Solbé, 1971) but principally in two batches, summer and winter. Although in *Eiseniella tetraedra* generations could not be traced as easily from month to month as in *Dendrobaena subrubicunda*, it was apparent that in both species the batch of cocoons deposited in winter did not develop. O'Connor (1958) showed that enchytraeid cocoons deposited in the soil in winter hatched together in spring, but in filters mass-hatching was not considered to occur because the filter temperatures were high enough to allow embryonic development to continue. The earthworms reached mature size in the filter about 5–6 months (*E. tetraedra*) or 4–5 months (*D. subrubicunda*) after hatching.

Terry (1951) gave a description of the leech *Haemopsis sanguisuga* in filters and found that leeches were observed in some humus tanks at all times of the year but especially in the vernal slough. This is also the case for enchytraeids and lumbricids.

Apart from fragments of insects, very little of the humus from percolating filters could be ascribed directly to animals (Solbé *et al.*, 1967). Animal debris, however, does settle more rapidly than non-animal particles, and the rates of settlement of both types of material varied seasonally with the varying concentration of solids discharged. Williams and Taylor (1968) found that at 15°C, filters with macro-invertebrates produced effluents whose solids settled at twice the rate (up to settlement periods of two hours) of effluents from filters with no animals. Presumably as numbers of animals change seasonally the rates of settlement of solids vary. Solbé *et al.* (1974) showed that the settlement properties of effluent solids improved with increases in temperature from 7 to 15°C.

In most percolating filters the most obvious seasonal phenomena are the accumulation of film in winter followed by the spring sloughing, after which there may be a period when the food supply for some of the grazing organisms is reduced. Superimposed on this rhythm in the annelids is the pattern of cocoon production, development and maturation. Depositions of enchytraeid cocoons which hatched successfully have been observed in spring and autumn and of lumbricid cocoons in summer. Lumbricid cocoons may also be found in the winter but do not necessarily hatch. It may be difficult to distinguish seasonal rhythms in population size unless samples are carefully taken from the full depth of the filter. Local migrations may temporarily eliminate the population near the surface at certain times of year while in fact the population as a whole is growing. The true reproductive capacity of a filter organism cannot be assessed without such samples and estimates of loss in the effluent. Even with such precautions the picture may be complicated by the continually changing rate of development of an organism, so that stages quickly passed through may appear under-represented in samples.

V. Spatial Distribution

A mature percolating filter does not present a single homogeneous habitat for the members of its community but changes horizontally according to the pattern of sewage flow produced by the distributors and, particularly, vertically as the materials in the sewage are progressively oxidized and modified by the filter organisms. In this respect the changes occurring vertically in a filter correspond to horizontal distance along the aeration tank of the activated-sludge process, but while in filters the animals are more or less static, in the aeration tank they are moving in the body of water as its composition changes.

Hawkes (1952) found that larvae of the window midge *Sylvicola* (*Anisopus*) *fenestralis* were more abundant in the area below the sewage distributor jet while pupae were more common in the drier inter-jet zones. Reynoldson (1947a) mentions that the area of medium surrounding the central column of a circular filter did not receive a direct supply of the toxic sewage but remained moist from condensation of water vapour, rain and occasional splashes of sewage. In this area *Enchytraeus albidus* and *Lumbricillus lineatus* were able to survive, although the remainder of the bed would not support enchytraeids.

Lumbricillus lineatus (Reynoldson, 1947a) and *Lumbricillus rivalis* Solbé *et al.*, 1967; Williams *et al.*, 1969) have been found principally in the upper medium of filters treating non-toxic sewage. At Cooper Bridge, however, Reynoldson found *L. lineatus* to be at a minimum near the surface. *L. rivalis* cocoons were also found with the adults near the top of a filter in spring 1965 (Williams *et al.*, 1969), but in the following year the cocoons were lower down, below 0·6 m (2 ft). Either cocoon production rates varied considerably with depth or cocoons were carried down the filter with the microbial film and debris which were being sloughed at the time. The large accumulation of cocoons near the base of the filter did not give rise to large numbers of juveniles in the region. The size structure of the population remained fairly constant with depth, so if hatching occurred the young worms may have died, have been discharged from the filter in the effluent or have migrated to the upper regions of the filter.

Enchytraeus coronatus and *Enchytraeus minutus* increased in numbers towards the base of the filter (Solbé *et al.*, 1967) so that below 0·6 m at least 50% of the enchytraeid population belonged to this genus and below 0·91 m (3 ft) more than 80% did so. Williams *et al.* (1969) found that *Enchytraeus* sp. cocoons were most abundant in the mid-depths of the filter and found an explanation for this discrepancy by examining the change in ratio of clitellate to non-clitellate worms with depth (Williams *et al.*, 1969, Fig. 3). Although the worm populations were relatively small at the site of maximum deposition of cocoons, the chances of a worm being mature were more than three times greater than towards the base of the filter. Possibly as a direct result of the lower position of *E. coronatus* in the filter, the wash-out rate was much greater than for *Lumbricillus rivalis*.

The depth-distribution of the lumbricids *Eiseniella tetraedra* and *Dendrobaena subrubicunda* was described by Solbé (1971). *E. tetraedra* was found in particular around the mid-depth of the filter, although the population in the upper half of the filter, being subject to the seasonal variation described above, could at times reach high densities. The percentage of the *D. subrubicunda* population increased towards the base of the filter, more than 30% of the worms being found between 1·2 and 1·5 m (4–5 ft). Reynoldson (1939a) found *Lumbricus rubellus* to be rare in the top 15 cm (6 in) but common in the deeper layers. At times, however, immature specimens were seen in the surface medium. Solbé (1971) also found anomalies in the size-class structure of earthworms in the top 15 cm, the largest *D. subrubicunda* (>4·5 cm) and *E. tetraedra* (3·51–7·5 cm) and the smallest *D. subrubicunda* (<1·25 cm) being less well represented than in the rest of the filter. *D. subrubicunda* of 2·51–4·5 cm, on the other hand, were much better represented.

Terry (1951) did not attempt depth-distribution studies but did observe that in filters with larger medium (5 cm, 2 in) earthworms were rarely found near the surface. Smaller media supported large populations of

Lumbricus rubellus and *Eisenia foetida.* In some filters the medium is deliberately stratified, larger grades being used at the surface in an effort to reduce winter ponding. Such a stratification would contribute to the pattern of distribution of earthworms and possibly other annelids in filters.

VI. The Activity of Annelids in Relation to Their Role in Percolating Filters

In examining the role of various species of annelids in the purification process, classic experiments have been set up in which animals have been introduced to one filter and prevented from entering another, the control. Until recently, replication of experiments was not attempted in such studies and the period of the experiment was of relatively short duration. Nevertheless the work of Reynoldson (1939a) and Terry (1951) showed clearly (in much the same way as Parkinson and Bell (1919) with Collembola) that enchytraeids and lumbricids respectively could cause clear effluents to be produced for some time from filters while the control filters deteriorated considerably.

Reynoldson applied heat-sterilized sewage to two filters 0·91 m long and 75 mm in diameter (3 ft × 3 in.) having a medium of marble chippings. After allowing the beds to mature for 5 weeks, 1500 *Lumbricillus lineatus* worms were introduced to bed "A". The worms deposited cocoons and by the 6th week after inoculation the dense growth of film was being reduced. There then followed a 5-week period of "sloughing" and the medium became almost free of deposits. Filter "B", however, had become choked with film so that by week 14 it was severely ponded and was rested for 5 days, although this had little effect on reducing the ponding. The effluent from bed "B" resembled the sewage in appearance while that from filter "A" was very clear. Later, worms were introduced to bed "B" but although ponding was eliminated by their activity they never managed to clear the bulk of the film and they died in the filter, presumably because the ventilation was reduced so much that the film became largely anaerobic. Chemical analyses were carried out on the unsettled effluents and the relatively high concentrations of suspended matter probably explain the poor results in terms of oxygen absorbed test (PV) from the effluent of filter "A". Nitrification was negligible in filter "B" but up to 30 mg nitrite/1 was found in the effluent of the filter with animals.

Terry (1951) set up two rather larger filters (1·8 m × 100 mm) (6 ft × 4 in.) and applied heat-sterilized sewage to the 19 mm ($\frac{3}{4}$ in.) coal medium. After 24 days in which the film became established he added 50 adult *Eisenia foetida* to filter "A", leaving filter "B" as the control. In both filters the purification, as given by the PV in four h at 80°C, increased to 70% by day 43, but the control filter then declined in performance at an increasing rate, reduction in PV values being less than 60% by day 72, 40% by day 102 and 10% by day 133. Filter "A", on the other hand, increased in performance to nearly 80% purification until around day 70, after which there was a slow reduction to 45% by day 133. As with the enchytraeids, the effluent from the filter containing lumbricids was bright and clear while that of the control filter remained dull and contained solid material.

Williams and Taylor (1968) prepared an artificial sewage (to eliminate fluctuations in the quality of the effluent) and applied it to twelve identical filters 0·91 m × 150 mm diam. (3 ft × 6 in.) in size containing 1·3–1·9 cm ($\frac{1}{2}$–$\frac{3}{4}$ in.) clinker. The filters were built and run in a constant temperature room at 15°C and were all inoculated with three species of fungus and six of protozoa. During the first 7 months of the study the degree of similarity between the performance of the filters was assessed. Four groups of three filters each were established by random selection; the difference in performance between these groups was normally insignificant (probabilities greater than 5%). The filters performed well for the first 4 months, without animals, at hydraulic loadings of 0·47 m^3/m^3 per day (80 gal/yd^3 per day) and organic loadings of 0·1 Kg BOD/m^3 per day) (0·16 lb BOD/yd^3 per day). The strength of the sewage was then increased to 0·21 Kg BOD/m^3 per day and the hydraulic loading to 0·6 m^3/m^3 per day, and this caused an increase in biomass of the microbial film and a decrease in the efficiency of the filters. Seven months from the beginning of the study some of the filters were inoculated with macroinvertebrates: the larvae of *Psychoda alternata* were introduced into three filters, another group received *Lumbricillus rivalis*, the third group received a mixture of the two species and the fourth remained uninoculated as a control.

While the controls continued to deteriorate in efficiency the filters with animals improved. The most efficient treatment was observed in the filters with a mixed fauna, and the filters with enchytraeids alone were the slowest of the animal filters to show improvement in the quality of the effluent, particularly in the production of oxidized nitrogen. As was suggested above, the worm populations developed more rapidly in the filters with fly larvae because the latter with their shorter life-cycle had conditioned the filter, destroying many areas of dense film growth. As Reynoldson (1939a) reported, high concentrations of solids are found in the effluents of filters with animals, but these solids were shown by Williams and Taylor (1968) to settle more rapidly than those from control filters. There was little difference between the settleability of the humus produced by the different animal groups. The solids concentration fluctuated in a regular manner, the peaks probably being related to periods when high numbers of animals were active in the film.

Solbé *et al.* (1974) used a similar experimental procedure to Williams and Taylor (1968) but worked at lower temperatures (7, 10 and 13°C) and with only two groups of three filters at each temperature, the controls and filters having both psychodid larvae and enchytraeid worms. The weight of each filter was noted each week and the animals were added after the filters had been operating for only 3 weeks. Improved settleability, BOD and organic carbon reduction were again established in the filters with animals and much of the improvement was thought to have been caused by the reduced accumulation of film leading to a reduction in channelling through the medium. Such improved flow-through characteristics and oxidative conditions might also be expected in the control filters before film had built up; the reduction of BOD and organic carbon support this. Nitrification, however, which should also have occurred at such times, did not occur unless animals were present, even at 13°C. Some direct effect of the inoculum may be responsible; perhaps the animals held nitrifying bacteria in their guts at the start of the experiment, the bacteria originating from the large experimental filter from which the animals were obtained. It is possible that colonization of the control filters by nitrifying bacteria may have been at a critically low rate. The benefit of the presence of animals to the metabolism of carbon in the filters was summarized as being equivalent to a 5°C rise in temperature.

In addition to the above studies other approaches have been made in attempting to quantify the role of annelids in filters. These approaches include the studies on settleability by Solbé *et al.* (1967), discussed in Section IV, and the work of Williams *et al.* (1969) and Solbé (1971) on the metabolism of enchytraeid and lumbricid worms respectively. In order to assess the metabolism of a population and thus its contribution to community metabolism, estimates of the size-class structure must be obtained and the relationships between oxygen consumption, temperature and body size must be elucidated. The choice of respirometer will influence the results, as Solbé (1971) showed: lumbricids held in respirometric chambers in which the worms were immersed in water consumed oxygen at ten times the rate of worms placed in thin films of water. The first procedure caused the worms to move actively during the whole experiment (1–2 h) while in the second situation the animals remained quiescent. Also it seems likely that the submerged worms were being forced to expend energy on the removal of water from their bodies against an osmotic gradient. The question arises as to which set of data is relevant to the calculation of rates of metabolism in filters, and the best solution would appear to be an average of active and resting rates.

Williams *et al.* (1969) used a Cartesian Diver micro-respirometer (Holter, 1943) for *Lumbricillus rivalis* and showed that the whole population of this species consumed a weight of oxygen equivalent (assuming a

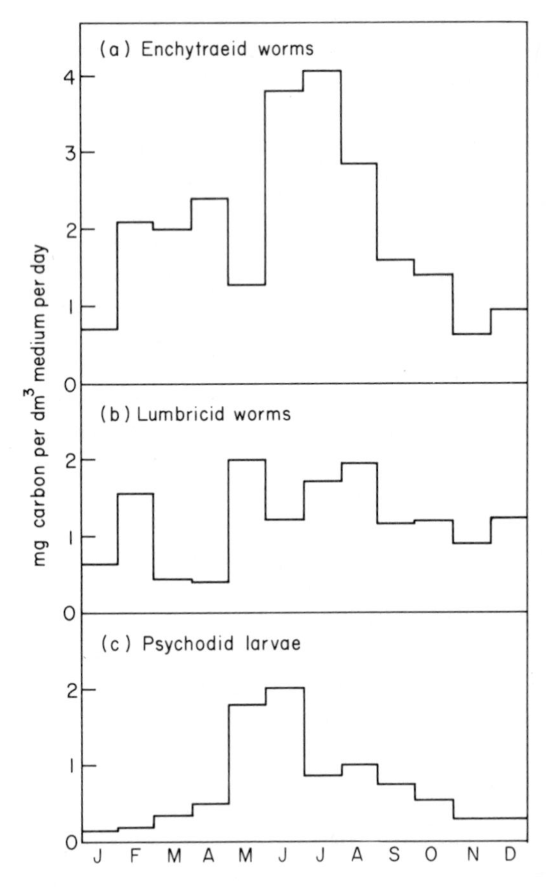

FIG. 3. Comparison of dissipation of carbon from percolating filters by macro-invertebrates.

respiratory quotient of unity) to 2·5% of the total organic carbon dissipation from the filter. If the *Enchytraeus coronatus* population consumed oxygen *pro rata*, taking into account the smaller average weight of *Enchytraeus* individuals, the total annual contribution to carbon dissipation of the enchytraeids averaged 2·9% (Fig. 3a). For the same period

(November 1965–October 1966), Solbé (1971) evaluated the oxygen consumption of both lumbricid species in the filter. Two methods were adopted: manometry, using the Warburg apparatus (Umbreit *et al.*, 1964), and polarography, in which the worms were totally immersed, using a dropping-mercury electrode (Briggs and Knowles, 1961). If a respiratory quotient of 0·8 is taken (Spector, 1956), *Eiseniella tetraedra* and *Dendrobaena subrubicunda* together averaged 0·2–2% of the total dissipation of carbon from the filter, their contributions being about equal (Fig. 3b). The third group of grazers in the filter, the psychodid larvae, were examined by Solbé and Tozer (1971) (Fig. 3c). Taking the respiration of the fly larvae into account, the macro-invertebrates were found to eliminate 5% of the influent carbon between January and April, a large part of this being due to the respiration of the enchytraeids. After April the lumbricids and psychodids became more active and by July the animals were responsible for a little over 10% of the filter's metabolic activity. The enchytraeids accounted for 60% and the lumbricids for 25% of this contribution.

Another direct effect of annelids in the purification of sewage may be in the consumption and/or conditioning of the micro-organisms in the film. *Eisenia foetida* is known to digest the bacteria *Serratia marcescens* (Day, 1950) and *Escherichia coli* (Brüsewitz, 1959), the latter being one of the key species in sewage whose elimination is an important criterion of successful treatment. Studies on the rate of feeding of soil-inhabiting lumbricids have been reviewed by Satchell (1967), who found rates to average 27 mg food (dry weight)/g worm (dry weight) per day. Solbé (1971) found that *Dendrobaena subrubicunda* in filters ingested 133 mg/g per day at 15°C. Ravera (1955) obtained intermediate values for aquatic oligochaetes, production of faeces being equivalent to 40–60% of their dry body weight per day. Solbé calculated that the whole lumbricid population ingested an amount of film equal to 55% of the daily input of carbonaceous material to the filter. The accumulated material in the filter could potentially have been ingested once over a period of 41 days. Jeuniaux (1969) has reviewed studies of the enzymes found in oligochaetes; earthworms are capable of digesting a wide range of substances. They produce proteolytic enzymes, amylases, oligo- and polysaccharidases. Naids can digest proteins, fats and starch; enchytraeids possess α-glucosidases (including invertase and trehalase), β-glucosidases, α- and β-galactosidases. Trehalase would probably allow the digestion of fungus spores, and there are times when these might be numerous in filters. Parle (1963) found that cellulase activity occurred even in the absence of bacteria in the earthworm gut. Chitin, common in filters having insect populations and possibly one of the materials most resistant to degradation, was digested by a combination of the activity of the earthworm and its gut flora. In a normally-feeding *Lumbricus terrestris* the food may take 20 h to pass through the gut, and this period is sufficient for cellulase activity to be effective.

Miles (1963) reviewed some of the controversial studies on the fate of algae, protozoa, bacteria and fungi in the earthworm. Diatoms, blue-green algae, desmids, yeasts and rhizopods have been found to be killed by passage through the gut (Aichberger, 1914). Bassalik (1913), Stöckli (1928), Dawson (1947) and Day (1950) found fungi to be unaffected and bacteria to be either unaffected or reduced, although their methods may have caused the worms to burrow actively, allowing only 11 or 12 hour's residence in the gut. Parle (1963) also found fungi to remain unchanged in the earthworm but bacteria and actinomycetes to be increased. The increase occurred progressively with passage along the gut and may have been caused partly by the increase of sugars released by enzymic degradation of organic material and partly by mechanical action creating finely divided material in the oesophagus and gizzard. Barley and Jennings (1959) showed that the earthworm *Allolobophora caliginosa* stimulated the microbial decomposition of a mixture of herbage litter and dung. The inoculum of 3 g worms/kg material was within the range encountered in the field but considerably below the density of earthworms in filters. Nitrate and ammonium nitrogen production was 20% higher than in the controls. Such decomposition would be valuable in filters, as would the function of tubificids in recycling material from detritus and sediments (Whitten and Goodnight, 1969).

In addition to digesting or multiplying the micro-organisms in filters, annelids may be important in forming water-stable aggregates as they do in soils. Such aggregates would tend to settle in the humus tanks more readily than finely divided matter. Swaby (1950) studied the mechanism of the formation of such aggregates in earthworms, and all the conditions he found to be necessary would appear to be met in filters. The aggregates are held together by gums produced by bacteria if the soil is sufficiently nutritious, as filter film would be. Mechanical reinforcement by fibres, calcium carbonate or the earthworms' own mucus was not considered important. Fungi are only valuable in maintaining the aggregate once it has left the limiting conditions of low dissolved oxygen within the gut.

The function of a percolating filter involves converting settled sewage into a clear effluent, easily settled solids and innocuous gases. The microbial film has the principal role in oxidizing and synthesizing material from the sewage. More than 90% of the influent carbon may be removed from the filter as carbon dioxide from the metabolism of the film.

However, micro-organisms in filters do not appear to be able to flocculate solids to the same extent as can micro-organisms in activated-sludge plants, neither can the micro-invertebrates restrict the accumulation of film. Thus filters without macro-invertebrates produce cloudy effluents incapable of being cleared by settlement and may become grossly overgrown with film.

The roles of annelids in percolating filters, while possibly involving all of the activities summarized below, will vary in extent from filter to filter. When large worms occur in small media Terry (1951) has suggested that the medium will be moved from time to time, impairing the complete adhesion of the surface film. It is doubtful whether this movement would occur with the necessary force to be of any significance in the maintenance of an "open" filter. Worms feed on film and affect it in several ways. In some filters they may be a prime cause of the spring unloading (Reynoldson, 1939a). They are certainly capable of preventing the accumulation of excess solids in filters. Within the annelid gut the more complex materials are broken down, some being absorbed, others being passed out as aggregates enriched with some species of bacteria and capable of relatively rapid settlement. Other bacteria, particularly *Escherichia coli*, may be completely eliminated by the worm's digestive system. In one filter annelid respiration contributed up to 8·5% towards the dissipation of carbon.

Only one form of nuisance is associated with the larger annelids in filters: on some works they may collect in the humus tanks and putrefy, possibly reducing the dissolved oxygen content of the final effluent and certainly not improving working conditions for the operators. Against this their benefit would seem to be considerable. In full-scale filters their numbers are not subject to the fluctuations found in the Diptera so that, for instance, a reasonably constant discharge of effluent solids may be expected. They are capable of feeding on bacterial film, breaking down and recirculating for further microbiological action material in the filter which might otherwise remain as solids (to the detriment of oxidative conditions), and of considerably increasing the rate of settlement of humus. Their respiration accounts for a small but useful proportion of the carbon dissipation by the filter, and in addition by competing with the diptera the worms reduce the fly nuisance from many filters.

Among the many species of macro-invertebrates to be found in filters the annelids do as much good and cause as little nuisance as any. It is fortunate that despite our present inability (with a few notable exceptions, e.g. Hawkes, 1955) to manipulate the macro-invertebrate community in filters, the niches available in the filter are sufficiently attractive to allow colonization by effective populations of annelids.

Acknowledgements

The author wishes to thank M. A. Learner of the University of Wales Institute of Science and Technology, Cardiff, and N. S. Thom of the Water Pollution Research Laboratory, Stevenage, for supplying data used in Sections II "Taxonomic Survey" and III A6 "Toxicity" respectively.

References

AICHBERGER, R. VON (1914). *Kleinwelt* **6,** 53–88.

ALSTERBERG, G. (1922). *Lunds Univ. Arsskr., Avd.* 2 (N.S.) **18,** 1.

BACKLUND, H. O. (1945). *Opusc. ent. Suppl.* **5,** 1–256.

BARLEY, K. P. and JENNINGS, A. C. (1959). *Aust. J. agric Res.* **10,** 364–370.

BASSALIK, K. (1913). *Z. Gär Physiol.* **2,** 1–32.

BAYLEY, R. W. and DOWNING, A. L. (1963). *J. Instn publ. Hlth Engrs* **62,** 303–332.

BOCH, J. (1960). *In* "Handbuch der Frischwasser- und Abwasserbiologie" (Ed. H. Liebmann), Vol. 2, pp. 820–822. Gustav Fischer, Jena.

BRIGGS, R. and KNOWLES, G. (1961). *Analyst,* Lond. **86,** No. 1026, 603–608.

BRINKHURST, R. O. (1971). "A Guide for the Identification of British Aquatic Oligochaeta." Freshwater Biological Association, Scientific Publication No. 22, 2nd edn (revised).

BRINKHURST, R. O. and KENNEDY, C. R. (1965). *J. anim. Ecol.* **34,** 429–443.

BRUCE, A. M. and MERKENS, J. C. (1970). *Wat. Pollut. Control,* 113–148.

BRÜSEWITZ, G. (1959). *Arch. Mikrobiol.* **33,** 52–82.

DAUSEND, K. (1931). *Z. vergl. Physiol.* **14,** 557.

DAWSON, R. C. (1947). *Proc. Soil Sci. Soc. Am.* **12,** 512.

DAY, G. M. (1950). *Soil Sci.* **69,** 175–184.

DEPARTMENT OF SCIENTIFIC AND INDUSTRIAL RESEARCH (1963). "Water Pollution Research, 1962." H.M. Stationery Office, London. p. 22.

EDEN, G. E. BRENDISH, K. and HARVEY, B. R. (1964). *J. Proc. Inst. Sew. Purif.,* 513–525.

EL-DUWEINI, A. K. and GHABBOUR, S. I. (1964). *Bull. zool. Soc. Egypt* **19,** 89–100.

FOX, H. M. and TAYLOR, A. E. R. (1955). *Proc. R. Soc.* B**143,** 214–225.

GERARD, B. M. (1964). The Linnean Society of London Synopses of the British Fauna, No. 6: Lumbricidae (Annelida).

GILLAR, J. (1962). *Sb. vys. Sk. chem.-technol. Prazi* **6** (1), 435–457.

GOLIGHTLY, W. H. (1940). *Ann. appl. Biol.* **27,** 406–421.

HARDING, W. A. (1910). *Parasitology* **3,** 130–201.

HAWKES, H. A. (1952). *Ann. appl. Biol.* **39,** 181–192.

HAWKES, H. A. (1955). *J. Proc. Inst. Sew. Purif.*, 48–58.

HAWKES, H. A. (1957). *J. Proc. Inst. Sew. Purif.*, 88–110.

HAWKES, H. A. (1963). "The Ecology of Waste Water Treatment." Pergamon Press, Oxford.

HAWKES, H. A. (1965). *Int. J. Wat. Air Pollut.* **9,** 693–714.

HERBST, H. V. (1967). *Gewäss. Abwäss.* **44–45,** 37–47.

HOLTER, H. (1943). *C. r. Trav. Lab. Carlsberg* (*Ser. chim.*) **24,** 399–478.

HOLTJE, R. H. (1943). *Sewage Wks J.* **15,** 14–29.

JEUNIAUX, C. (1969). *In* "Chemical Zoology". (Eds M. Florkin and B. T. Scheer), Vol. IV: Annelida. Echiura and Sipuncula. Pp. 69–91. Academic Press, London and New York.

KANAGINA, A. V. and LEBEDEVA, M. P. (1957). *Vodosnab. sanit. Tekh.* No. 1, 15–17.

LAVERACK, M. S. (1963). "The Physiology of Earthworms." Pergamon Press, Oxford.

LEARNER, M. A. and EDWARDS, R. W. (1963). *Proc. Soc. Wat. Treat. Exam.* **12,** 161–168.

LIEPOLT, R. and WEBER, E. (1958). "Wässer und Abwässer" (Ed. R. Liepolt), pp. 335–353. Winkler, Wien.

LLOYD, L. (1943a). *Ann. appl. Biol.* **30,** 47–60.

LLOYD, L. (1943b). *Ann. appl. Biol.* **30,** 358–364.

LLOYD, L. (1945). *J. Proc. Inst. Sew. Purif.*, 119–139.

LLOYD, L., GRAHAM, J. F. and REYNOLDSON, T. B. (1940). *Ann. appl. Biol.* **27,** 122–150.

MANN, K. H. (1954). "A Key to the British Freshwater Leeches." Freshwater Biological Association, Scientific Publication No. 14.

MANN, K. H. (1958). *J. exp. Biol.* **35,** 314–323.

MILES, H. B. (1963). *Soil Sci.* **95,** 407–409.

MITCHELL, N. T. and EDEN, G. E. (1963). *Wat. & Waste Treat.* **9,** 366–370.

NIELSEN, C. O. and CHRISTENSEN, B. (1959). *Natura Jutlandica* **8–9,** 1–160; **10** (1961), 1–23.

O'CONNOR, F. B. (1958). *Oikos* **9,** 11, 272–281.

O'CONNOR, F. B. (1963). *In* "Soil Organisms" (Eds J. Doeksen and J. van der Drift), pp. 32–48. North-Holland, Amsterdam.

PALMER, M. F. (1966). *J. Zool.* **148,** 449–452.

PALMER, M. F. (1968). *J. Zool.* **154,** 463–473.

PARKINSON, W. H. and BELL, H. D. (1919). "Insect Life in Sewage Filters." Sanit. Pub. Co., London.

Parle, J. N. (1963). *J. gen. Microbiol.* **31,** 1–11.
Ralph, C. L. (1957). *Physiol. Zoöl.* **30,** 41–55.
Ravera, O. (1955). *Mem. Ist. Ital. Idrobiol.*, suppl. 8, 247–264.
Reynoldson, T. B. (1939a). *Ann. appl. Biol.* **26,** 138–164.
Reynoldson, T. B. (1939b). *Ann. appl. Biol.* **26,** 782–799.
Reynoldson, T. B. (1941) *J. Proc. Inst. Sew. Purif.*, 109–128.
Reynoldson, T. B. (1942). *J. Proc. Inst. Sew. Purif.*, 116–139.
Reynoldson, T. B. (1943). *Ann. appl. Biol.* **30,** 60–66.
Reynoldson, T. B. (1947a). *J. Anim. Ecol.* **16,** 26–37.
Reynoldson, T. B. (1947b). *Ann. appl. Biol.* **34,** 331–345.
Reynoldson, T. B. (1948). *J. Anim. Ecol.* **17,** 27–38.
Richards, A. G. and Cutkomp, L. K. (1946). *Biol. Bull.* **90,** 97–108.
Rudolfs, W. (1924). *Bull. N. J. agric. Exp. Stn* No. 403.
Rudolfs, W. (Chairman), Barnes, G. E., Edwards, G. P., Heukelekian, H., Hurwitz, E., Renn, C. E., Steinberg, S. and Vaughan, W. F. (1950). *Sew. ind. Wastes* **22,** 1157–1191.
Satchell, J. E. (1967). *In* "Soil Biology" (Eds A. Burges and F. Raw), pp. 259–322. Academic Press, London and New York.
Schweiger, G. (1957). *Arch. Fisch Wiss.* **8,** 54–78.
Seibold, A. (1955). *Vom Wass.* **22,** 90–166.
Solbé, J. F. de L. G. (1971). *J. appl. Ecol.* **8,** 845–867.
Solbé, J. F. de L. G. and Tozer, J. S. (1971). *J. appl. Ecol.* **8,** 835–844.
Solbé, J. F. de L. G., Williams, N. V. and Roberts, H. (1967). *Wat. Pollut. Control* **66,** 423–448.
Solbé, J. F. de L. G., Ripley, P. G. and Tomlinson, T. G. (1947). *Wat. Res.* **8,** 557–573.
Spector, W. S. (1956). "Handbook of Biological Data." p. 255. W. B. Saunders Co., Philadelphia, U.S.A.
Stammer, H. A. (1953). *Vom Wass.* **20,** 34–71.
Stöckli, A. (1928). *Landw. Jb Schweiz* **42,** 1.
Swaby, R. J. (1950). *J. Soil. Sci.* **1,** 195–197.
Terry, R. J. (1951). *J. Proc. Inst. Sew. Purif.*, 16–23.
Tomlinson, T. G. (1946). "Animal Life in Percolating Filters." Department of Scientific and Industrial Research, Water Pollution Research Technical Paper No. 9. H.M. Stationery Office, London.
Tynen, M. J. (1969). *Oikos* **20,** 41–53.
Umbreit, W. W., Burris, R. H. and Stauffer, J. F. (1964). "Manometric Techniques" (4th edn). Burgess Publ. Co., Minneapolis, U.S.A.
Walter, G. (1966). *Wiss. Z. Karl-Marx-Univ. Lpz.* **15,** 247–269.
Whitley, L. S. (1968). *Hydrobiologia* **32,** 193–205.
Whitten, B. K. and Goodnight, C. J. (1966). *J. Wat. Pollut. Contr. Fed.* **38,** 227–235.

WHITTEN, B. K. and GOODNIGHT, C. J. (1969). *Proc. 2nd Natn. Symp. Radioecol.* 1967, Ann Arbor, 1969, 270–277.
WILLIAMS, N. V. and TAYLOR, H. M. (1968). *Wat. Res.* **2,** 139–150.
WILLIAMS, N. V. SOLBÉ, J. F. DE L. G. and EDWARDS, R. W. (1969). *J. appl. Ecol.* **6,** 171–183.
ZELINKA, M. (1957). *Voda* **36,** 242–244.

9

Insecta

M. A. Learner

Department of Applied Biology
University of Wales
Institute of Science and Technology
Cathays Park
Cardiff CF1 3NU
Wales

I. Introduction

The importance of insects in the treatment of sewage is principally associated with their successful colonization of the percolating filter, although they do occur to some extent in other kinds and stages of treatment as well. Published lists of insect species inhabiting filters and investigations aimed at elucidating the factors affecting the occurrence of different species have been largely restricted to beds receiving sewage containing a high proportion of industrial waste, situated in either the West Riding of Yorkshire or Birmingham (England) (Lloyd, 1945; Terry, 1951; Crisp and Lloyd, 1954; Hawkes, 1963, 1965a). Only Usinger and Kellen (1955) in California and Solbé *et al.* (1967) at Stevenage (England) have published lists of insect species found in filters receiving domestic sewage. Tomlinson (1946a) has described some seventeen insect species found in filters receiving sewages of varied compositions, but he made no attempt to assess the factors affecting their occurrence. In view of the restricted nature of the information concerning the geographical distribution and ecology of the insect inhabitants of filters, the author (ms in prep.) in 1966–1967 surveyed the fauna of filters at 48 sewage works situated throughout Great Britain, in relation to a number of physical and chemical parameters. A total of 186 species of insect, belonging to 38 families, was recorded. Several of these species were of infrequent occurrence and unlikely to have any significant effect upon the filter ecosystem, and only the principal species occurring are listed in Table I. These insects were collected from the top 35 cm (12 in.) of the filter, the filter effluent, and from sticky-traps of the pattern described by Solbé *et al.* (1967).

II. Principal Insect Inhabitants of Percolating Filters

A. Collembola

Christiansen (1964) reviewed the diet of spring-tails, showing them to ingest a wide variety of organic substances including fungal hyphae, fungal spores, bacteria, dead or decaying plants, and algae, all of which are available in percolating filters. However, only five species of springtail were found in more than 10% of the filters sampled (Table I). Of these, *Hypogastrura viatica* and *Tomocerus minor* have frequently been recorded from filters, the former species from North America as well as from Europe (Parkinson and Bell, 1919; Dreier, 1946; Tomlinson, 1946a). Two other species, *Proisotoma* (*Ballistura*) sp. and *Anurida tullbergi*, although of infrequent occurrence, reached densities of about 1500 and 250 per litre of medium respectively. Lawrence (1961) also records the latter species as reaching considerable numbers in a filter. Figures and brief descriptions of all the principal species occurring in filters are given by Lawrence (1970).

B. Coleoptera

Members of four beetle families (Chrysomelidae, Ptiliidae, Hydrophilidae and Staphylinidae) occurred frequently in the filter samples, but of these families only the latter two were common, being found on over 70% of the sticky-traps. Nearly all the hydrophilid species found frequently in the filters belonged to the Sphaeridiinae, a group commonly associated with dung. The most common species was *Cercyon ustulatus* and larvae of this species have been found in filters in July and August. Terry (1951) and Hawkes (1963) also list *C. ustulatus* as occurring frequently in filters, although other hydrophilids are clearly locally dominant (Table I). Usinger and Kellen (1955) also record a species of *Cercyon* inhabiting filters in California.

The occurrence of staphylinid species in filters is probably also associated with the preference of a number of them for decaying animal and vegetable matter, and the two species most common in filter samples, *Platystethus arenarius* and *Oxytelus tetracarinatus*, are both associated with dung. Although adult staphylinids are quite numerous in filters their larvae are only occasionally found, and the conditions are probably not suitable for these insects to breed successfully in filters.

C. Hymenoptera

Many of the hymenopteran species found during the survey are known parasites of dipteran larvae and pupae. A few species have been reared from dipteran pupae collected from filters; Baines and Finlayson (1949) recorded *Phygadeuon cylindraceus* (Ichneumonidae) as parasitizing *Spathiophora hydromyzina* pupae, and Mr M. B. Green (pers. comm.) has obtained *Spalangia erythromera* (Pteromalidae) from pupae of *Themira*

TABLE I. The frequency and abundance of the commoner species of insect found during a survey of the fauna of filters at forty-eight sewage-works. Species occurring in less than 10% of the filters sampled or in less than 20% of the traps are not included unless their abundance was greater than 50/l of medium or 500/m^2 per year. (A) = adults only

Species	Synonyms used in previous literature on filter-bed fauna	Occurrence in survey (%)		Numbers found (maximum)	
		In filter	On insect-traps	In filter (no./l of medium)	On insect-traps (no./m^2 per year)
COLLEMBOLA					
Entomobryidae					
Heteromurus nitidus (Templeton)		12	—	—	—
Tomocerus minor (Lubbock)		28	—	57	—
Isotomidae					
Proisotoma minuta (Tullberg)		21	—	57	—
Proisotoma (*Ballistura*) sp.		1	—	1488	—
Poduridae					
Anurida granaria (Nicolet)	*Anurida* = *Aphoromma*	16	—	56	—
A. tullbergi Schött		7	—	240	—
Hypogastrura viatica (Tullberg)	*Achorutes subviaticus*	51	—	3800	—
COLEOPTERA					
Chrysomelidae		—	29	—	—
Hydrophilidae		—	82	—	2637
Cercyon atricapillus (Marsh)		—	12	—	2035
C. ustulatus Preys		6 (A)	41	—	—
Cryptopleurum minutum (F.)		—	24	—	—
Helophorus brevipalpis Bed.		—	29	—	—

Ptiliidae		—	29	—	6555
Staphylinidae		10	71	—	14 919
Oxytelus tetracarinatus Block		—	29	—	—
Philonthus umbratilis (Grav.)		—	12	—	750
Platystethus arenarius (Fourc.)		—	70	—	750
P. cornutus (Grav.)		—	12	—	975
Trogophloeus pusillus (Er.)		—	18	—	521
T. rivularis (Mots)		—	6	—	595
HYMENOPTERA		—	71	—	3391
DIPTERA					
Trichoceridae					
Trichocera maculipennis Mg.		4	12	57	9534
Anisopodidae					
Sylvicola fenestralis (Scop.)	*Sylvicola* = *Anisopus* (has been confused with *A. cinctus*)	49	47	1680	2640
Psychodidae					
Psychoda alternata Say.		89	94	44 700	3 838 026
Psychoda cinerea Banks		4	18	120	62 615
P. severini Tonn.		72	82	1240	609 429
Chironomidae					
Hydrobaenus minimus Mg.	*Spaniotoma minima*; *Limnophyes minimus*	61	76	3460	328 149
H. perennis Mg.	*Orthocladius perennis*; *Spaniotoma perennis*	24	23	1400	180 990
Metriocnemus hirticollis Staeg.		6	18	10 850	526
M. hygropetricus Kieff.	*M. longitarsus*	54	82	1892	15 522

TABLE I. *(Continued)*

Species	Synonyms used in previous literature on filter-bed fauna	Occurrence in survey (%)		Numbers found (maximum)	
		In filter	On insect-traps	In filter (no./l of medium)	On insect-traps (no./m^2 per year)
Ceratopogonidae					
Culicoides vexans (Staeg.)		—	6	—	975
Forcipomyia bipunctata (L.)		—	29	—	—
Scatopsidae					
Swammerdamella brevicornis Mg.		—	23	—	—
Cecidomyiidae		—	29	—	—
Sciaridae			88	—	5651
Bradysia sp.		—	53	—	—
Scatopsciara sp.		—	47	—	—
Phoridae			53	—	—
Megaselia brevicostalis (Wood)		—	24	—	—
Triphleba nudipalpis (Becker)		—	24	—	—
Ephydridae		16	—	—	—
Scatella silacea Lw		8 (A)	18		6524
S. stagnalis Fall.		—	35	—	529

Sepsidae		—	35	—	1205
Themira putris L.		—	24	—	1205
Sphaeroceridae		24	—	—	—
Leptocera appendiculata Vill.		—	18	—	979
L. aterrima Hal.	*Leptocera = Limosina*	—	23	—	749
L. atoma Radi.		—	23	—	971
L. caenosa Radi.		4 (A)	47	—	1278
L. cambrica Richards		—	23	—	3217
L. coxata Stenh.		—	59	—	15 957
L. curvinervis Stenh.		1 (A)	41	—	32 995
L. ferruginata Stenh.		—	29	—	—
L. fontinalis Fall.		1 (A)	18	—	601
L. heteroneura Hal.		1 (A)	47	—	—
L. humida Hal.		—	65	—	1945
L. limosa Fall.		—	23	—	1117
L. mirabilis Collin		1 (A)	82	—	23 350
L. puerula Radi.		—	29	—	1734
L. spinipennis Hal.		—	71	—	4745
L. sylvatica Mg.		—	6	—	979
L. vagans Hal.		—	59	—	828
Sphaerocera pusilla Fall.		—	29	—	1424
S. subsultans L.		—	12	—	1122
Cordyluridae					
Spathiophora hydromyzina Fall.	*Spaziphora hydromyzina*	36	6	—	—

putris and *Apheareta tenuicornis* (Braconidae) from *Leptocera* pupae. The author has found *Scatella silacea* pupae parasitized by a *Kleidotoma* sp. (Figiditae).

D. Diptera

It is clear from the survey carried out by the author (Table I) that this order contributes most of the principal insect species which inhabit filters. Twenty-eight species from eleven dipteran families occurred frequently, often in considerable numbers, especially *Psychoda* spp., *Hydrobaenus* spp. and *Leptocera* spp. Some of these dipteran species and genera are very widespread, occurring in filters both in Europe and North America (Usinger and Kellen, 1955), and *Psychoda alternata* has also been reported from South Africa (Murray, 1939). Virtually all the principal filter species have larvae which feed on decaying vegetable matter in more or less wet conditions, and many feed on dung, especially larvae of anisopodid, psychodid, sepsid and sphaerocerid species (Laurence, 1954, 1955).

III. Identification

A large number of the species found to be numerically important in the survey are described by Tomlinson (1946a), and references to the descriptions of most of the larvae, pupae and adults of the remainder may be obtained from Kerrich *et al.* (1967).

IV. Factors Affecting the Presence and Abundance of Insects in Percolating Filters

It is clear from a comparison of filters where sticky-traps were sited during the author's survey (Table II) that the conditions in some filters favoured a much greater diversity of insect species than in others; thus the total number of species caught during a year's trapping ranged from 18 from one filter to 54 from another. It is also clear that the status of particular species varied very much from filter to filter; thus *Psychoda cinerea* was more successful than other species in the Minworth filters but was not caught on sticky-traps at any other sewage-works. There are many factors in filter operation and design which could account for such differences in the insect communities, but even where apparently identical filters have been operated under identical conditions (Bruce *et al.*, 1967) the macro-invertebrate populations have differed, suggesting that the kind of animal community developing in a filter may be only partly due to the environmental conditions provided by that filter. The filter fauna themselves may modify their environment such that the climax community established in each filter is different, owing possibly to differences in the order in which the dominant species colonize the filters.

The most obvious factors likely to affect the kind of insect community established in a filter will be the physical and chemical nature of the applied sewage, the hydraulic and organic loading, the frequency of dosing, the type and size of the medium, the climate, and biotic interactions. For a specific filter the insect species most successful at adapting to the prevailing conditions will dominate, and should these conditions become more rigorous there will be a tendency towards fewer abundant species and a lower species diversity. The filter is a three-dimensional habitat and there will be changes, especially in the chemical and physical quality of the sewage, in relation to depth, so that the environment extant at the different levels within the filter may be suited to different species. The majority of the insects inhabiting filters are grazers feeding on the film which develops on the medium, and the factor most significant in determining the abundance and composition of the insect fauna is the quantity of film present within the filter. Many of the factors mentioned earlier directly affect the quantity and distribution of film within the filter.

A. Organic Loading

The strength of the sewage, the rate at which it is applied and the proportion and type of industrial waste treated all affect the insect community inhabiting a filter. Of particular importance is the organic loading, which is the product of the rate of application of the sewage and the concentration of degradable organic matter (BOD value), since this determines the rate at which nutrients are being supplied to the filter which in turn determines the rate of film development; the amount of film generally increases with increased organic loading. Tomlinson and Stride (1945) found that the density of *Psychoda* species, estimated by insect traps, increased with organic loading up to 0·34 kg BOD/m^2 per day (0·62 lb BOD/yd^2 per day), the densities being sometimes depressed by loadings in excess of this (Fig. 1). In general the chironomid species were more numerous in lightly loaded filters and almost non-existent in filters receiving a loading greater than 0·22 kg BOD/m^2 per day (0·41 lb BOD/yd^2 per day). The densities of both *Hydrobaenus perennis* and

TABLE II. Proportional representation of the principal insect species caught on sticky-traps in 1967, the total number of species caught and the estimated number of insects emerging per m² of filter-bed surface in the year

Sewage-works	Industrial waste (% of average flow)	Species; % of total annual catch						Total no. of insect species	Total insect emergence (no./m² per year)
		P. alternata	*P. cinerea*	*P. severini*	*H. minimus*	*H. perennis*	*M. hygropetricus*		
Ashford	13	93	—	1	2	P	P	38	1 497 228
Aviemore	0	—	—	37	34	20	P	45	903 351
Aycliffe (new bed)	0	93	—	P	P	—	6	8*	—
Bedford (primary filter)	45	87	—	12	P	P	P	26	928 424
Bedford (secondary filter)	—	8	—	4	45	17	7	54	235 660
Galashiels	40	30	—	34	34	—	P	41	961 389
Kendal	18	79	—	1	20	—	P	18	1 478 574
Langley Mill (self-propelled dosing)	0	97	—	P	P	—	P	7*	(21 630)
Langley Mill (controlled dosing 1 rev/ 20 min)	0	65	—	P	P	—	21	20*	(10 840)
Lincoln	30	98	—	P	—	—	P	14*	—
Minworth B	60	6	89	—	—	—	—	22	70 137
Minworth C	60	39	58	—	—	—	—	21	98 402
Minworth D	60	4	84	—	—	—	P	24	42 539
Norwich	30	98	—	1	P	—	—	24	489 605
Saltford (Bath)	0·3	97	—	P	P	—	P	18	739 231
Stevenage (5 min dosing cycle)	0	86	—	14	P	—	P	29	4 460 912
Wrexham	10	99	—	P	P	—	P	20	1 099 855

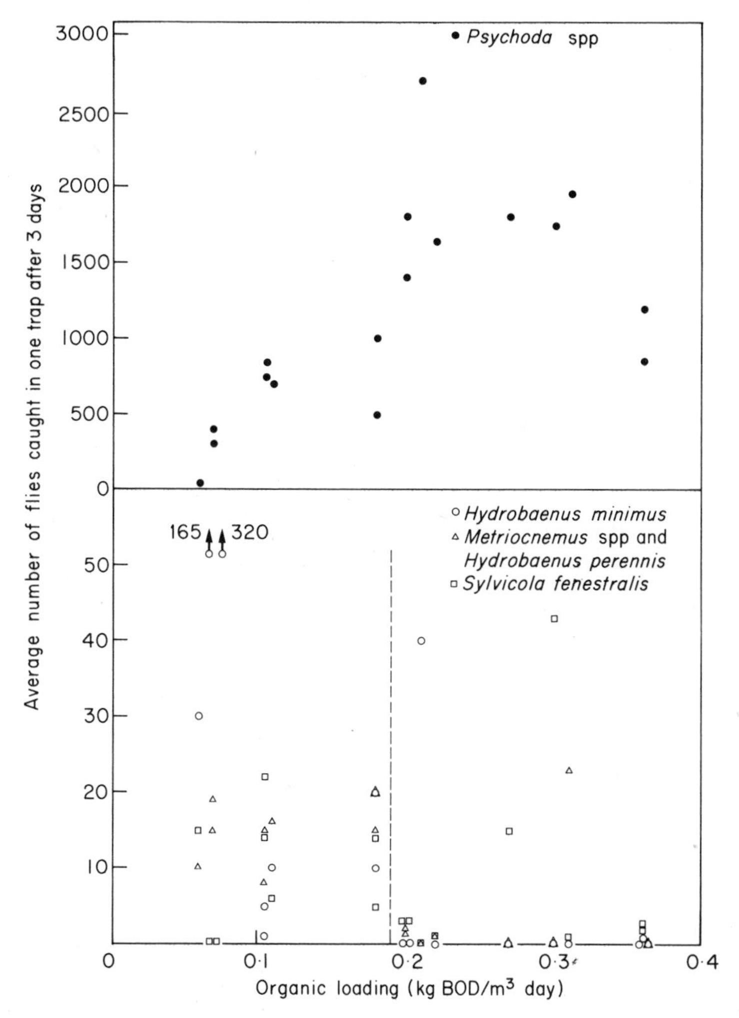

FIG. 1. Average number of flies trapped on different filters in relation to filter loading (data from Tomlinson and Stride, 1945).

Sylvicola fenestralis appeared unrelated to filter loading, but Hawkes (1952a) found twice as many larvae of *S. fenestralis* in a filter receiving an organic loading in excess of 0·32 kg BOD/m^2 per day (0·6 lb BOD/yd^2 per day) as in filters receiving 0·2 kg BOD/m^2 per day (0·4 lb BOD/yd^2 per day), and he showed that trapping the adults of this species at the filter surface gave inaccurate estimates of the number of larvae within the filter.

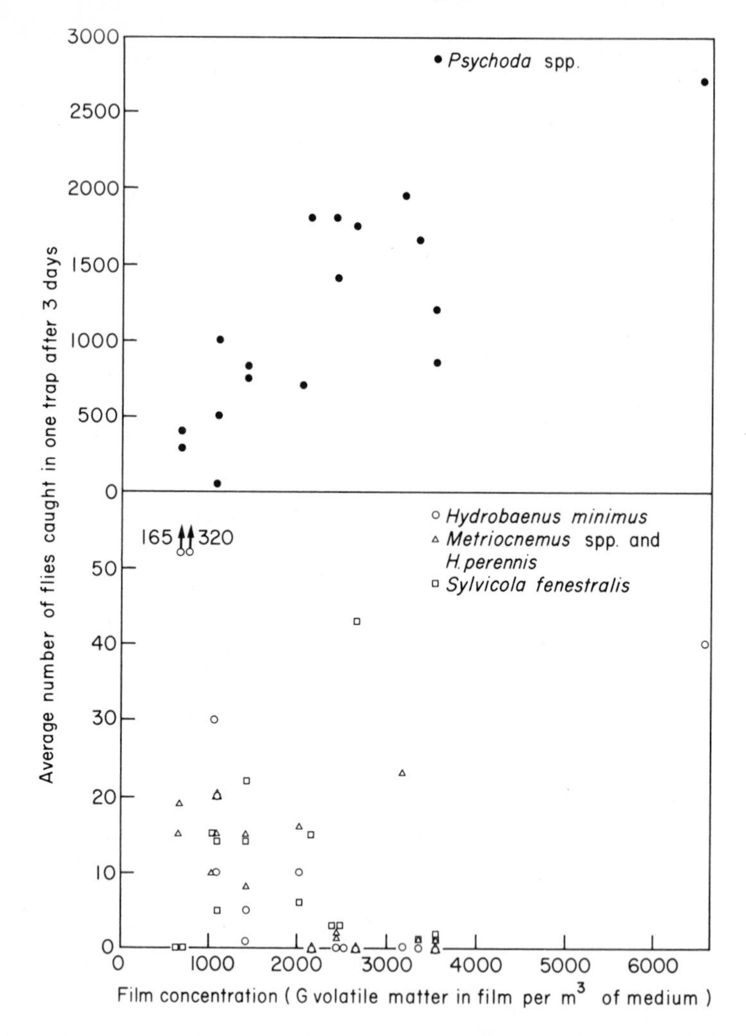

FIG. 2. Average number of flies trapped on different filters in relation to the concentration of film (data from Tomlinson and Stride, 1945).

Figure 2 shows the relationship between the numbers of individuals trapped of some insect species and the amount of film in the filter. It is clear that the numbers of *Psychoda* generally increased with increasing amount of film, whereas the chironomid species generally showed an inverse relationship with increasing amount of film. It is possible that the larvae of *Psychoda* spp. and *Sylvicola*, all of which bear respiratory siphons at the posterior end, enabling them to draw in oxygen direct from the air, are better able to cope with a thick film than are the chironomid larvae which have a closed tracheal system and can only obtain oxygen by diffusion through the body surface.

B. Hydraulic (Volumetric) Loading

Tomlinson and Hall (1950) demonstrated the effect of increasing hydraulic loading on the densities of *Psychoda* larvae, *Sylvicola* larvae and *Hypogastrura viatica* (Fig. 3). The densities of the latter two species

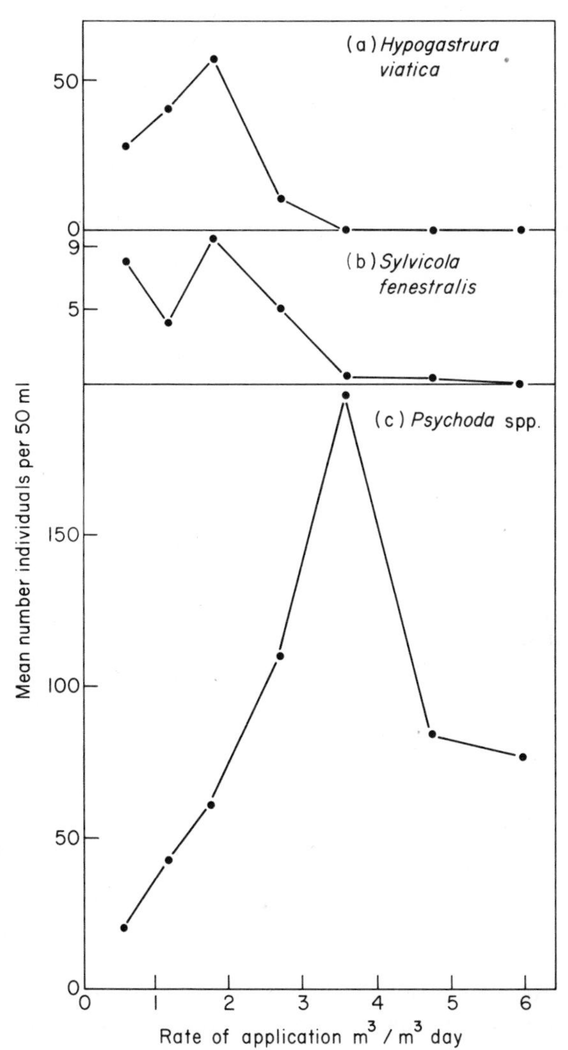

FIG. 3. Mean number of insects in filters receiving different hydraulic loadings (data from Tomlinson and Hall, 1950).

declined when the rate of application exceeded 1·8 m^3/m^3 per day (△ 300 gal/yd^3 per day), while the density of *Psychoda* did not decline until the rate of application exceeded 3·6 m^3/m^3 per day (△ 600 gal/yd^3 per day). The differences in density shown in Fig. 3 may not be entirely due to a direct effect of differences in the hydraulic loading, since both Thompson and Watson (1944) and Tomlinson and Hall (1950) showed that increased hydraulic loading generally resulted in an increased amount of film, and this has already been shown to affect the density and occurrence of filter insects (Fig. 2). Bruce and Merkens (1970) considered a hydraulic loading in excess of 3 m^3/m^3 per day (△ 500 gal/yd^3 per day) to be an acceptable definition of "high-rate" filtration, and it is clear (Fig. 3) that *Psychoda* can maintain densities at loadings up to 6 m^3/m^3 per day (△ 1000 gal/yd^3 per day), which is at least four times those customarily used (0·3–1 m^3/m^3 per day; △ 50–170 gal/yd^3 per day). However, these authors noted that no *Psychoda* adults were observed to leave the surface of high-rate filters operated at 6 m^3/m^3 per day even though there were plenty of larvae within the filter. This was probably because the almost continuous application of sewage customary at this loading prevents the adults from readily gaining access to the filter surface and from flying.

C. Nature of Sewage (Presence of Toxic Substances)

In general, the presence of industrial effluents in the amounts normally found in sewage does not have a markedly adverse effect on the filter fauna (Bruce, 1969). Where industrial effluents have been known to affect the filter fauna there tends to be a reduction in the diversity of the fauna such that only the most resistant species survive; owing to the reduced competition these species may become very abundant. Thus Reynoldson (1942) considered that the acidity of the sewage treated at Huddersfield (England), along with unspecified toxic substances in the sewage, accounted for the very impoverished insect fauna (only *Psychoda alternata* and *P. severini* were found there). Golightly (1940) showed that *P. alternata* in the filters at Huddersfield were ten times more numerous for most of the year than at Knostrop (Leeds) where the sewage contained a lower proportion of industrial waste and the filters had a more diverse fauna. However, during 1940 to 1942 the proportion of industrial waste in the Huddersfield sewage gradually increased and even *P. alternata* was ultimately severely affected (Reynoldson, 1942).

The application of a specific industrial waste to a filter can give rise to an insect fauna of unusual composition; thus the collembolan *Folsomia* sp., which rarely occurs in filters, was the only insect found in a filter treating gas liquor (Hawkes, 1963), and another insect rare to filters, *Trichocera maculipennis*, was the dominant fly in a filter treating a distillery waste (Karandikar, 1931).

D. Operational Procedures

Many single-pass filters are overloaded with sewage, resulting in a considerable accumulation of film especially in the winter. Various operational procedures have been adopted, such as double filtration, alternating double filtration, recirculation, and controlled periodicity of dosing, all of which are intended to reduce the amount of film in the filters or alter its distribution and so permit increased hydraulic and organic loadings (Hawkes, 1963). Hawkes (1961a) showed that reduced film accumulation tends to change the animal community in the filter from being fly-dominated to worm-dominated, this being due partly to the reduction in the amount of film and partly to the increased flow of sewage through the filter-bed. Hawkes believed that the chaetigerous prehensile worms were better able to withstand the higher velocities associated with higher instantaneous rates of doseage than were the fly larvae. Apart from reducing the numbers of insects relative to the other components of the filter community, reduced film accumulation also results in an insect fauna having an increased diversity. Thus the secondary filter where insect-trapping was carried out at Bedford (Table II) had twice as many insect species as the primary filter, and three times as many insect species were trapped leaving the Langley Mill filter which had a controlled frequency of dosing as were trapped leaving the filter with a self-propelled distributor. Furthermore the total numbers of insects emerging were lower in those filters having a reduced film accumulation. Tomlinson and Stride (1945) observed that, in proportion to the amount of sewage treated, fewer flies emerged from filters operating on ADF than with a single-pass system.

E. Method of Application of Sewage

Hawkes (1959) compared the effect of six kinds of distributor (Fig. 4), each discharging equal volumes of sewage, on the filter fauna, and found that the distributors bearing either splash-plates or fish-tails gave a more even distribution of sewage; this resulted in the greatest accumulation of film and generally low densities of *Psychoda* and *Sylvicola* larvae and of *Hypogastrura*, but high densities of the enchytraeid worm *Lumbricillus*. The distributors bearing one, two or four nozzles every 60 cm (24 in.) produced jets of sewage with a greater impinging force and resulted in

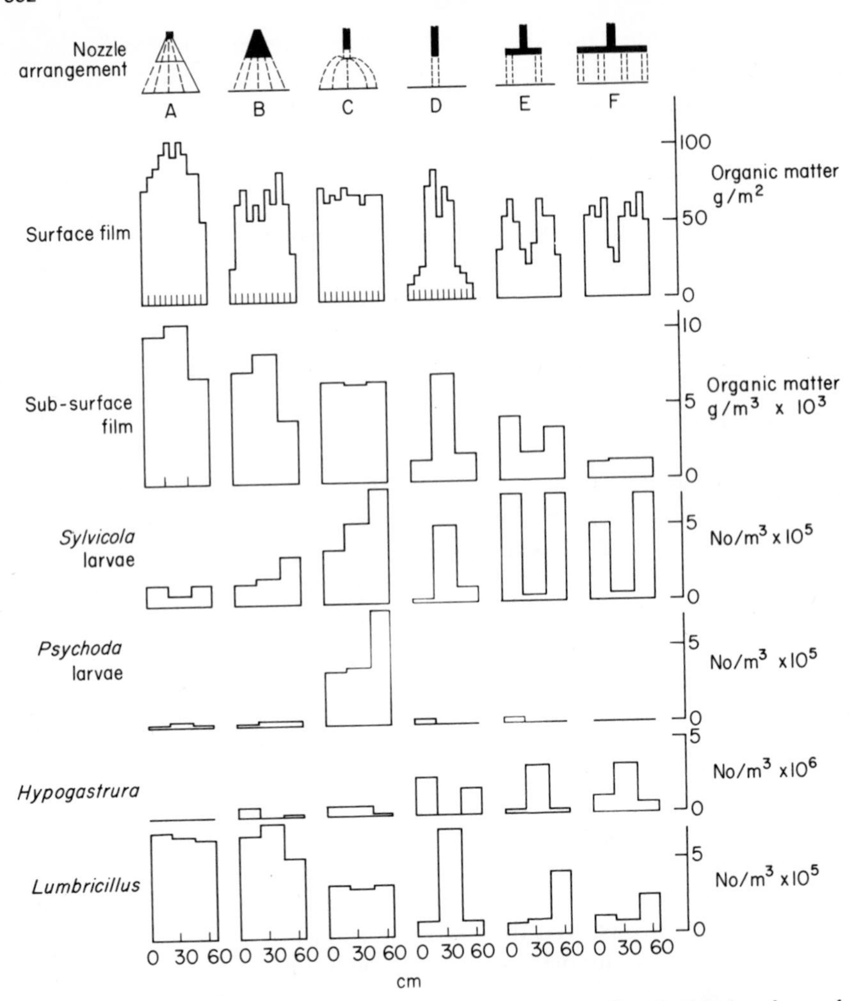

FIG. 4. Comparison of film and fauna occurring under six kinds of nozzle arrangement, each discharging equivalent volumes of sewage. (A) Open fish-tail. (B) Closed fish-tail; sewage discharged with greater force. (C) Splash-plate; sewage discharged on to a circular disc 6·4 cm (2·5 in.) diam. producing a circular sheet of spray. (D) Unmodified nozzle, 2·6 cm (1 in.) diam. (E) Two nozzles, 2 cm (0·75 in.) diam. (F) Four nozzles, 1·3 cm (0·5 in.) diam. (After Hawkes, 1963.)

less sub-surface film which was differentially distributed. This influenced the horizontal distribution of the macro-invertebrates such that fly-larvae and *Lumbricillus* were more abundant in the wetter areas where there was most film and *Hypogastrura viatica* was more abundant in the drier inter-jet areas. It was concluded from this experiment that a distributor bearing simple jets spaced 15·2 cm (6 in.) apart was best for the maintenance of a diverse fauna.

F. Type and Size of Medium

As with the other factors so far considered, the kind of medium and its size will strongly influence the amount of film developing. This is because of differences in the surface area of medium available and the suitability of the medium for film attachment, and differences in the size of the interstices, which may readily block if small. Hawkes and Jenkins (1955) studied four grades of media (3·2 to 6·4 cm—$1\frac{1}{4}$ to $2\frac{1}{2}$ in—diam.) and found that film accumulation and the mean numbers of *Sylvicola* and *Psychoda* larvae increased with decreasing particle size. Hawkes and Jenkins (1951) found that a medium of 1·6 cm ($\frac{5}{8}$ in.) or less diam. inhibited the egress of adult *Psychoda* and *Sylvicola* from filters. The topping of a filter with 1·9 cm ($\frac{3}{4}$ in.) medium to a depth of 23 cm (9 in.) was shown by Tomlinson and Stride (1945) to reduce the emergence of *Sylvicola* by 80% but to have much less effect upon the emergence of *Psychoda*. This effect upon adult emergence by media of small size may be partly a physical effect in that the interstices readily block, so making the egress of the adult flies difficult or impossible, and partly physiological in that the small medium plus film reduces the ventilation of the filter and so the amount of oxygen available becomes limiting to larval development near the filter surface. Certain insect species are favoured by a small medium, e.g. *Hypogastrura viatica* (Tomlinson and Stride, 1945).

The actual size at which a medium becomes limiting to a particular species is undoubtedly related to the sewage strength or to the organic loading. Terry (1956) showed that *Psychoda severini* and *Hydrobaenus minimus* were absent from filters containing a 1·9 cm ($\frac{3}{4}$ in.) medium fed with a strong sewage but abundant in such a medium if it received a weak sewage; *P. alternata* was not so restricted.

G. Temperature

Figure 5 shows the effect of temperature on the length of the life-cycle of some of the principal filter-inhabiting insects. Clearly *Psychoda alternata* has a higher rate of development than any of the other species at temperatures above 10°C, but below this temperature higher rates are achieved by *P. cinerea*, *P. severini* and *Hydrobaenus minimus*. The latter two species, as well as *Metriocnemus hygropetricus*, can complete their life-cycles at very low temperatures. However, *H. minimus* will not mate

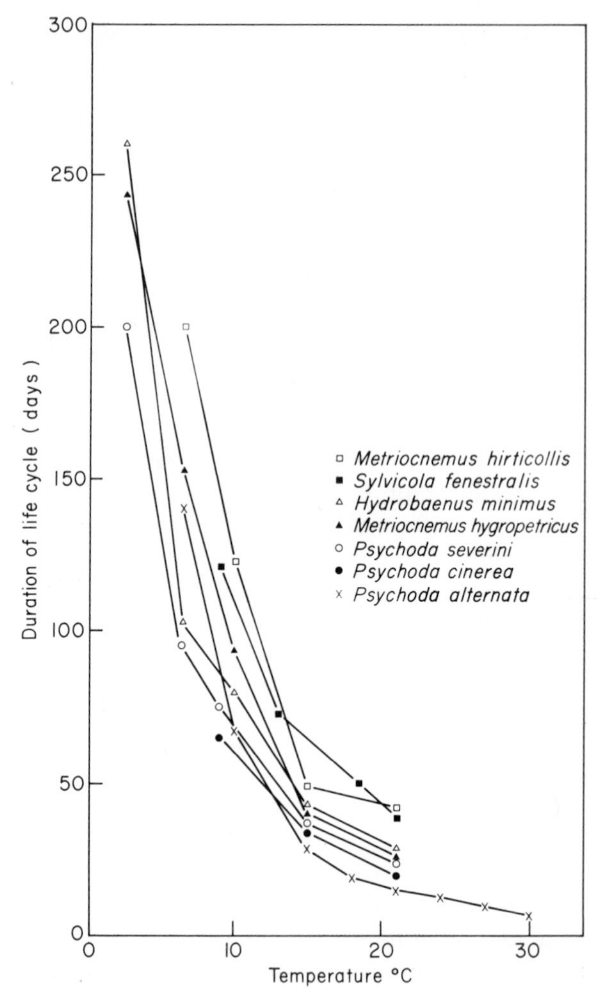

FIG. 5. Duration of the life-cycles at different temperatures of some of the principal insect species which inhabit filters. (Data from Fair, 1934; Lloyd, 1943; and T. G. Tomlinson, unpublished.)

at temperatures much below 10°C; the temperature-threshold for mating of *M. hygropetricus* must be considerably lower, since it and the parthenogenic *P. severini* have been observed to complete their life-cycles in an ice-box (Lloyd, 1943). It would be expected that the development of populations of those flies which do better at lower temperatures would be favoured during the winter and early spring, while those such as *P. alternata* would be favoured during the summer, and this generally holds true in practice (Lloyd, 1937). *M. hirticollis* and *Sylvicola fenestralis* develop at a slower rate than the other species shown in Fig. 5 and may only become successful in special conditions favouring their survival.

The relationship between the development period (D) and the temperature (T) can be expressed as

$$D = \frac{C}{T - K}$$

where C is the thermal constant (day degrees) and K is the minimum threshold temperature for development (Uvarov, 1931). This relationship assumes that once the temperature has reached a value at which development can occur, the amount of heat required by the insect for its development is a constant quantity. Table IV gives the thermal constants and threshold temperatures for the total life-cycle and for various developmental stages of several insect species. These data show that *Psychoda severini*, *Metriocnemus hygropetricus* and *Sylvicola fenestralis* have lower thresholds for development but require a greater number of day degrees for full development than *P. alternata*, *P. cinerea* and *Hydrobaenus minimus*.

Temperature not only affects the rate of development but may be lethal above and below an optimum range. Thus Lloyd *et al.* (1940) gave the temperature ranges within which there was no detectable mortality for several species (Table III), and Table IV compares the mortality during various stages of development of three *Psychoda* species at different temperatures. These figures confirm that *P. severini*, *Metriocnemus hygropetricus* and *Sylvicola fenestralis* are "low-temperature" species, while *P. alternata* is a "high-temperature" species. *P. cinerea* did less well in laboratory culture than did *P. alternata* at higher temperatures and *P. severini* at lower temperatures (Table IV), and the rarity of this species (see Table I) suggests that it is not so well adapted to conditions which favour the other two species even though its temperature requirements seem well suited to the filter ecosystem.

V. Vertical Distribution of the Insect Fauna

The amount of film in a filter has already been shown to affect the occurrence and density of the various insect species, and the vertical distribution of film might also be expected to influence the vertical distribution of the insect fauna. In filters operated as single-pass systems with frequent dosing the bulk of the film occurs in the top 60 cm (24 in.), and where the amount of film is sufficient to suppress any chironomid

TABLE III. Thermal constants and threshold temperatures for the life-cycle and various stages of development of several filter-inhabiting flies

Species	Egg incubation		Larva		Pupa		Maturation of female		Total for life-cycle		Temperature range at which there is no detectable temp. mortality (°C)	Mating bar (°C)
	Threshold temp. (°C)	Thermal constant (day degrees)	°C	Day degrees	°C	Day degrees	°C	Day degrees	°C	Day degrees		
Hydrobaenus minimus	4·3	38	3·5	383	5·5	37	0·0	63	4·5	450	9–20	10
Metriocnemus hygropetricus	1·0	63	1·7	440	2·5	32	—	—	1·0	540	5–18	very low
M. hirticollis	4·5	46	3·0	660	3·7	26	—	—	4·5	610	10–20	10
Psychoda alternata	5·9	20	5·5	244	7·7	31	2·0	46	6·0 5·0**	315 240**	15–28	6
P. severini	2·2	48	0·6	353	1·5	77	0·2	35	1·7 1·0**	480 480**	8–18	partheno-genetic
P. cinerea	—	—	—	—	—	—	—	—	3·5**	350**	—	—
Sylvicola fenestralis	—	—	—	—	—	—	—	—	2·1* 3·5**	814* 680**	<9–18**	very low

All data from Lloyd (1937) and Lloyd *et al.* (1940), except* Khalsa (1948) and** T. G. Tomlinson (unpubl.)

TABLE IV. The effect of temperature on egg and larval mortality of three *Psychoda* species (data from Golightly, 1940)

Species	*P. alternata*			*P. cinerea*			*P. severini*	
Temperature °C	10–12	17–19	27	9–12	13–15	17–20	11–14	17–20
Percentage eggs hatching	89	84	75	52	71	73	87	43
Percentage larvae successfully developing into adults	39	68	58	43	36	47	73	54
Percentage eggs developing into adults	38	53	51	23	31	37	85	18

population, the *Psychoda* and *Sylvicola* larvae are also concentrated in the top 60 cm of the filter (Table V; Fig. 6). However, if the chironomids develop successfully, competition between them and the *Psychoda* larvae for the available food results in a reduced proportion of the *Psychoda* population occupying the top 60 cm of the filter (Fig. 6) and may even result in a lowering of the zone of maximum abundance of the *Psychoda* larvae to a depth greater than 60 cm (Lloyd *et al.*, 1940). The chironomid larvae, especially those of *Hydrobaenus minimus* and *Metriocnemus hygropetricus*, may also actively eat the *Psychoda* larvae (Lloyd *et al.*, 1940). *Sylvicola*, however, appears unaffected by competition from chironomids (Fig. 6); it is distributed throughout the filter but seems always to occur in greatest numbers in the top 60 cm irrespective of organic and hydraulic loading (Tomlinson and Hall, 1950), although this distribution may be

TABLE V. Vertical distribution of insects at Minworth, Filter-block B, August 1968

Species	*Anurida granaria*	*Hypogastrura viatica* + *H. purpurescens*	*Cercyon ustulatus*	*Sylvicola fenestralis*	*Psychoda alternata*	*P. cinerea*	*P. severini*
Depth (ft)	No. per litre of medium						
0–0·5	18	18	—	—	89	89	—
0·5–1·0	71	18	36	107	36	214	53
1·0–1·5	142	89	36	18	18	231	18
1·5–2·0	1175	53	53	71	—	303	18
2·0–2·5	552	125	18	—	36	71	—
2·5–3·0	445	71	—	—	—	89	—
3·0–3·5	677	142	18	—	—	36	—
3·5–4·0	944	71	—	—	18	36	18
4·0–4·5	427	53	—	—	—	53	18
4·5–5·0	374	—	—	—	—	18	—
5·0–5·5	214	71	—	—	—	—	—
5·5–6·0	641	89	—	—	—	—	—

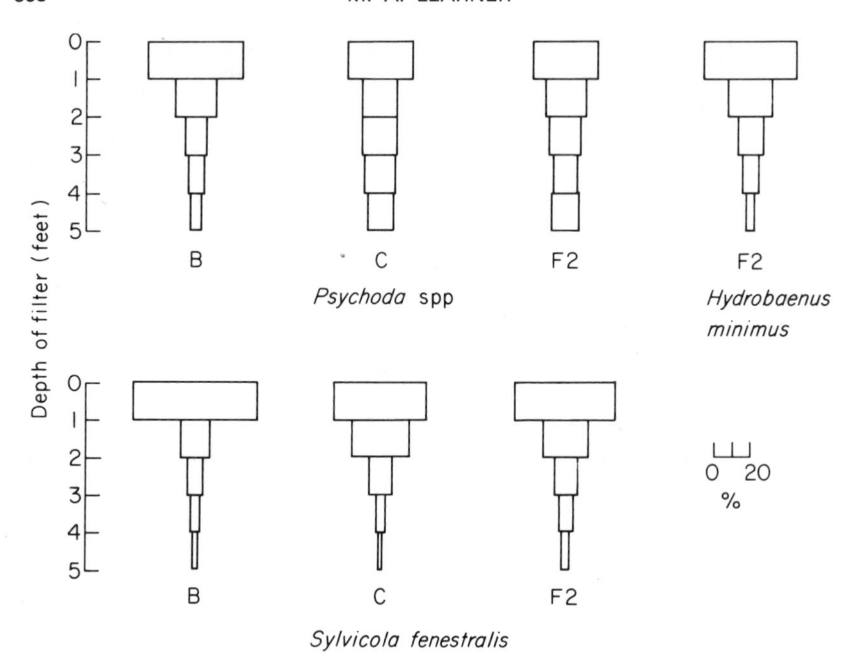

FIG. 6. Vertical distribution of flies in filters at Minworth, Birmingham; the numbers of insects found at each level are expressed as a percentage of the total count for three years (1945–1948). Filter B was operated as a single-pass system treating an effluent from a bio-flocculation plant at an applied rate of 0·9 m^3/m^3 per day (150 gal/yd^3 per day) from a rotary distributor. Filter F2 dealt with the same effluent, also as a single-pass system, but at an applied rate of 0·6 m^3/m^3 per day (100 gal/yd^3 per day) from a travelling distributor. Filter C treated settled sewage by alternating double filtration at an applied rate of 0·9 m^3/m^3 per day (150 gal/yd^3 per day) from a rotary distributor. (Unpublished data from T. G. Tomlinson.)

altered downwards by a high instantaneous rate of sewage application (Hawkes, 1963). Although Hawkes (1952a) considered the occurrence of *Sylvicola* to be closely correlated with film distribution, this may not be the overriding factor determining its vertical distribution; this species requires a drier environment for successful pupation than that tolerated by the larvae. There is, therefore, an upward and horizontal migration by the larvae before pupation into drier areas of the filter such as the inter-jet zones or the edge of the filter. It seems that *Sylvicola* is particularly successful in rectangular filters where widely spaced distributor jets and relatively low concentrations of film provide the environmental conditions necessary for successful pupation. Similar conditions may also be required by *Spathiophora hydromyzina* for successful pupation (Lloyd *et al.*, 1940). An upward migration has also been recorded for larvae of *Metriocnemus hygropetricus* (Dyson and Lloyd, 1936), so that pupation principally occurs close to the filter surface. Many chironomid adults require to swarm before successful mating occurs, and it is therefore an advantage if the adults hatch near the surface. *M. hygropetricus*, however, will mate successfully within the filter, and no such migratory movement of the larvae of *Hydrobaenus minimus* and *Metriocnemus hirticollis*, the adults of which require a mating swarm, has been reported. Adults of *Hydrobaenus perennis* probably also need to produce a mating swarm, although mating can proceed in confined spaces. The larvae of this species have been observed to burrow into the substrate to pupate when kept in laboratory culture (Lloyd *et al.*, 1940), but whether this results in a general downward migration in the filter, producing a high proportional wash-out from the base of the filter, is not known. *Psychoda* spp., like *Sylvicola*, can mate in the confines of the filter, but *Psychoda* pupae, unlike those of *Sylvicola*, seem able to tolerate similar conditions to the larvae and there is no evidence that larval migration occurs in relation to pupation, although there may be a nett downward movement of larvae in the filter as they develop; the median level of the pupal population is lower than that of earlier stages in the life-cycle (Table VI).

The vertical distribution of some other filter-bed insects is shown in Table V. The collembolans *Hypogastrura purpurescens*, *H. viatica* and *Anurida granaria* all occurred throughout the filter examined, with the greater proportion of the population occurring below the top 60 cm of the filter; beetle larvae (*Cercyon ustulatus*) were largely confined to the top 60 cm. Tomlinson and Hall (1950) also found *H. viatica* throughout the filter, and their data suggest that the depth of maximum abundance as well as overall abundance (see Fig. 3) may be more dependent upon hydraulic loading than on the amount of film present; this species appears to be readily flushed through the filter at hydraulic loadings in excess of 2 m^3/m^3 per day (336 gal/yd^3 per day). Nothing has so far been published on the vertical distribution of the other major filter-inhabiting insects, e.g. sphaerocerids, ephydrids and staphylinids. There is also little information

TABLE VI. Median depths (cm) of *Psychoda* larvae and pupae in a filter-bed (data from Solbé and Tozer, 1971)

Species	Larval instars I	II	III	IV	Pupae
P. alternata	9·1	15·3	12·2	10·7	15·3
P. severini	18·3	19·0	27·4	26·0	35·0

available on seasonal changes in the vertical distribution of the various insect species. Hawkes (1963) showed the vertical distribution of film to alter with season, and he considered (Hawkes, 1965b) that in general the incidence of film (perhaps fungal film rather than bacterial film) was more important in determining the incidence of grazers than vice versa. However, Solbé *et al.* (1967) showed that though the level of greatest concentration of *Psychoda* larvae did vary with season, being low down in the filter in summer but near the top in the winter, this level was not necessarily associated with the level at which the greatest concentration of film occurred.

The vertical distribution of film is more even in those filters where operational procedures (e.g. alternating double filtration) have been geared to reduce the rate of film accumulation (Hawkes, 1963). This change in film distribution may also reduce the variation of abundance with depth of the *Psychoda* population (Fig. 6), but it has little observable effect on the distribution of *Sylvicola*; this is further evidence that the vertical distribution of *Sylvicola* is not dependent upon the distribution of film.

VI. Seasonal Periodicity of Emergence

The numbers of each species in a filter fluctuate considerably with season, so that different species tend to dominate at different times during the year. The majority of filter insects have winged adults, and the emergence of these from the filter also shows seasonal fluctuations in numbers. It has already been established (Section IV G; Table III) that for many of the principal filter inhabitants there is a direct relationship between development and temperature, and thus the seasonal variations in numbers will be determined initially by the different responses of the various species to temperatures within the filter (Lloyd, 1945), which are largely dependent upon the temperature of the applied sewage (Bayley and Downing, 1963). Thus in Britain *Metriocnemus hygropetricus*, *Psychoda severini* and *Sylvicola fenestralis*, all with low thresholds for development (Table III), are able to develop relatively rapidly during the winter months and generally produce a peak adult emergence early in the year (April–May), while *Hydrobaenus minimus*, *M. hirticollis* and *P. alternata*, having higher threshold temperatures, reach their peak later in the year (June–September). *P. cinerea*, with a threshold between those of the above two groups, produces a peak emergence from May to September. Less is known about the temperature requirements of *H. perennis*, but it develops successfully at low temperatures and maximum output of adults occurs during April and May. A pre-pupal diapause occurs in this species from July to November (Lloyd *et al.*, 1940) and this may be a factor limiting the success of this species in filters, particularly if the medium is coarse and a large proportion of the population is washed out. However, despite the possible disadvantage of a diapause for this species in the filter ecosystem, it can become very abundant (see Table I) (Lloyd *et al.*, 1940). The other filter insects of numerical importance, the ephydrids and sphaerocerids, emerge chiefly in July, August and September (Terry, 1952; Solbé *et al.*, 1967; Green, 1970).

Temperature is not the only factor affecting the periodicity of adult emergence. The thermal constant of 480 day degrees for *Psychoda severini* (Table III) means that in Britain this species is, in theory, capable of producing a generation each month throughout the summer (Lloyd, 1937), and Satchell (1947) has shown that *P. severini* populations occurring in other habitats *do* produce generations throughout the summer, yet the species is usually almost entirely absent from filters from July to January. This intense summer–autumn depression may be due to the longer period spent by this species in the egg and pupal stages compared with *P. alternata* or the chironomids (Table III), so that it is particularly vulnerable to both predation and wash-out. Similarly, the establishment of *Sylvicola* in a sewage-works near Birmingham, England, limited the *Psychoda* emergence period, previously April–November, to August–October (Hawkes and Jenkins, 1951).

Differences in the summer and winter temperatures from year to year may also be sufficient to alter the pattern of emergence by favouring one species at the expense of another (Lloyd, 1943). Hawkes (1951) has shown that a mild winter favouring the earlier development of *Psychoda alternata* can result in a lowered *Sylvicola* emergence.

VII. Factors Affecting Flight

A. Temperature

Once the flies have emerged from the filter, their subsequent flight and dispersal appear to be largely dependent upon climatic conditions. Insects will not fly unless the temperature is above a threshold value. This value is about 10°C for *Psychoda* spp. (Scouller and Goldthorpe, 1932), 7·8°C for *Hydrobaenus minimus* (Gibson, 1945) and near 4°C for *Metriocnemus* spp. and *Sylvicola* (Gibson, 1945; Hawkes, 1961b). However, Lloyd (1943) considered that *Metriocnemus hirticollis* would not form swarms at temperatures much below 10°C. Increasing temperature above the threshold produces an increase in flight activity; Hawkes (1961b) found that every 1·2°C rise in temperature above the threshold value up to about 24°C doubled the number of *Sylvicola* in flight.

B. Light Intensity

Light intensity appears to have an important influence upon the diel rhythms displayed by filter insects. Figure 7 shows a typical rhythm for *Psychoda alternata*; this species is diurnal with a peak emergence in the early afternoon. The emergence data were obtained from two filters receiving an equivalent hydraulic loading of 0·475 m^3/m^3 per day (80 gal/yd^3 per day); sewage was not applied to one of them for a period of 8 h during daylight. The pattern of emergence was unaffected by the prolonged resting period. It is also clear from Fig. 7 that while the absence of light considerably reduced the number of individuals emerging it did not prevent emergence. Experiments in which a filter was kept in darkness for 5 days, by means of black polythene sheeting, showed that while darkness suppressed the oscillations in the normal diel rhythm it had no effect upon the numbers of *P. alternata* emerging from the filter-bed (Ministry of Technology, 1966). Therefore it would appear that the existence of daily light and dark periods serves to synchronize the flight periodicity of the adult *P. alternata* but otherwise does not influence whether they fly or not. *P. severini* has a very similar daily emergence pattern to *P. alternata* (Fig. 8) and the chironomids *Hydrobaenus minimus* and *Metrocnemus* spp. are also diurnal; Gibson (1945) showed that these chironomids fly much less readily at light intensities below 108 lx (10 foot-candles). *Sylvicola*, however, tends to be crepuscular with a peak emergence at dusk and a smaller peak around dawn; some flying occurs at all light intensities although the flies are much less active during darkness (Hawkes, 1961b).

C. Wind Velocity

The *Psychoda* spp. are normally considered weak fliers, generally flying for no more than a few seconds at a time, but so far as the author is aware no information exists upon the influence of wind velocity on flight. The chironomids and *Sylvicola* are much stronger fliers; however, Gibson (1945) found that *Hydrobaenus minimus* virtually ceased flying at wind velocities above 0·5 m/sec (1 mph), although *Metriocnemus* swarms were observed at a velocity of 1·4 m/sec (3·2 mph). The optimum velocity for flight and swarming was about 0·3 m/sec (0·6 mph), the size of the swarms

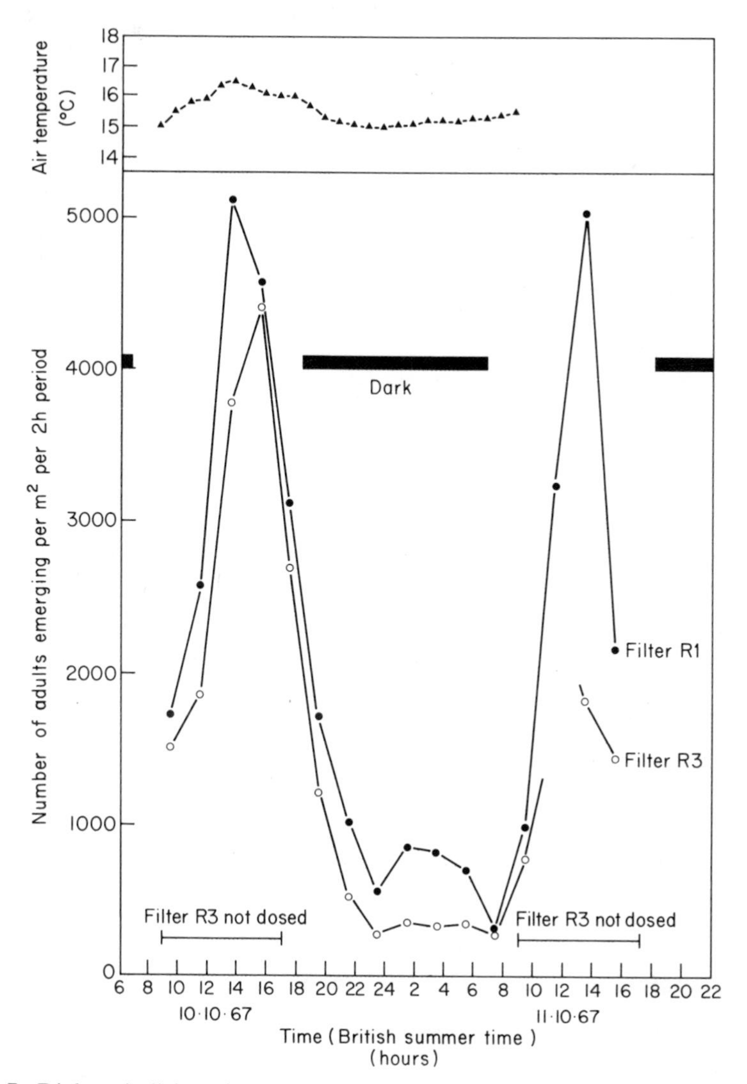

FIG. 7. Diel periodicity of emergence of *Psychoda alternata* adults from filters receiving a hydraulic loading of 0·475 m^3/m^3. Filter R3 was not dosed between 0900 and 1700 h.

progressively decreasing as wind velocities increased above this value. Hawkes (1952b, 1961b) showed that for *Sylvicola* every 0·67 m/sec (1·5 mph) increase in wind velocity from 0 to 4·5 m/sec (0–10 mph) halved the number of adults in flight, although some individuals flew in wind velocities up to about 6·7 m/sec (15 mph).

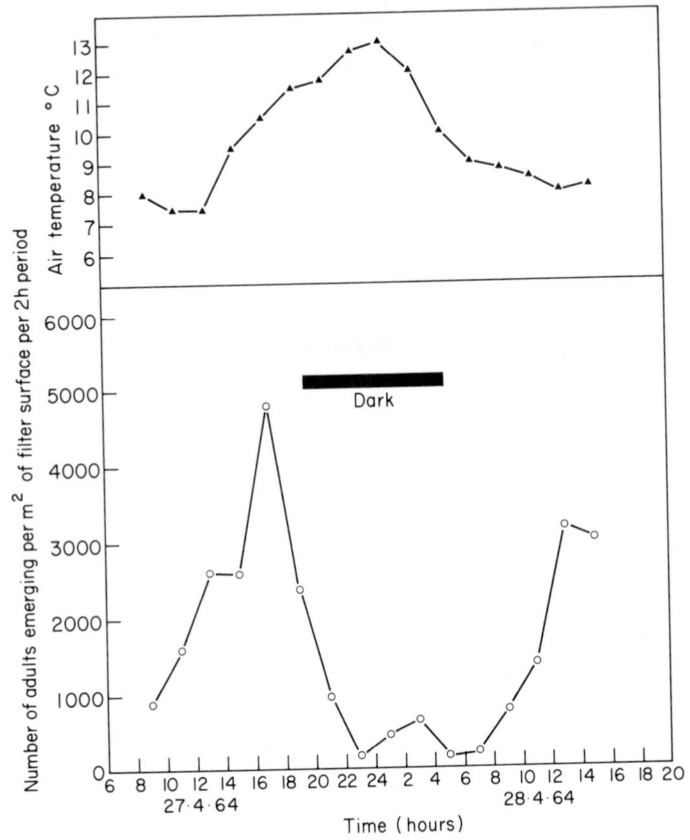

FIG. 8. Diel periodicity of emergence of *Psychoda severini* adults from pilot-scale filters (redrawn from Ministry of Technology, 1965).

VIII. Dispersal and Causes of Nuisance

A mating flight is only necessary as a prelude to mating for certain of the chironomid inhabitants of filters, principally *Hydrobaenus minimus* and *Metriocnemus hirticollis*. However, most chironomid species swarm and these swarms continually seek a balance between optimum light intensity and the minimum shelter necessary for "dancing". Perhaps because of this the swarms are frequently orientated with respect to some fixed raised object, e.g. wall, building, line of trees (Gibson, 1945). For other insect species of filters flight is only a means of dispersing the species, enabling the rapid colonization of suitable habitats as they occur. In general, however, the flies do not disperse far from their particular source.

Satchell (1947) recorded that *Psychoda* derived from filters were found up to about 1·6 km (1 mile) away in the direction of the prevailing wind. He was unable to give any indication either of the proportion of the emerging flies that were dispersed or of the numbers involved, because the species found in the sewage-works were also breeding in a number of other places near his sampling points. Other authors (Headlee, 1919; Simpson, 1933; Holtje, 1943) also considered the range of *Psychoda* to be about 1·6 km. The range of *Sylvicola* is similar, for T. G. Tomlinson (unpublished data) made observations over a number of years at Minworth (Birmingham) and found considerable numbers at distances of 0·8 to 1·2 km (0·5–0·75 mile) from the filters but none at a distance greater than 2·4 km (1·5 mile). Although the dispersal of the foregoing species is very restricted they have frequently proved to be a considerable nuisance both within the sewage-works, where they are apt to be drawn into the mouths and nostrils of the employees, and outside, where they readily enter nearby houses or factories. None of these species bites; indeed among those insects occurring in filters only three species, *Culicoides impunctatus*, *C. pulicaris* and *C. vexans*, are known to bite man. All three are rare, only *C. vexans* being trapped in numbers greater than $500/m^2$ per year (Table I). There is also no evidence to show that any of the flies carries pathogens likely to harm man. However, as they have been associated with sewage, it is undesirable both from the public-health and from aesthetic viewpoints for these insects to enter homes or factories in large numbers. Additionally, in factories they may actually adversely affect certain manufacturing processes. Where filters are in urban areas, therefore, control measures have frequently been used (see Section X).

Much less is known about the dispersal of the chironomids, sphaerocerids and ephydrids, but they have rarely been cited as a nuisance in dwellings near sewage-works.

IX. Role of Insects in Percolating Filters

The Collembola and the majority of fly-larvae are grazers feeding on the film or components of it, and there is clear evidence that under normal operational procedures practised in Britain their activities are extremely important in preventing the filters from blocking (Parkinson and Bell, 1919; Reynoldson, 1939; Tomlinson, 1941; Lloyd, 1945; Hawkes, 1965b; Williams and Taylor, 1968). They are most effective during the summer when temperatures promote greater activity, since although the rate of film growth (at least over 10–15 days) is about twice that recorded at winter temperatures (Tomlinson, 1946b), "ponding" of the filter rarely occurs during the summer period. Even during the winter when "ponding" is frequently reported, the presence of large populations of certain genera, e.g. *Sylvicola,* which are relatively active at low temperatures may limit film accumulation (Hawkes, 1965b).

By grazing on the film, insects perform two functions. Firstly, they directly convert some of the film to carbon dioxide and other simple substances which are excreted, and heat. It has been estimated for one filter (Solbé, 1971) that total macro-invertebrate respiration may account for between 2 and 10% of the total organic carbon oxidized in the filter, although only a small proportion of this was due to insects. Secondly, they alter the character of the film by converting much of it into faecal pellets which may aid bacterial action. These activities enable the filter to remain operational at higher organic loadings than if the insects and other macro-invertebrates were absent. Hawkes (1957), however, did not consider the activity of macro-invertebrates to be important in preventing the winter accumulation of film when this is dominated by fungi rather than bacteria, although their activity may assist the vernal sloughing. It seems that the operational procedures used at any sewage-works and the kind of sewage treated may well determine whether the insects play an important role in limiting film growth or not. Thus, although Usinger and Kellen (1955) in the U.S.A. considered insects to be of importance, most workers in the U.S.A. (e.g. Holtje, 1943; Cooke and Hirsch, 1958) have suggested otherwise; this may be because in the U.S.A. the medium is generally larger, the sewage weaker, and the hydraulic loadings greater than in Britain. Brink (1967) in Sweden, using experimental filters, and Bruce and Merkens (1970) in England, investigating the use of synthetic media in high-rate filters, likewise believed that the insect grazers were not of importance in affecting the degree of purification in their systems.

There is no evidence that the insect grazers remove so much film that the filter no longer operates efficiently; in fact, the film need be no more than about 0·2 mm thick to operate at maximum effectiveness (Hawkes, 1963). However, reduced availability of food in filters with little film has a clear effect upon insects such as *Sylvicola* (Hawkes, 1951), *Psychoda severini* and *P. alternata* (Golightly, 1940), *Hydrobaenus minimus* and *Metriocnemus hirticollis* (Lloyd, 1945), resulting in an increase in the duration of the larval phase, smaller pupae and adults and much higher mortalities among pupating larvae. Increasing temperature also reduces the final size of the adult of at least *Psychoda* and chironomids (Todd, 1936; Golightly, 1940; Lloyd, 1945), the shortening of the duration of the larval phase resulting in smaller pre-pupal larvae. During the summer months both these factors operate, the overall effect being fewer adults than theoretically expected and smaller than those from winter populations. Small adults lay fewer and smaller eggs than large adults and these responses help to regulate the population at times of food shortage.

The quality of the food available is also important in determining the duration of larval development, and it is possible that film containing different dominant fungi or bacteria will influence the rate of development of, and favour the occurrence of, different species. T. G. Tomlinson (unpublished data) compared the rate of development of *Psychoda alternata*, *P. cinerea* and *Sylvicola* reared on three different substrates, the fungus *Sepedonium*, filter film, and sludge from a bio-flocculation plant, and showed that development was quickest on *Sepedonium* for all three insect species. Dyson and Lloyd (1936), Lloyd and Turner (1936) and T. G. Tomlinson (unpublished data) have all noted an association, though not exclusive, between the occurrence of the alga *Phormidium* and the presence of *Metriocnemus* spp.

Insects and other macro-invertebrates in filters have another effect on sewage treatment in that they may indirectly promote a more rapid settlement of solids in the humus tanks. Williams and Taylor (1968) found that the solids in effluents from filters containing *Psychoda* larvae only, or *Psychoda* and the worm *Lumbricillus*, settled considerably more quickly, 65–70% in one hour, compared with solids in effluents from filters without macro-invertebrates, 34% in one hour. This is probably because of the greater density of the solids from the filters containing macro-invertebrates, resulting from the presence of animal fragments and faecal material.

X. The Control of Filter Flies

It has frequently been necessary to adopt measures to control the numbers of adult flies emerging from filters, for reasons discussed in Section VIII. Several methods have been tried with varying success, but whatever control method is used it should be specific to the species causing the nuisance and it must not reduce the efficiency of the filter. Control can be long-term or short-term but must in any case be as economical as possible.

A. Physical Control

The flooding of filters for about 24 h has frequently been advocated in the U.S.A. (Headlee, 1919; Fair, 1934) but the degree of success varies (Dreier, 1946); the construction of the filter must be suitable otherwise the necessary alterations are likely to be expensive. This method is only effective for short periods. A more permanent method is the topping of a filter with a fine medium (1·3 to 1·9 cm–$\frac{1}{2}$ to $\frac{3}{4}$ in.) to a depth of about 23 cm (9 in.). This reduces the number of adult *Psychoda* and *Sylvicola* emerging but may give rise to "ponding" in the winter. A small medium favours *Hypogastrura viatica* and populations of this species may help to reduce ponding by grazing within the superficial layer. Complete enclosure of filters has sometimes been adopted but this is expensive and only partially successful (Murray, 1939).

B. Chemical Control

Before the introduction of organo-chlorine pesticides, a considerable range of chemicals was tried (Fair, 1934; Lloyd, 1945; Tomlinson and Jenkins, 1947), but none was specific to flies and affected both other components of the grazing fauna and the film. The introduction of organo-chlorine pesticides, particularly DDT and BHC, provided a much more effective way of dealing with the fly nuisance (Brothers, 1946; Tomlinson and Jenkins, 1947; Tomlinson and Muirden, 1948; Jenkins *et al.*, 1949; Tomlinson *et al.*, 1949). These chemicals have been employed largely against the larvae of *Psychoda*, *Sylvicola*, *Hydrobaenus minimus* and *Scatella silacea*. Such pesticides can be added directly to the sewage feed: this facilitates effective dispersion within the filter and continuous filter operation. It has the disadvantage, like all methods of direct control on filters, that larvae which may aid in filter efficiency are destroyed. Such adverse effects are minimized by restricting application to periods immediately before fly emergences.

Concentrations of DDT and BHC effective against fly-larvae apparently have little effect upon other macro-invertebrate grazers such as *Hypogastrura viatica* and the worm *Lumbricillus* (Tomlinson and Jenkins, 1947), and Hawkes (1957) found that insecticide treatment reduced the density of *Sylvicola* and enabled *H. viatica* to increase its density, so maintaining the purifying efficiency of the filter. However, when pesticide control ceased, the population of *Sylvicola* recovered because the environment was basically more suitable for *Sylvicola* than for *H. viatica*.

An extremely important disadvantage in the use of organo-chlorine compounds is that when DDT and BHC are used at concentrations suitable for fly-control, concentrations of these substances in the filter effluent are sometimes sufficiently high to be lethal to fish such as rainbow trout (*Salmo gairdneri*) (Tomlinson and Jenkins, 1947). BHC is more labile than DDT, the use of which has already been banned in several countries and much restricted in others, but other fairly labile compounds such as malathion (Watson and Fishburn, 1964) and the carbamates may prove to be of value in fly control.

There are now many examples of insect species which have developed populations resistant to specific insecticides (Cherrett *et al.*, 1971). In general an insecticide acts as a "selective agent" on a population in that a gene/genes already present in some individuals is/are able to protect them from the lethal effects of the pesticide. The continued use of the pesticide concentrates the gene(s) in subsequent generations and the population contains an ever increasing proportion of resistant individuals. This may be happening among some filter populations regularly receiving insecticide treatment. Watson and Fishburn (1964) and Clifford (discussion in Watson and Fishburn, 1964) believed that *Sylvicola* larvae were becoming increasingly resistant to BHC at Keighley (Yorkshire) and Leeds (Yorkshire) sewage-works respectively.

A similar adaptation among insect larvae may occur where the filter receives waste from factories using or formulating pesticides. The author found larvae of *Psychoda alternata* drawn from a population inhabiting such a filter (Lowden *et al.*, 1969) to be twice as resistant to DDT as larvae which had not been exposed to DDT.

C. Biological Control

Because of the disadvantages of pesticides, some form of biological control appears an attractive alternative, particularly as it is unlikely to lead to the total extinction of species which are of benefit to the filter. However, there are few natural predators in the filter ecosystem; *Spathiophora hydromyzina* (Table I) occurs in many filters but never in numbers large enough to make a significant impact upon fly populations. *Cercyon* larvae are omnivorous but only occur in small numbers. The omnivorous midge larvae, especially *Hydrobaenus minimus* and *Metriocnemus hygropetricus*, do occur in large enough numbers to reduce the numbers of other fly-larvae (Lloyd, 1945) but they may in turn become a nuisance. Apart from the insects a few other predators, e.g. spiders and some mites (see Chapter 10: Arachnida), may exert some small influence. In other fields parasites have provided the highest proportion of successes in biological control possibly because they can remain effective at low host densities (Cherrett *et al.*, 1971). However, although various hymenopterans are known to parasitize many of the sphaerocerid and ephydrid species in filters, none has been found parasitizing *Psychoda*, *Sylvicola* or the chironomids.

D. Nutritional Control

Since the flies which cause a nuisance all feed on the film, the most logical way of reducing the number of flies is to operate the filter so that the amount of food (film) available is restricted throughout the year. Where filters have been operated to reduce the amount of film and to produce a fairly homogeneous distribution of film, the filter fauna has become more diverse with a greater number of species and fewer flies. It would be still more useful if the operating conditions were such as to favour a worm- rather than a fly-dominated community, but there is still insufficient information to provide a basis for such control of community structure. It should eventually be possible to do this, at least for the important species, and then operating conditions could be altered to produce a particular community. An additional complication to the idea of controlling the fly population by altering the conditions of operation is that the design of some sewage-works may not allow any latitude in operating procedures. Under such circumstances additional fly-control methods, either physical or chemical, will be necessary during the main periods of fly emergence.

XI. Occurrence of Insects in Other Used-water Treatment Processes

The sewage-works, whether using the filter or activated-sludge method of treatment, provides few other habitats for the development of large insect populations, but a range of species successfully colonizes sludge-drying beds and the numbers of sepsid flies arising from such beds have given rise to complaints (Green, 1970; Volume 2, Chapter 6). Occasionally chironomids have successfully invaded activated-sludge systems, causing a considerable reduction in the degree of purification (Anderson, 1943; Jackson, 1943; Pillai and Subrahmanyan, 1944), but generally this method of treatment does not provide an environment favourable for the build-up of large numbers of insect larvae. Chironomid larvae, usually *Chironomus dorsalis* or *C. riparius*, also occur frequently in any open channels carrying filter effluent, and larvae of *Eristalis* spp. occur in humus and sludge tanks; the densities of all these species are low.

Oxidation ponds are frequently used as a secondary treatment process in areas of the world where the climate is suitable; these may provide a habitat for a range of insect species either wholly aquatic or with aquatic stages. The composition of such communities is largely dependent upon the organic loading received by such ponds and the detention period (Usinger and Kellen, 1955).

Tubicolous chironomid larvae often reach high densities (up to 45 000/m^2) in oxidation ponds and are important in removing algae and promoting mixing at the mud–water interface.

XII. Summary

The secondary treatment of sewage using filters provides an environment suitable for a range of insect species, some of which occur in considerable numbers. The principal members of this community (*Psychoda* spp., *Sylvicola fenestralis*, several species of chironomid midges, and the collembolan *Hypogastrura viatica*) all feed on the film and, depending upon the operational methods employed, may play a very important part in the sewage purification process. The amount of film present is the principal factor determining the kind of community occurring in any particular filter; a large amount of film produces an insect community having a low species diversity dominated by *Psychoda* spp., while a small amount of film produces a community having a higher species diversity dominated by chironomid midges. A community with a high species diversity is likely to have species which dominate at different times of the year, and this increases the usefulness of such a community in that the growth of film is restricted for much of the year. Filters inhabited by a community having a high species diversity do not generally give rise to complaints of fly nuisance, partly because the lower film growth reduces fly production and partly because, with the wide variety of species, emergences at any time are not unduly high.

The various operational procedures aimed at reducing the amount of film in the filter will also favour the development of a community with a high species diversity, but it is unlikely that any procedural changes can entirely eradicate the fly problem where filters are concerned, except perhaps where high rate filtration is adopted. For this reason it is advisable to site new sewage-works (if filters are to be used) at least 1·6 km (1 mile) from houses or dwellings and to restrict housing development close to existing works. Should a fly nuisance arise, there may at present be no alternative to short-term chemical control using highly labile pesticides such as malathion or the carbamates. More information is needed about the rates of application needed and the effects of such insecticides on the other components of the filter fauna.

References

ANDERSON, R. A. (1943). *Sewage Wks J.* **15,** 929.

BAINES S. and FINLAYSON, L. H. (1949). *Entomologist's Mon. Mag.* **85,** 150–151.

BAYLEY, R. W. and DOWNING, A. L. (1963). *J. Instn publ. Hlth Engrs* **62,** 303–332.

BRINK, N. (1967). *Int. Revue ges. Hydrobiol. Hydrogr.* **52,** 51–122.

Brothers, W. C. (1946). *Sewage Wks J.* **18,** 181.
Bruce, A. M. (1969). *Process Biochem.* **4,** 19–23.
Bruce, A. M. and Merkens, J. C. (1970). *Wat. Pollut. Control* **69,** 113–139.
Bruce, A. M., Truesdale, G. A. and Mann, H. T. (1967). *J. Instn publ. Hlth Engrs* **66,** 151–172.
Cherrett, J. M., Ford, J. B., Herbert, I. V. and Probert, A. J. (1971). "The Control of Injurious Animals." English Universities Press, London.
Christiansen, K. (1964). *A. Rev. Ent.* **9,** 147–178.
Cooke, W. B. and Hirsch, A. (1958). *Sewage ind. Wastes* **30,** 138–156.
Crisp, G. and Lloyd, L. (1954). *Trans. R. ent. Soc. Lond.* **105,** 269–314.
Dreier, D. E. (1946). *Sewage Wks J.* **18,** 704–729.
Dyson, J. E. B. and Lloyd, L. (1936). *Proc. Leeds phil. lit. Soc.* **3,** 174–176.
Fair, G. M. (1934). *Sewage Wks J.* **6,** 966–979.
Gibson, N. H. E. (1945). *Trans. R. ent. Soc. Lond.* **95,** 263–294.
Golightly, W. H. (1940). *Ann. appl. Biol.* **27,** 406–421.
Green, M. B. (1970). *Wat. Pollut. Control* **69,** 399–408.
Hawkes, H. A. (1950). "The chemical control of *Anisopus fenestralis* in percolating sewage filters." M.Sc. Thesis, University of Leeds, England.
Hawkes, H. A. (1951). *Ann. appl. Biol.* **38,** 592–605.
Hawkes, H. A. (1952a). *Ann. appl. Biol.* **39,** 181–192.
Hawkes, H. A. (1952b). *Proc. Birmingham Nat. Hist. Phil. Soc.* **18,** 41–53.
Hawkes, H. A. (1957). *J. Proc. Inst. Sew. Purif.*, 88–102.
Hawkes, H. A. (1959). *Ann. appl. Biol.* **47,** 339–349.
Hawkes, H. A. (1961a). *J. Proc. Inst. Sew. Purif.*, 105–127.
Hawkes, H. A. (1961b). *Ann. appl. Biol.* **49,** 66–76.
Hawkes, H. A. (1963). "The Ecology of Waste Water Treatment." Pergamon Press, London and Oxford.
Hawkes, H. A. (1965a). *In* "Ecology and the Industrial Society" (Eds G. T. Goodman, R. W. Edwards and J. M. Lambert), pp. 119–148. Blackwell, London.
Hawkes, H. A. (1965b). *Int. J. Air Wat. Pollut.* **9,** 693–714.
Hawkes, H. A. and Jenkins, S. H. (1951). *J. Proc. Inst. Sew. Purif.*, 300–318.
Hawkes, H. A. and Jenkins, S. H. (1955). *J. Proc. Inst. Sew. Purif.*, 352–357.
Headlee, T. J. (1919). *J. econ. Ent.* **12,** 35–41.
Holtje, R. H. (1943). *Sewage Wks J.* **15,** 14–29.
Jackson, R. B. (1943). *Sewage Wks J.* **15,** 1253–1254.
Jenkins, S. H., Baines, S. and Hawkes, H. A. (1949). *J. Proc. Inst. Sew. Purif.*, 178–184.
Karandikar, K. R. (1931). *Trans. ent. Soc. Lond.* **79,** 249–260.

KERRICH, G. J., MEIKLE, R. D. and TEBBLE, N. (1967). "Bibliography of Key Works for the Identification of the British Fauna and Flora." The Systematics Association, London.

KHALSA, H. G. (1948). "Biology of *Anisopus fenestralis* Scop., a new invader of sewage filter-beds at Knostrop, Leeds." Ph.D. Thesis, University of Leeds, England.

LAURENCE, B. R. (1954). *J. Anim. Ecol.* **23,** 234–260.

LAURENCE, B. R. (1955). *J. Anim. Ecol.* **24,** 187–199.

LAWRENCE, P. N. (1961). *Wat. Waste Treat.* **8,** 497.

LAWRENCE, P. N. (1970). *Wat. Waste Treat.* **13,** 106–109.

LLOYD, L. (1937). *J. Proc. Inst. Sew. Purif.*, 150–165.

LLOYD, L. (1943). *Ann. appl. Biol.* **30,** 47–60.

LLOYD, L. (1945). *J. Proc. Inst. Sew. Purif.*, 119–139.

LLOYD, L. and TURNER, J. N. (1936). *Proc. Leeds phil. lit. Soc.* **3,** 177–188.

LLOYD, L., GRAHAM, J. F. and REYNOLDSON, T. B. (1940). *Ann. appl. Biol.* **27,** 122–150.

LOWDEN, G. F., SAUNDERS, C. L. and EDWARDS, R. W. (1969). *Proc. Soc. Wat. Treat. Exam.* **18,** 275–287.

MINISTRY OF TECHNOLOGY (1965). "Water Pollution Research, 1964." H.M. Stationery Office, London.

MINISTRY OF TECHNOLOGY (1966). "Water Pollution, 1965." H.M. Stationery Office, London.

MURRAY, K. A. (1939). *J. Proc. Inst. Sew. Purif.*, 350–352.

PARKINSON, W. H. and BELL, H. D. (1919). "Insect Life in Sewage Filters." Sanitary Publishing Company, London.

PILLAI, S. C. and SUBRAHMANYAN, V. (1944). *Nature, Lond.* **154,** 179–180.

REYNOLDSON, T. B. (1939). *J. Proc. Inst. Sew. Purif.*, 158–167.

REYNOLDSON, T. B. (1942). *J. Proc. Inst. Sew. Purif.*, 116–134.

SATCHELL, G. H. (1947). *Ann. appl. Biol.* **34,** 611–621.

SCOULLER, W. D. and GOLDTHORPE, H. H. (1932). *J. Proc. Inst. Sew. Purif.*, 202–206.

SIMPSON, W. J. (1933). *Sewage Wks J.* **5,** 103.

SOLBÉ, J. F. DE L. G. (1971). *J. appl. Ecol.* **8,** 845–867.

SOLBÉ, J. F. DE L. G. and TOZER, J. S. (1971). *J. appl. Ecol.* **8,** 835–844.

SOLBÉ, J. F. DE L. G., WILLIAMS, N. V. and ROBERTS, H. (1967). *Wat. Pollut. Control* **66,** 423–448.

TERRY, R. J. (1951). "Some ecological studies of the relations between the bed medium and the fauna in percolating sewage filters." Ph.D. Thesis, University of Leeds, England.

TERRY, R. J. (1952). *Proc. Leeds phil. lit. Soc.* **6,** 104–111.

TERRY, R. J. (1956). *J. Anim. Ecol.* **25,** 6–14.

THOMPSON, J. T. and WATSON, H. J. (1944). *J. Proc. Inst. Sew. Purif.*, 65–71.

TODD, J. P. (1936). *J. Proc. Inst. Sew. Purif.*, 201–204.

TOMLINSON, T. G. (1941). *J. Proc. Inst. Sew. Purif.*, 39–56.

TOMLINSON, T. G. (1946a). "Animal life in Percolating Filters." Water Pollution Research Technical Paper No. 9. H.M. Stationery Office, London.

TOMLINSON, T. G. (1946b). *J. Proc. Inst. Sew. Purif.*, 168–178.

TOMLINSON, T. G. and HALL, H. (1950). *J. Proc. Inst. Sew. Purif.*, 338–363.

TOMLINSON, T. G. and JENKINS, S. H. (1947). *J. Proc. Inst. Sew. Purif.*, 94–114.

TOMLINSON, T. G. and MUIRDEN, M. J. (1948). *J. Proc. Inst. Sew. Purif.*, 168–177.

TOMLINSON, T. G. and STRIDE, G. O. (1945). *J. Proc. Inst. Sew. Purif.*, 140–148.

TOMLINSON, T. G., GRINDLEY, J., COLLETT, R. and MUIRDEN, M. J. (1949). *J. Proc. Inst. Sew. Purif.*, 127–136.

USINGER, R. L. and KELLEN, W. R. (1955). *Hilgardia* **23,** 263–321.

UVAROV, B. P. (1931). *Trans. ent. Soc. Lond.* **79,** 1–246.

WATSON, W. and FISHBURN, F. (1964). *J. Proc. Inst. Sew. Purif.*, 464–465.

WILLIAMS, N. V. and TAYLOR, H. M. (1968). *Wat. Res.* **2,** 139–150.

10

Arachnida

R. A. Baker

Department of Pure and Applied Zoology
University of Leeds
Leeds
England

I. Introduction

Although considerable work has been done on the meso- and macrofauna of percolating filters, the arachnids have largely escaped attention. The paucity of information on this group forms one of the major gaps in our knowledge of the biology of sewage purification processes. No work has been carried out on their seasonal incidence or spatial distribution and little is known about the factors determining their presence and abundance. They are not found in activated-sludge plants.

II. Classification

The class Arachnida forms a major division of the phylum Arthropoda. It is represented in sewage percolating filters by the mites (Acari) and the spiders (Araneae). The mites, though sometimes occurring in vast numbers, are smaller and less well known than spiders, insects and worms, and often go unnoticed.

Arachnids have four pairs of walking legs in the adult stage and their mouth appendages consist of paired chelicerae and pedipalps. They are morphologically quite distinct from the insects, lacking the mandibles and antennae which typify that group. Arachnids form a vast assemblage of animals and include, in addition to the mites and spiders, the scorpions, pseudoscorpions, harvestmen and sun spiders.

Before discussing their biology and role in sewage purification, some mention must be made of the classification of the Arachnids, in particular those groups which are represented in the filter fauna. Bristowe (1939) and Locket and Millidge (1951, 1953) for spiders, and Evans *et al.* (1961) for mites, give good accounts of the morphology and taxonomy of these groups; the latter's interpretation of the systematics is adopted here.

The spider sub-class (Araneae) is divided into 24 families, the largest of which is the Linyphiidae which has more than 250 British species; all the filter species are included in this family. The Linyphiidae, or hammock-web building spiders, produce a web in the form of a sheet and its members are distinguished by the form and position of an accessory reproductive structure called a paracymbium. In male spiders the distal segment of the pedipalp is greatly modified into a secondary sexual organ and contains an apparatus for taking up sperm. Part of the tarsus is hollowed out for this purpose and is called the cymbium and the accessory branch of this is called the paracymbium. These complex structures are collectively referred to as the palpal or "sexual" organs and are the principal morphological features upon which the identification of members of the family is based (Locket and Millidge, 1953).

The sub-class Acari is divided into seven orders of which two, the Astigmata and Mesostigmata, are represented in the filter fauna. This classification is largely based on the nature of the respiratory system. In the Mesostigmata the tracheal system opens by a pair of laterally situated stigmata in the region of coxae II–IV. Their bodies are covered by a number of chestnut-brown sclerotized plates and their mouthparts are normally held in a recess at the anterior end of the body. The chelicerae of many Mesostigmata are chelate–dentate. The majority of members are found free-living in the soil or decaying organic matter and many are ectoparasites.

By contrast adult astigmatid mites are weakly sclerotized and lack a tracheal system and stigmata. Their mouthparts are not held in a recess, being clearly visible from above. They are often colourless but may have a whitish colour due to the presence of stored guanine in the body. Each pedipalp is reduced to two segments only. In general they are more slow-moving animals than mesostigmatids.

III. Taxonomic Survey

A. List of Species Recorded from Percolating Filters

Phylum ARTHROPODA
Class ARACHNIDA
Sub-class ARANEAE
Family Linyphiidae

Araeoncus humilis (Bl.)
Diplocephalus cristatus (Bl.)
Erigone arctica (White)
Erigone atra (Bl.)
Erigone dentipalpis (Wid.)
Erigone promiscua (O.P.—Camb.)
Lepthyphantes ericaeus (Bl.)
Lepthyphantes leprosus (Ohl.)
Lepthyphantes nebulosus (Sund.)
Lepthyphantes tenuis (Bl.)
Leptorhoptrum robustum (Westr.)
Lessertia dentichelis (Sim.)
Meioneta rurestris (C.L.K.)
Oedothorax fuscus (Bl.)
Ostearius melanopygius (O.P.—Camb.)
Porrhomma convexum (Westr.)

Sub-class ACARI
Order Astigmata

Histiogaster carpio (Kramer)
Histiostoma feroniarum (Dufour)
Rhizoglyphus echinopus (Fumouze & Robin)

Order Mesostigmata

Eugamasus sp.
Macrocheles glaber (Müller)
Parasitus fucorum (Degeer)
Platyseius italicus (Berlese)

In addition the larval stage of *Ixodes ricinus* (L.), the castor bean or sheep tick, and an unidentified species of water mite have been recorded once only.

B. General Biology and Distribution of Species Listed

Perhaps the first person to record the presence of spiders in filters was Johnson (1914). He collected 109 mature spiders from three sewage works near Wakefield and found three species to be dominant. These were *Lessertia dentichelis, Erigone arctica* var. *maritima* and *Porrhomma convexum*. They constituted 55%, 17·4% and 13·6% respectively of the total spider population. These three species thus constituted 86% of the total spiders observed. Johnson also noted a number of relatively rare species.

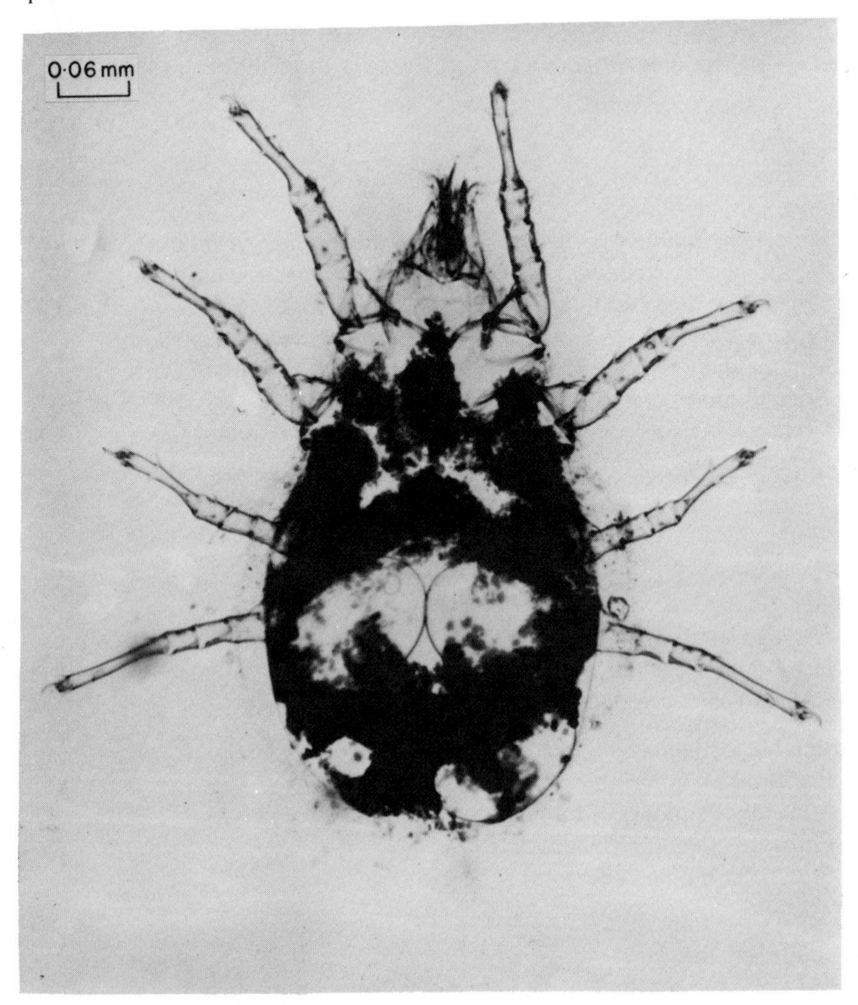

FIG. 1. *Histiostoma feroniarum* (Dufour) ♀. Ventral view.

Bristowe (1939) listed 15 species, all belonging to the Linyphiidae, which he had collected from sewage-works in the West Riding of Yorkshire. The most common species again appeared to be *Lessertia dentichelis*, *Erigone arctica* and *Porrhomma convexum*. Lloyd *et al.* (1940) thought that spiders represented the only important predators of worms and insects in percolating filters.

Among mites, the astigmatids are numerically the most abundant forms. The filter species require a high humidity or very damp surroundings in which to exist. The commonest mite in filters is *Histiostoma feroniarum*, which has been recorded from a large number of sites both in England and on the continent. Figure 1 illustrates an adult female of this species. The life-cycle, according to Scheucher (1957), takes 3–3½ days at 25–30°C, whilst Cooreman (1944) gives figures of 2–4 days at 20–25°C. Hughes (1961) gave a good description of this species and in particular an account of the hypopus stage.

The Astigmata are noted for the production of an optional hypopial or "wandernymph" stage in their life cycle. Robinson (1953), Perron (1954), Wallace (1960) and Griffiths (1969) have examined hypopus production in different species, but the factors influencing its formation remain incompletely understood and may vary from species to species. The hypopus of *Histiostoma feroniarum* is heteromorphic, having a reduced gnathosoma and a well sclerotized cuticle. It possesses a sucker plate, consisting of eight suckers, on the posterior ventral surface of the body, and is therefore well adapted for survival and dispersal attached to the bodies of other animals living in the same environment. The hypopus of this species is illustrated in Fig. 2. Hughes (1961) stated that the hypopus of *H. feroniarum* is commonly found clinging to the bodies of myriapods and insects. Baker (1964b) collected the hypopus of *H. feroniarum* from a number of other arthropods living in filters; these arthropods may be listed as follows:

Linyphiid spiders—including *Leptorhoptrum robustum* (Westr.)
Porcellio scaber (Latreille)
Cercyon ustulatus Preys
Lithobius forficatus (L.)
Anisopus fenestralis Scop
Parasitus fucorum (Degeer)
Staphylinid beetles

Baker also recovered the hypopus from the surface of the active film surrounding the filter medium. The hypopus did not show a preference for any specific arthropod but rather associated with these organisms in relation to their relative incidence in the habitat. It is noteworthy that the

FIG. 2. *Histiostoma feroniarum* (Dufour). Hypopus. Ventral view.

hypopus was never recovered from psychodids. Although large numbers of these "moth flies" were collected and examined, no hypopi were found, perhaps because the fine coating of hairs over the bodies of these insects forms an unsuitable surface for their attachment.

Rhizoglyphus echinopus, though not so frequently found and not in such large numbers as *Histiostoma feroniarum*, is also a common member of the mesofauna in many filters. Its body is smooth, colourless and glistening, and the appendages are reddish brown in colour (Hughes, 1961). Figure 3 illustrates the general appearance of this animal. The life cycle from egg hatch to adult is completed in 17–27 days at 13·3–23·9°C and in 9–13 days at 20–26·7°C, according to Zachvatkin (1941), while Woodring (1969) gave the following figures: 11·2 days at 93–98% relative humidity and 23°C. Garman (1937) quoted the duration of the egg stage as 6·5–7 days at 15–24°C and 4 days at 21–26·5°C. Baker's (1961) figures for the life-cycle of this species agree fairly closely with those of other workers; at room temperatures the life-cycle took approximately 18 days and the egg, on average, required 5 days for incubation. Although this mite is known to produce a hypopus, Baker (1961) was

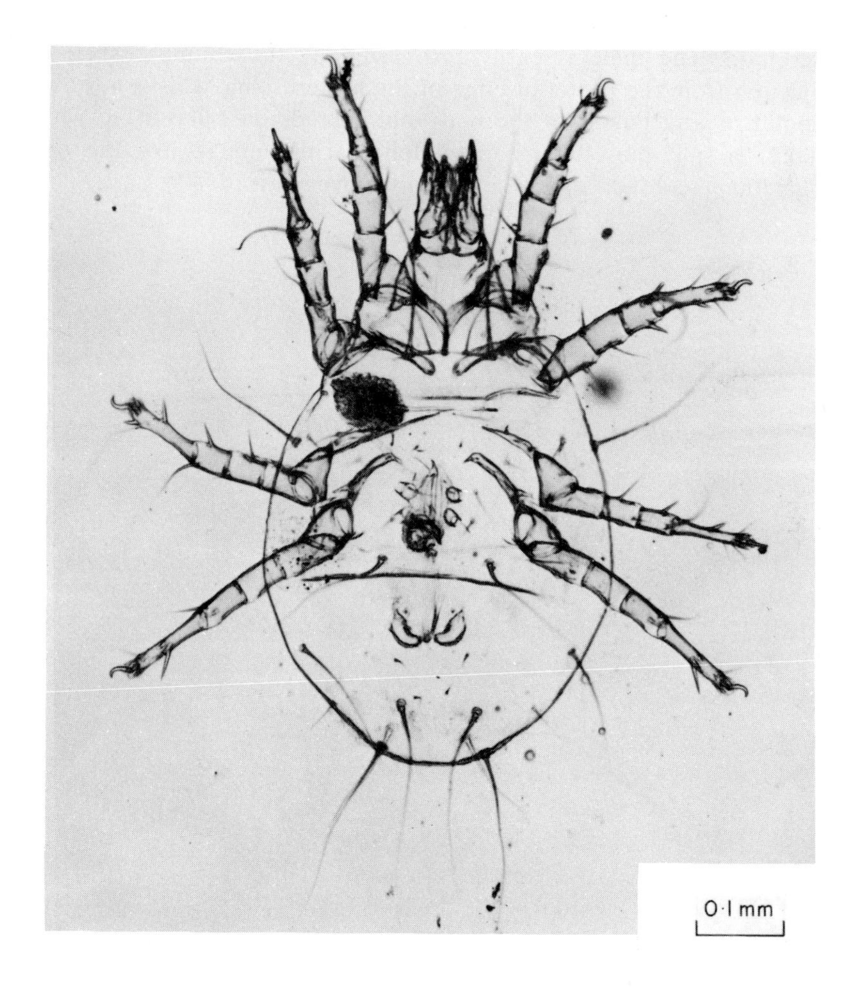

FIG. 3. *Rhizoglyphus echinopus* (F. and R.) ♂. Ventral view.

unable to obtain this stage in laboratory cultures. It is significant, therefore, that Woodring (1969) found that hypopi never formed from isolated groups of fewer than 20 eggs, larvae or protonymphs, but that in stock cultures 1–2% consisted of hypopi.

Histiogaster carpio was found by Learner (pers. comm.), Baker (1975) and recorded by Solbé *et al.* (1967) in the filter, effluent and traps of an experimental filter on the site of the former Water Pollution Research Laboratory at Stevenage. It is clearly more commonly distributed than was originally thought but does not occur so regularly as the other two astigmatid species. Michael (1903) listed the species as having been found abroad but not in Britain; Woodring (1966) gives a good taxonomic description. The male is easily recognized by the chitinous shelf-like projection from the posterior edge of the hysterosoma. This is narrow at its point of attachment to the body but spreads out, like a fan which, distally, is split into four rounded lobes. This animal, like the other astigmatids recorded, is very slow in its movements.

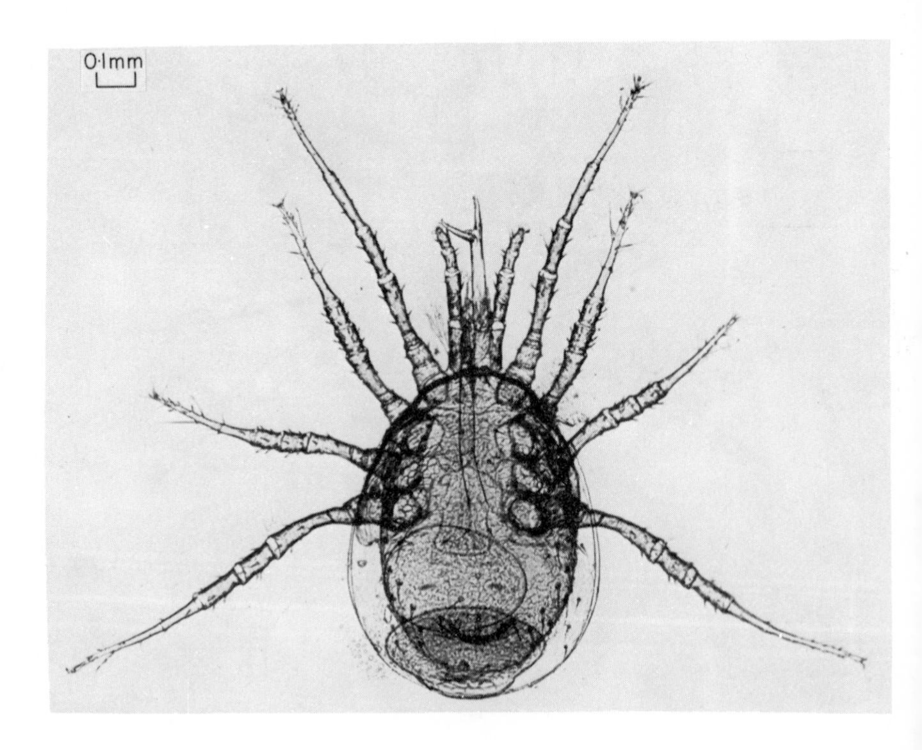

FIG. 4. *Platyseius italicus* (Berlese) ♀. Ventral view. The chelicera on one side is fully extended, the other withdrawn.

Of the Mesostigmata, *Platyseius italicus* (Fig. 4) is the most frequently recorded species in filters both in this country and abroad (Baker, 1964b). They are commonly found in clusters as brown specks on the surface of the bed medium (Tomlinson, 1946a). Hawkes (1965) also lists *P. italicus*, under its synonym *Platyseius tenuipes* Halbt, as a commonly recorded member of the mesofauna.

Other Mesostigmata are sometimes found; *Macrocheles glaber*, belonging to the family Macrochelidae, are commonly found in dung where they probably feed on the eggs, larvae and pupae of insects and worms. *Parasitus fucorum* belongs to the family Parasitidae; Hughes (1959) believes that members of this genus are coprophagus and detritus feeders. The other mesostigmatid recorded by Baker (1964b) was an undetermined species of *Eugamasus*.

There can be little doubt that the arachnid list is not an exhaustive one and that with further collection and identification other species will be found in filters.

IV. Other Habitats in which Filter Species are Found

Lloyd (1944) considered that the insect fauna of percolating filters derived from freshwater mud-flats. Although in the latter environment he found eight out of the ten species of insects normally found in sewage filters, the two environments are not much alike. The insects are the only group of the mud-flat fauna to have received detailed study and they appear to correspond closely to the filter fauna. No spiders or mites, however, were recorded from the mud-flats, and one has to look elsewhere for the natural setting of the arachnid fauna of percolating filters.

Baker (1961) noted that a large number of the filter meso- and macrofauna, including spiders, has been recorded from caves, and the same author (1964b) indicated that a number of filter mites also occurred in dung and in soil.

Bristowe (1939) considered alternative habitats of the filter spiders, and recorded that a number of them could also be found in cellars, rubbish heaps, caves, mines and on the seashore. Wolf (1934, 1938) lists the spiders recorded from caves, and these include the following filter species: *Erigone atra*, *Erigone dentipalpis*, *Lepthyphantes leprosus*, *Lepthyphantes nebulosus*, *Lepthyphantes tenuis*, *Diplocephalus cristatus*, *Lessertia dentichelis* and *Porrhomma convexum*.

Reynoldson (1947) stressed the resemblance of the filter environment to that of the decaying seaweed zone at high water-mark, but he was concerned essentially with the enchytraeid worms and did not refer to the arachnid fauna. By reference to Backlund (1945) who worked on the wrack fauna of Sweden and Finland, Reynoldson (1948) was able to demonstrate further the similarity of the two environments. All the major faunistic genera of the filter, except chironomids, occur in wrack. Egglishaw (1965) listed the following spiders as members of the wrack fauna: *Erigone arctica*, *Erigone dentipalpis*, *Lepthyphantes tenuis*, *Ostearius melanopygius* and *Centromerita* sp. It is noteworthy that *E. arctica* var. *maritima* is found in filters. The presence of decaying vegetation, the periodic washings and the alternate "wet" and "dry" nature of the surroundings are features common to the two environments.

Various workers have mentioned the following habitats for the most commonly occurring filter Acari. *Histiostoma feroniarum* is widely distributed in many kinds of decaying vegetation; Hughes and Jackson (1958) recorded it from the exudations of trees, on mushrooms and other fungi and in stable manure and compost heaps. *Rhizoglyphus echinopus* is commonly found in the bulbs and roots of many plants; it feeds on decaying plant or animal matter (Woodring, 1969). Sheals (1957) records it in uncultivated soil, while Dhillon and Gibson (1962) noted that it was restricted to the upper layers of undisturbed grassland. Hughes (1961) lists it as being present in all kinds of vegetable matter, bulbs, roots of wheat, oats, vines, mushrooms and decaying grain. Evans and Hyatt (1960) list *Platyseius italicus* as having been found in moss, under dead rotting leaves, on a stone in a mountain stream and also at Minworth Sewage Works, Birmingham.

Many of the species described are each found in a variety of habitats and are often widespread and common. There is no suggestion, therefore, that the arachnid fauna of filters has been directly derived from one of these habitats. It appears more likely that the arachnids have colonized filters from a number of habitats which share a number of features in common with filters, the chief of which appears to be the presence of large amounts of decaying organic matter. The wrack fauna or stranded seaweed zone at high water-mark, compost heaps, and caves are of particular significance in this respect.

The dispersal of mesostigmatids is probably effected by adult females clinging onto the bodies of flying insects, and astigmatids are likely to be transported mainly via the hypopus stage.

V. Possible Factors Determining Presence and Abundance

There are many uncontrollable factors in the filter environment which make it hazardous to attempt to identify those which determine the presence and abundance of the individual species. The few observations which have been made on mites, involving different types of sewage treated by different methods, have shown variations in the distribution of the individual species, and although no controlled experiments have been carried out, and therefore no clear-cut results obtained, some general trends are apparent.

A. Type of Sewage

Observations have been made by the staff of the Water Pollution Research Laboratory on the filter fauna at 48 sewage-works in England, Scotland and Wales (Ministry of Technology, 1968). The works were classified on the basis of the proportion of industrial wastes treated, 29 of them receiving less than 10% of industrial wastes by volume. Mites were found in every filter at each of the works visited, although the population density varied considerably from site to site; no attempt to classify them into categories was attempted. The mites and other invertebrates did not appear to be adversely affected by sewages containing more than 10% of industrial wastes. Baker (1961) has also found large variations in the populations of mites at different sewage-works. In some filters mesostigmatids predominate and few, if any, astigmatids are present, whereas at other sites, astigmatids are the predominant forms. In some filters, e.g. Esholt (Bradford), although seasonal variations have not been analysed, both groups exist side by side in large numbers. Some works in the West Riding of Yorkshire treat strong sewages which contain 40–60% industrial wastes by volume (at Bradford and Huddersfield). The filters nevertheless contain large numbers of astigmatids and this suggests that the group is not markedly affected by the strength of the sewage (Baker, unpublished observations).

Reynoldson (1948) demonstrated that under these conditions the fauna is limited to a few successful species which can tolerate the strong sewages, and there is evidence that some astigmatids, notably *Histiostoma feroniarum*, should be included in this restricted faunal list.

B. Biological Factors

Insufficient attention has been given to biological factors, which may play an important role in determining the abundance of certain filter species. Hawkes (1951) stressed the importance of interspecific competition in the case of the insects. The presence of many species may result in such strong interspecific competition for food that reduction in numbers of any one species might occur. With a restricted fly fauna, on the other hand, competitive forces may be more selective and a single species may dominate the few represented. Competition for food and space may also be an important factor in the case of *Histiostoma feroniarum*. This species may be unable to compete successfully when other grazers are present in a mixed population but may become established as a dominant member of a more restricted filter fauna.

Mesostigmatids may be found in close association with enchytraeid worms at some works and, since it has been demonstrated that the latter form the staple diet of some mesostigmatids, their appearance together in dense masses may in part be accounted for as a predator–prey relationship.

Hawkes (1952) has demonstrated that different stages in the life-cycle of certain filter flies are also distributed differently within the filters, larvae preferring the sub-jet and pupae the inter-jet zones. It is therefore important that observations are made on all stages of the life-cycle. In the case of *Histiostoma feroniarum*, Baker (1964b) noted very large numbers of the deutonymph or hypopus in filters which had become dry after being out of commission for several days. The resting of sewage filters for short periods, though a common practice at some works, is not thought to be advantageous, often resulting in inadequate purification (Ministry of Technology, 1968). When practised it is apparent that one of the changes observed is the production of resistant, distributive stages, such as the hypopus. In another paper Baker (1964a) was able to show that the hypopus of the same species, when provided with a film of water, moulted to the tritonymphal stage in 2 to 3 days at temperatures ranging from 16 to 22°C, whereas control hypopi did not develop further. It may be argued that hypopi are readily produced when environmental conditions become unsuitable to the species, due to the drying out of the film in this case, and that when favourable conditions are restored, the hypopi continue their developmental cycle. Since *H. feroniarum* appears to feed on micro-organisms in fluid or semi-fluid media, it is perhaps not surprising to find that if this fluid film is absent, a resistant stage is produced to withstand the temporary adverse conditions created.

C. Method of Distribution of Sewage

Hawkes (1952, 1959) has shown that the method of distribution of sewage on to a filter using travelling distributors has a marked effect on the horizontal distribution of the fauna. Between the jets is a zone, referred to as the inter-jet zone, which is kept moist only by splashing and condensation. The sub-jet zone immediately below the jet is indicated at some works by a striped appearance at the surface due to unicellular algae and at others, e.g. Esholt (Bradford), by the additional presence of a fungus. This fungus, referred to as *Fusarium aqueductum*, is especially abundant in winter and is evidently suited to the acid conditions prevailing at certain works (Butcher, 1932). In the inter-jet zone, according to Hawkes (1952), the algae are grazed away. In beds where distribution methods favour the creation of permanently "dry" inter-jet zones, these zones and the "wet" zones each establish a characteristic fauna. Thus collectively the filters show a more varied and mixed grazing fauna. Splash plates are said to give a more even distribution of sewage but a more restricted fauna. Some animals, such as *Hypogastrura viatica* (*Achorutes*) are unable to exist under the jet of sewage, with the result that this collembolan tends to congregate in the drier areas of filter not subjected to a strong flow of sewage (Hawkes, 1959).

At Esholt (Bradford) there are rectangular filters with reciprocating distributors having travelling arms fitted with splash plates; these produce an umbrella spray as the sewage impinges on the plates. In spite of this there are "drier" areas between the jets and between the occasional unmodified nozzles. These "drier" areas contrast markedly in amount of film and grazing fauna with the areas immediately below the splash plates. In the sub-jet zone there is, in winter, a thick growth of *Fusarium aqueductum* and algae are present throughout the year. In this zone the fauna is restricted and dominated by enchytraeid worms. In the "drier" areas, between the jets, which receive only a gentle spray of sewage, the filter medium is clean and supports a variety of grazing organisms including midge larvae, enchytraeids and large numbers of *Histiostoma feroniarum*, *Rhizoglyphus echinopus Histiogastio carpio* and mesostigmatid mites. The same phenomenon has been observed in other filters. It would appear that the mite fauna at the surface of the filters has a distinct horizontal distribution, the nature of which depends upon the method of distribution of sewage on to the filter. Mites are less frequently found in the very wet areas subjected to a strong flow of sewage but are abundant, at the surface, in the drier inter-jet zones which receive only a gentle spray of sewage and where the film is thinner (Baker, unpublished observations).

Mites occur in large numbers throughout the year, in filters treating sewages of different strength, with large differences in the rate of application, periodicity of dosing and method of distribution; thus the situation is far from clear and positive statements cannot yet be made: further work is required to verify the tentative conclusions reached so far. From the few observations which have been made to date, it would appear that the method of distribution of sewage on to the bed is an important factor and that strong chemical and industrial sewages do not eliminate the astigmatid mite fauna. The size and nature of the medium seems to be less important. Biological factors such as interspecific competition and predator–prey relationships probably also play a part in determining the presence and abundance of each acarine species. Much more work needs to be carried out both in the laboratory and in the field in order to determine the precise factors responsible for the presence, abundance and distribution of mites in filters.

VI. Role in Purification Process

It is necessary to understand the feeding habits of the arachnids before their role in the purification process can be critically assessed.

A. Spiders

Linyphiid spiders are predators feeding mainly on insects and worms. The chelicerae contain a poison gland and bear teeth; unlike many other spiders, they chew their prey until the exoskeleton is completely disrupted. Egglishaw (1965) observed that *Erigone arctica*, in the laboratory, fed on oligochaetes, fly larvae and adult insects, and stated that its natural prey is probably oligochaetes. Spiders may reduce the fly and worm populations but so far as is known they play no direct role in the purification process in filters. Tomlinson (1939) indicated that, when abundant, spiders must play a considerable part in controlling the insect life in filters, and probably with the mesostigmatid mites they form the chief predators.

B. Mesostigmata

Mesostigmatids are often large, active and predatory. Many have piercing, sucking and tearing mouthparts and feed by piercing the integument of the prey, inserting their chelicerae and sucking in body fluids (Wallwork, 1967). It is widely believed that external digestion takes place prior to imbibition.

Direct microscopical observations on the surface of filter media indicate that *Platyseius italicus* is predatory. Figure 4 illustrates the nature of the chelicerae in this species. In culture, this mite was presented with various living organisms commonly found in filters and from its responses it was concluded that enchytraeid worms were probably its chief prey. Whole living worms were attacked as well as chopped up pieces, but the cocoons of *Lumbricillus* sp. were not eaten. *P. italicus* also fed on fly larvae and when starved became cannabalistic, but living specimens of *Histiostoma feroniarum* and Collembola did not appear to be used as food (Baker, 1961). These results agree closely with the findings of other workers who have described the food and feeding in related mesostigmatid mites. Wallwork (1967) states that some of the large mesostigmatids such as *Pergamasus* spp. feed on their own young in culture, and Leitner (1946), investigating the mite communities of dung, considered that the majority of *Parasitus* and *Macrocheles* in this habitat were predatory; *Macrocheles glaber* refused decaying straw but fed on enchytraeid worms. Although living fly larvae and collembola were not attacked, parts of their dead bodies were consumed.

C. Astigmata

The feeding habits of astigmatids are not well known, and in the table included in Wallwork (1967) (an abridged form of which is reproduced here: Table I), there are a number of unconfirmed and conflicting reports on the feeding habits of the soil-dwelling members of this group. According to Wallwork (1967) *Rhizoglyphus echinopus* feeds on the decaying plant material present in soils. Karg (1963) considers many of the Anoetidae to be filter feeders on liquids rich in micro-organisms, and Hughes (1953) demonstrated that the mite *Anoetus sapromyzarum* (Dufour) and probably other anoetids also, fed by "combing" the microfauna from a fluid medium using specially modified mouthparts. The mouthparts of *Histiostoma feroniarum* are shown in Fig. 5.

Baker and Barker (unpublished observations) have kept both *H. feroniarum* and *R. echinopus* on agar plates inoculated with different strains of bacteria, and both species have completed several life-cycles

TABLE I. Feeding habits of selected groups of soil mites (modified from Wallwork, 1967)

	Carnivore	Herbivore/Decomposer			
Group	Predator	Plant detritus	Moss and lichens	Wood	Fungi and algae
Macrochelidae	X				
Parasitidae	X				
Acaridae					
Histiogaster				X*	X
Rhizoglyphus		X			X*
Anoetidae		X			X

* Unconfirmed or conflicting reports

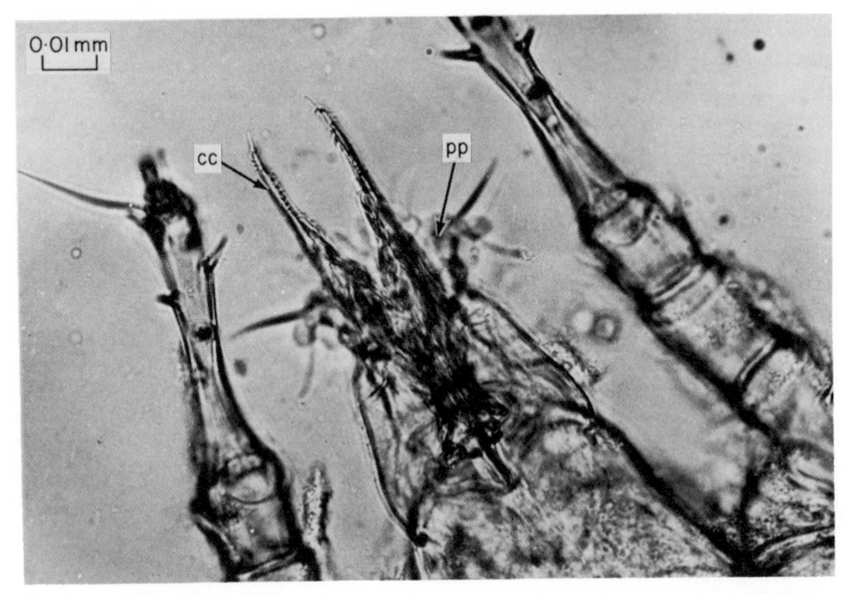

FIG. 5. Mouthparts of *Histiostoma feroniarum* (Dufour), showing chelicerae (c.c.) and pedipalps (p.p.).

under these conditions. The latter species also appears capable of feeding at the surface of filters on the algal film covering the medium. Baker (1961) maintained cultures of both species on *Drosophila* food medium inoculated with yeast solution and suggested (1964b) that the astigmatids in filters feed on the zoogloeal film which surrounds the filter medium.* There seems little doubt that the filter astigmatids feed mainly on liquids rich in micro-organisms or on plant material. They thus form part of the scouring group of organisms in the filter.

British workers (Johnson, 1914; Lloyd, 1945; Reynoldson, 1939; Tomlinson, 1946b; Hawkes, 1957; Williams and Taylor, 1968) believe that the role of the grazing fauna in film removal is of primary importance. Without their activity filter efficiency would quickly become reduced due to the filter becoming clogged by too much film. The water-logging thus created and the impairment of air circulation would obviously contribute to the formation of a more anaerobic and therefore less beneficial gaseous environment. However, most American workers are of the opinion that the microbiological activity of the film and physical scouring are most important in film removal, and that the grazers play only a minor role (Hawkes, 1965).

From the evidence available concerning the Arachnida it would appear that the astigmatid mites of filters play a similar role to that of the fly

* See note added in proof.

larvae and worm populations and that film removal is their primary role. The mesostigmatids and spiders, on the other hand, by removing fly larvae and worms, must reduce the effectiveness of the grazing fauna, but possibly help to some extent in combatting the fly nuisance, especially during the summer months, in areas adjacent to percolating filters.

References

BACKLUND, H. O. (1945). *Opusc. ent.* (*Suppl.*) 1–237.

BAKER, R. A. (1961). "A preliminary survey of the mite fauna of sewage percolating filters." M.Sc. Thesis, University of London.

BAKER, R. A. (1964a). *A. Mag. nat. Hist.* **7,** 693.

BAKER, R. A. (1964b), *A. Mag. nat. Hist.* **7,** 707.

BRISTOWE, W. S. (1939, 1941). "The Comity of Spiders" Vols I, II. Ray Soc., London.

BUTCHER, R. W. (1932). *Trans. Br. mycol. Soc.* **17,** 112–124.

COOREMAN, J. (1944). *Bull. Mus. Hist. nat. Belg.* **20** (9), 1–16.

DHILLON, B. S. and GIBSON, N. H. E. (1962). *Pedobiologia* **1,** 189–209.

EGGLISHAW, H. J. (1965). *Trans. Soc. Br. Ent.* **16,** 191–216.

EVANS, G. O. and HYATT, K. H. (1960). *Bull. Br. Mus. nat. Hist. Zool.* **6** (2). 27–101.

EVANS, G. O., SHEALS, J. G. and MACFARLANE, D. (1961). "The Terrestrial Acari of the British Isles", Vol. 1. British Museum of Natural History, London.

GARMAN, P. (1937). *Conn. agric. Exp. Stn Bull.* **402,** 889–907.

GRIFFITHS, D. A. (1969). *Proc. Int. Congr. Acarol.* England (1967), 419–432.

HAWKES, H. A. (1951). *Ann. appl. Biol.* **38,** 592–605.

HAWKES, H. A. (1952). *Ann. appl. Biol.* **39,** 181–192.

HAWKES, H. A. (1957). *J. Proc. Inst. Sew. Purif.* (2), 88–112.

HAWKES, H. A. (1959). *Ann. appl. Biol.* **47,** 339–349.

HAWKES, H. A. (1965). "Ecology and the Industrial Society" (5th Symp. Brit. Ecol. Soc.), pp. 119–148. Blackwell, Oxford.

HUGHES, A. M. (1961). "The Mites of Stored Food Products." H.M. Stationery Office, London.

HUGHES, R. D. and JACKSON, G. G. (1958). *Va J. Sci. N.S.* **9,** 5–198.

HUGHES, T. E. (1953). *Proc. Acad. Sci. Amst.* **56C**(2), 278–287.

HUGHES, T. E. (1959). "Mites or the Acari." Athlone Press, University of London.

JOHNSON, J. W. H. (1914). *J. econ. Biol.* **9,** 105–124.

KARG, W. (1963). *In* "Soil Organisms" (Eds J. Doeksen and J. van der Drift), pp. 305–315. North-Holland, Amsterdam.

LEITNER, E. (1946). *Zentbl. Gesamtgeb. Ent.* **1**(3), 75–95; **1**(5/6), 129–156.

LLOYD, L. (1944). *Nature, Lond.* **154,** 397.
LLOYD, L. (1945). *J. Proc. Inst. Sew. Purif.* (2), 119–139.
LLOYD, L., GRAHAM, J. F. and REYNOLDSON, T. B. (1940). *Ann. appl. Biol.* **27,** 122–150.
LOCKET, G. H. and MILLIDGE, A. F. (1951, 1953). "British Spiders" Vols 1, 2. Ray Soc., London.
MICHAEL, A. D. (1901, 1903). "British Tyroglyphidae" Vols 1, 2. Ray Soc., London.
MINISTRY OF TECHNOLOGY (1968). "Water Pollution Research, 1967." H.M. Stationery Office, London.
PERRON, R. (1954). *Acta zool., Stockholm* **35,** 71.
REYNOLDSON, T. B. (1939). *J. Proc. Inst. Sew. Purif.* (1), 158–172.
REYNOLDSON, T. B. (1947). *J. Anim. Ecol.* **16,** 26–36.
REYNOLDSON, T. B. (1948). *J. Anim. Ecol.* **17,** 27–38.
ROBINSON, I. (1953). *Proc. zool. Soc. Lond.* **123,** 267.
SCHEUCHER, R. I. (1957). *Beitr. Syst. Ökol. Acarina* **1**(1), 233–284.
SHEALS, J. G. (1957). *J. Anim. Ecol.* **26,** 125–134.
SOLBÉ, J. F. DE L. G., WILLIAMS, N. V. and ROBERTS, H. (1967). *Wat. Pollut. Control* **66,** 423–448.
TOMLINSON, T. G. (1939). *J. Proc. Inst. Sew. Purif.* (1), 225.
TOMLINSON, T. G. (1946a). Department of Scientific and Industrial Research, Water Pollution Research Technical Paper No. 9. H.M. Stationery Office, London.
TOMLINSON, T. G. (1946b). *J. Proc. Inst. Sew. Purif.* (1), 168–178.
WALLACE, D. R. J. (1960). *J. Insect Physiol.* **5,** 216.
WALLWORK, J. A. (1967). *In* "Soil Biology" (Eds A. Burges and F. Raw), pp. 363–395. Academic Press, London and New York.
WILLIAMS, N. V. and TAYLOR, H. M. (1968). *Water Res.* **2,** 139–150.
WOLF, B. (1934, 1938). "Animalium Cavernarum Catalogus." W. Junk, 's- Gravenhage *1–3*.
WOODRING, J. P. (1966). *Proc. Louisiana Acad. Sci.* **29,** 113–136.
WOODRING, J. P. (1969). *Ann. ent. Soc. Am.* **62**(1), 102–108.
ZACHVATKIN, A. A. (1941). *Inst. Zool. Acad. Sci. Moscow*, N.S. No. 28, 1–475.

Note added in proof: Baker (1975) has recently studied the alimentary canal of *Histiogaster carpio* and found large numbers of bacteria in the lumen of the midgut. Reference: Baker, R. A. (1975). *Acarologia* **17,** in press.

11

Crustacea and Mollusca

M. A. Learner

Department of Applied Biology
University of Wales
Institute of Science and Technology
Cathays Park
Cardiff CF1 3NU
Wales

I. Introduction

Crustaceans and molluscs have been found in sewage-works only occasionally (Gurney, 1932; Tomlinson, 1939; Lloyd *et al.*, 1940; Tomlinson, 1941; Fryer, 1955; Brink, 1967; Solbé *et al.*, 1967). The results of the author's survey of percolating filters (see Chapter 9 for details) confirm the infrequent occurrence of species of both these groups except for two species of copepod, *Paracyclops fimbriatus* and *Bryocamptus pygmaeus* (Table I). The species found during this survey are in general aquatic and three appear largely restricted to the more liquid systems in sewage-works. Fryer (1955) recorded *Cyclops strenuus* from the settling tanks at several sewage-works in Yorkshire and this species, *Eucyclops serrulatus* and *Radix*(=*Lymnaea*) *peregra* were found only in filter under-drains during the author's survey; however the lower levels of the filters studied were not sampled. The other species listed in Table I have so far been reported only from the percolating-filter stage of sewage treatment.

TABLE I. Frequency and abundance of crustacean and molluscan species found during a survey of the fauna at 48 sewage-works in 1966–1967. Species found in the effluent only are marked "E"

	Occurrence in survey (%)	Maximum no. found (no./litre of medium)
CRUSTACEA		
Copepoda		
Cyclopoida		
Cyclopidae		
Cyclops strenuus (Fisch.)	1	E
Eucyclops serrulatus (Fisch.) = *E. agilis* (Koch, Sars)	1	E
Paracyclops chiltoni (Thomson)	18	619
P. fimbriatus (Fisch.)	73	1222
P. poppei (Rehb.)	12	458
Speocyclops demetiensis (Scourf.)	1	17
Harpacticoida		
Canthocamptidae		
Bryocamptus pygmaeus (Sars)	61	7790
B. zschokkei (Schmeil)	1	6
Moraria varica (E. Graet.)	1	17
Phyllognathopodidae		
Phyllognathopus viguieri (Maup.)	9	120
Isopoda		
Asellidae		
Asellus aquaticus L.	18	80
MOLLUSCA		
Gastropoda		
Arionidae		
Arion subfuscus (Drap.)	12	1
Limacidae		
Agriolimax reticulatus (Müll.)	8	1
Lymnaeidae		
Galba palustris Müll.	1	19
Radix peregra Müll.	6	E
Zonitidae		
Oxychilus sp.	1	<1

II. Identification and Biology

A. Crustacea

Cyclopoid copepods can be identified using Gurney (1933), Harding and Smith (1960) and Dussart (1969), harpacticoid copepods using Gurney (1932) and Dussart (1967), and asellid species using Hynes *et al.* (1960).

All the copepod species recorded in Table I, except *Cyclops strenuus* and *Eucyclops serrulatus*, are known to occur in temporary and in subterranean waters (Illies, 1967), the latter habitat having many environmental features resembling those of the percolating filter (Baker, 1961). Even so, the two species most successful in filters, *Paracyclops fimbriatus* and *Bryocamptus pygmaeus*, are highly adaptable, having been found in a wide variety of habitats, and both are able to live in a thin film of water (Gurney, 1932, 1933).

Relatively little is known about the diet of most species of copepod in filters. *Cyclops strenuus* is a carnivore feeding on dipteran larvae and oligochaetes and *Eucyclops serrulatus* is a herbivore feeding on a variety of algae (Fryer, 1957). Neither of these species has been found actually in filters and those species that do occur are probably mainly detritus feeders. It has been found that both *Paracyclops fimbriatus* and *Bryocamptus pygmaeus* live, breed and develop successfully in small watch-glasses when supplied with dewatered activated sludge for food; such sludge, however, contained protozoans (Learner, in prep.). The isopod *Asellus aquaticus* feeds principally on decaying plant material and filamentous algae (Williams, 1962) but is capable of living in a wide variety of habitats including subterranean ones, and can undoubtedly exist as a detritus feeder.

The functional significance of the crustaceans in sewage purification processes is, because of their generally low numbers, unlikely to be of great importance. Occasionally, however, the *Paracyclops* group and *Bryocamptus pygmaeus* occur in high numbers in some filters and probably contribute to the purification process in much the same way as other grazers.

The factors affecting the occurrence and distribution of crustaceans in filters have still to be determined. The amount of film is clearly important with respect to some species; Gurney (1932) reported *Phyllognathopus viguieri* as being most numerous in parts of filters where sewage purification was most complete, and Brink (1967) found *Cyclops strenuus* in his "horizontal" filter only where there were small amounts of film. Figure 1 shows that in one filter studied by the author, *Paracyclops fimbriatus*

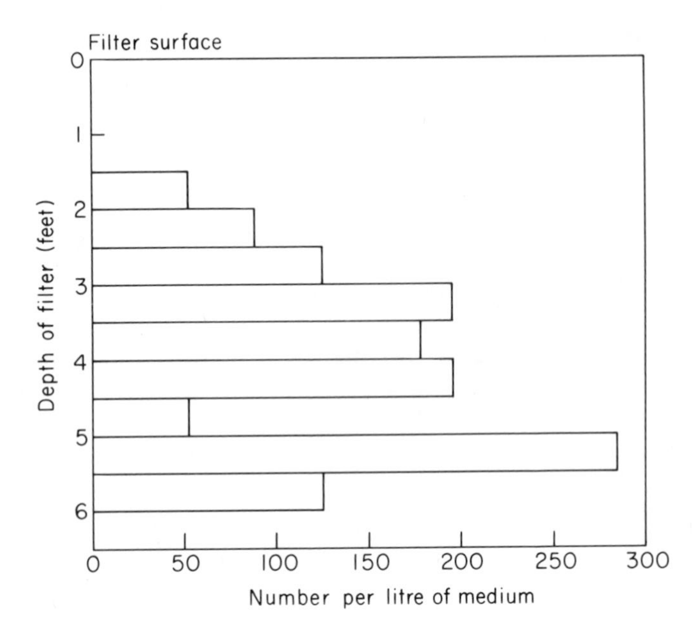

FIG 1. Vertical distribution of *Cyclops fimbriatus* in a filter, August 1968. Minworth sewage works, Birmingham (England).

occurred principally in the lower three feet (≃1 metre) of a filter sampled in August, i.e. where the amount of film was likely to be small and where purification was almost complete. The pH of the sewage may also limit the occurrence of some crustaceans, particularly in areas where there are acid wastes, since Roen (1957) found *P. fimbriatus* only in localities having a pH within the range 6·2 to 9·0.

Virtually nothing is known about seasonal changes in the population densities of crustaceans in filters. Solbé *et al.* (1967) found *Paracyclops* in a filter effluent throughout the year with maximum numbers occurring during the spring sloughing period; the occurrence of harpacticoids (probably *Bryocamptus pygmaeus*) in the effluent was spasmodic. The numbers of animals present in effluents from filters may not reflect population densities within filters, however. *B. pygmaeus* reproduces throughout the year with peak periods in March, June and October–November (Gurney, 1932) and so its spasmodic appearance in filter-effluents is somewhat unexpected. Solbé *et al.* (1967) believed that the harpacticoids were early colonizers of filters and might only survive during the early stages of filter colonization (about 18 months), but later work by Solbé (unpublished) and the author, showed that these copepods are not restricted to these early stages.

B. Mollusca

Descriptions and keys to the freshwater and terrestrial molluscs are provided by Quick (1949), Janus (1965), McMillan (1968) and Macan (1969).

There are very few records of molluscs occurring in sewage treatment processes in Britain, although Lloyd *et al.* (1940) referred to the presence of *Agriolimax agrestis* (probably *A. reticulatus*) on the surface of filters and to *Lymnaea* (= *Galba*) *glabra* as being common within the filters at Knostrop (Leeds, England). Hawkes (1963) reported *Lymnaea* (= *Radix*) *peregra* as being common in filters (Harrogate, England) receiving a high hydraulic loading and recirculated effluent.

Of the species of snail reported from British sewage-works, *Galba glabra* and *G. palustris* tend to be amphibious, living in very shallow water or temporary water (Janus, 1965), conditions similar to those they would experience in a filter, but in fact both species are rare in filters. The species most frequently found in filters in Britain is the more adaptable *Radix peregra*, which has been recorded from virtually every kind of watery habitat.

Snails of the genera *Physa* and *Physilla* have been reported from filters in the U.S.A. and *Physa integra* has colonized both standard-rate (loading of $4{\cdot}8\ m^3/m^2$ per day = 890 gal/yd^2 per day) and high-rate (loading of $15{\cdot}1\ m^3/m^2$ per day = 2800 gal/yd^2 per day) filters (Higgins 1948, Ingram *et al.*, 1958).

All the previously mentioned snail species are pulmonates and are not dependent upon the dissolved oxygen content of the sewage for their survival. They probably all feed on the film, and like most of the other macro-invertebrate inhabitants of filters, aid the purification process by this grazing activity. Their shells can become a serious nuisance, however, by blocking pipes and damaging pumps (Ingram *et al.*, 1958).

The distribution of snails within filters, their seasonal occurrence, and the factors affecting their distribution and density, have been little studied.

The slugs associated with filters seem largely confined to the surface of the filter. *Arion subfuscus* feeds on fungi and carrion and requires a damp environment, while *A. reticulatus* is omnivorous and is found in a wide variety of habitats. It is unlikely that either species is important in the filter ecosystem.

References

BAKER, R. A. (1961). "A preliminary survey of the mite fauna of sewage percolating filters." M.Sc. Thesis, University of London.
BRINK, N. (1967). *Int. Revue ges. Hydrobiol. Hydrogr.* **52,** 51–122.
DUSSART, B. H. (1967). "Les Copepodes des Eaux Continentales d'Europe Occidentale. Tome 1, Calanoides et Harpacticoides." Boubée et Cie, Paris.
DUSSART, B. H. (1969). "Les Copepodes des Eaux Continentales d'Europe Occidentale. Tome 2, Cyclopoides et Biologie." Boubée et Cie, Paris.
FRYER, G. (1955). *Naturalist, Hull.*, 101–126.
FRYER, G. (1957). *J. Anim. Ecol.* **26,** 263–286.
GURNEY, R. (1932). "British Fresh-Water Copepoda." **2.** Ray Society, London.
GURNEY, R. (1933). "British Fresh-Water Copepoda." **3.** Ray Society, London.
HARDING, J. P. and SMITH, W. A. (1960). *Scient. Publs Freshwat. biol. Ass.* **18,** 53 pp.
HAWKES, H. A. (1963). "The Ecology of Waste Water Treatment." Pergamon Press, London and Oxford.
HIGGINS, T. E. (1948). *Sewage Wks Engng* **19,** 143.
HYNES, H. B. N., MACAN, T. T. and WILLIAMS, W. D. (1960). *Scient. Publs Freshwat. biol. Ass.* **19,** 35 pp.
ILLIES, J. (1967). "Limnofauna Europaea." Gustav Fischer, Stuttgart.
INGRAM, W. M., COOKE, W. B. and HAGERTY, L. T. (1958). *Sewage ind. Wastes* **30,** 821–825.
JANUS, H. (1965). "The Young Specialist Looks at Land and Freshwater Molluscs." Burke, London.
LLOYD, L., GRAHAM, J. F. and REYNOLDSON, T. B. (1940). *Ann. appl. Biol.* **27,** 122–150.
MACAN, T. T. (1969). *Scient. Publs Freshwat. biol. Ass.* **13,** 44 pp.
MCMILLAN, N. F. (1968). "British Shells." Warne, London.
QUICK, H. E. (1949). "Slugs (Mollusca)." Synopsis of the British Fauna No. 8. Linnean Society, London.
ROEN, U. (1957). *Biol. Skr. Dan. Vid. Selsk.* **9,** 1–101.
SOLBÉ, J. F. DE L. G., WILLIAMS, N. V. and ROBERTS, H. (1967). *Wat. Pollut. Control* **66,** 423–448.
TOMLINSON, T. G. (1939). *J. Proc. Inst. Sew. Purif.* 225–235.
TOMLINSON, T. G. (1941). *J. Proc. Inst. Sew. Purif.* 39–56.
WILLIAMS, W. D. (1962). *Hydrobiologia* **20,** 1–30.

Subject Index

S

T

V

W

Y

Z

Species Index

Page numbers indicate where species are mentioned throughout the text, tables (except species lists) and illustrations. Illustrations are indicated in bold type.

T

U

V

X

Z